Cooperative Guidance & Control of Missiles
Autonomous Formation

Sentang Wu

Cooperative Guidance & Control of Missiles Autonomous Formation

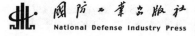

National Defense Industry Press

Sentang Wu
Beihang University
Beijing, China

ISBN 978-981-13-4539-5   ISBN 978-981-13-0953-3   (eBook)
https://doi.org/10.1007/978-981-13-0953-3

Jointly published with National Defense Industry Press, Beijing, China

The print edition is not for sale in China Mainland. Customers from China Mainland please order the print book from: National Defense Industry Press, Beijing.

Translation from the Chinese language edition: 导弹自主编队协同制导控制技术 by Sentang Wu, © National Defense Industry Press 2015. All Rights Reserved.
© National Defense Industry Press, Beijing and Springer Nature Singapore Pte Ltd. 2019
Softcover re-print of the Hardcover 1st edition 2019

Printed on acid-free paper

This Springer imprint is published by the registered company Springer Nature Singapore Pte Ltd.
The registered company address is: 152 Beach Road, #21-01/04 Gateway East, Singapore 189721, Singapore

Preface

The monograph mainly illustrates the rationale, design, and technical realization/verification for the Cooperative Guidance & Control System (CGCS) of Missiles Autonomous Formation (MAF). According to the structure of the seven functions and five major compositions of CGCS by the way of combining theory method and engineering practice, the book systematically explains the concepts and definitions of MAF, the structure and integrated approach of CGCS, information acquisition system(IAS), decision and management system (DMS), flight control system of MAF (FCSM), member flight control system (MFCS), support networks system (SNS), simulation and verification for the CGCS, etc.

In view of the demand for the autonomous formation of cruise missiles, and combining bionics theory and practical project, this book illustrates the research content on the basic theory, synthesis design method, technology implementation, and technical identification by use of figure and figure lines as well as the application of a large number of wind tunnel test data digital simulation analysis, hardware-in-the-loop (HIL) system simulation test, and embedded system test results.

This book consists of eight chapters. Chapter 1 outlines the seven functions of CGCS and its structure which is made up of five major compositions. Chapter 2 offers the concepts and definitions of MAF and discusses the structure of CGCS and the five major compositions in a thorough way. Chapter 3 introduces information acquisition system (IAS) and discusses space-time registration and relative navigation and other related concepts. Chapter 4 explains decision and management system (DMS) and illustrates formation principles, combat effectiveness indication, task planning, weapon-target assignment, route planning, cooperative guidance, management of leaving and joining formation, etc. Chapter 5 is about formation generation and guidance methods, focusing on flight control under the framework of loose formation and dense formation. Chapter 6 pays special attention to member flight control system (MFCS), including its modeling and design for the system. Chapter 7 defines support networks system (SNS) of MAF and explains networks support protocols and other related concepts. Finally, Chapter 8, respectively, talks about mathematical simulation analysis for the CGCS, the system simulation and test based on hardware-in-the-loop (HIL), the embedded platform for equivalent flight test verification of the CGCS.

This monograph is well-suited to be read by the engineers working on design of guidance and control system, as well as to be referred by the undergraduate and graduate majoring in guidance, navigation, and control. It should prove useful for both undergraduate and graduate courses in control theory and engineering.

The author wishes to acknowledge many helpful discussions with colleagues from Beihang University and University of Chinese Academy of Sciences, in particular with Mr. Zhongbo Wu and also with Ph.D. Xing Liu, Ph.D. Xiaomin Mu, Ph.D. Chen Peng, Ph.D. Jian Sun, Ph.D. Yang Du, Ph.D. Nanxi Hu, Ph.D. Xiang Jia, Ph.D. Xiaolong Wu, Mr. Da Cai, Yongming Wen, Zheng Yao, Wenlei Liu, and Ms. Hongbo Zhao. Special thanks are due to Ms. Jia Deng and associate professor Miao Zhang for many hours of proofreading. The author also wishes to acknowledge the support of National Defense Industry Press.

Beijing, China   Sentang Wu

Contents

Chapter 1
Introduction

1.1 Application Background and Significance

In 1997, the United States put forward the concept of "network centric warfare", then it expanded to Cyberspace, which is regarded as the most important revolution in the military field and the core of the new global military strategy. "Network centric warfare" refers to applying the powerful computer information network to synthesis all of the distributive various sensors and various types of weapons to a unified and efficient big system, achieving battlefield situation and weapons collaborative interworking operation. In this mode, combat units can be integrated a variety of operational information in real time obtained by the reconnaissance satellite, early warning airplane, UAV/spy plane, surface warship, intelligence system and ground reconnaissance unit troops; Combat personnel can fast, comprehensive, reliable insight into the whole battlefield, co-command of the platform or other platform's weapons, implement of continuous operations at a faster speed command and the damage probability of higher, in order to obtain more comprehensive synthesized combat effectiveness.

Under the framework of "network centric warfare", various types of weapons systems can be divided into land, sea, air and space-based four categories from the platform type which is corresponding to weapon equipments and installations of the land, sea, air and space deployment; The unmanned combat air vehicle, (UCAV) is the most important part of the network centric warfare; At present, the application of the unmanned combat air vehicle (UCAV) in the military mainly includes the following aspects: military surveying and mapping, reconnaissance and detection, electronic warfare, information warfare, precision strikes, battlefield evaluation and cooperation with other platforms and so on.

The missile studied in this book is an important member of the unmanned combat air vehicle (UCAV) family, it is an indispensable important weapon equipment to realization cooperative combat under the framework of "network centric warfare" winged missile (WM) (referred to as cruise missile). This type of

© National Defense Industry Press, Beijing and Springer Nature Singapore Pte Ltd. 2019
S. Wu, *Cooperative Guidance & Control of Missiles Autonomous Formation*,
https://doi.org/10.1007/978-981-13-0953-3_1

winged missile is classified by the characteristics of the aerodynamic layout and the trajectory and the characteristics of the target, that is a type of missile mainly used to attack ground or water targets by producing an aerodynamic lift in the atmosphere. According to the type of launch platform and target, there are two sorts of winged missile. One is a missile that can attack ground targets,such as ground-to-ground missile, submarine-to-ground missile, air-to-ground missile. And the other is a missile that can attack targets of the sea, such as shore-to-ship (submarine) missile, air-to-ship (submarine) missile, ship-to-ship (submarine) missile, submarine-to-submarine missile [1].

This type of winged missile includes US army "Tomahawk" cruise missile, "harpoon" anti-ship missiles category, similar to the LOCAAS and LAM low cost autonomous patrol attack missile, like the Russian X-59 M type air to surface TV guided missile, and hypersonic winged missile equipped with an air breathing scramjet with "liftingbody" or "waverider" aerodynamic layout, even as Israel reconnaissance/attack type UAV "Hobbit" [1].

Due to the rapid development of modern sea air and space integration defense technology, especially the area air defense and a short range point defense forces of the high value of military targets (such as aircraft carrier battle group and strategic command center) to form a multi-layer anti-missile air defense system, so the penetration capability and attack effect of the guided weapon is greatly reduced. In order to break through the enemy's multi anti-missile air defense system, enhance the comprehensive combat-effectiveness of the guided weapon equipment, World military powers raced to do research and development of integrated intelligent guidance capability of winged missile, and used of multiple types of coordination, multi batch and scale operation of the winged missile autonomous cooperative penetration combat mode, which can make full use of low cost guided weapons the advantages of scale, but also can make use of the high technical advantages of new guided weapons, the cooperative electronic countermeasure, echelon joint penetration, scale saturation attack and other means to enhance the ability of the new tactical system deterrence and effective system confrontation capability.

The cooperative guidance control (CGC) of missile autonomous formation (MAF) refers to the principles, methods and technology which ensure the missile members form formation with situational awareness and group cognitive ability through the support network according to the task requirements, and according to the principle of the maximum comprehensive combat effectiveness, can autonomously the formation decision-making and management, and guide, control formation to complete combat task. That is to say, the CGC technology can form the formation through wireless self-organizing network support system by all kinds of missiles launched by various platforms, which with the cognitive ability of situation perception and analysis of battlefield, the members and the whole formation, and according to principle of the maximum missile team comprehensive combat effectiveness, the implementation of decision-making and formation independent management, and control the formation of high quality and complete the task. The CGC technology can better meet the task requirements, to support more efficient implementation of missile formation cooperative combat firepower equipped with echelon coordination, task planning and target assignment through the formation decision-making and management system, cooperative route planning and cooperative guidance, electronic countermeasure and synergistic penetration and saturation attack, significantly to improve the comprehensive efficiency of the whole missile formation.

With the rapid development of computer technology and wireless network technology, the CGC technology of MAF will play a more and more important role, will become the key technology to support the new operation style and means of modern warfare. Collaborative operation mode of MAF, which means the target with a high defense capability and high military value, especially the aircraft carrier battle group and a large range of major military facilities, in the high strength electronic warfare conditions, the implementation of an important means of sudden and intensive high precision strike, is the main attack form of the future combat burst of war in the region. At present, every military power of the world has invested a lot of manpower, material resources and financial resources, to strengthen the overall technology research for cooperative attack system of MAF, and constantly improve the combat effectiveness of cooperative attack weapon system. With the rapid development and improvement of the CGC technology, weapon system with autonomous formation cooperative combat ability will undoubtedly become one of the killer weapons of modern war process.

The CGC-of-MAF (cooperative guidance control of missile autonomous formation) technology has the following significance:

(1) To promote a new model of precision guided weapons which can be implemented sudden, high density, strong confrontation, full saturation of accurate combat to the objective with high defense capability and high military value
(2) It can't only give full play to the advantages of low cost guided weapons and equipment, but also make use of the advantages of new guided weapons, To form collaborative electronic warfare, echelon joint penetration, scale saturation attack means, effectively restrain the enemy of high-tech weapons and equipment advantages, and implement the tactical level of deterrence and effective system against;
(3) Under the framework of "network centric warfare", it can provide a more flexible and more tactical application space for the new type of guided weapons, which can greatly improve the overall combat effectiveness of the guided weapon system.

1.2   Summary

The CGC-of-MAF is a new research field which involves many fields and many subjects. The research work in this field is still at the stage of starting and accelerating. The following will introduce some basic theories, methods and research status and development trend of the technique's application related to this field.

1.2.1   Wireless Mobile Ad Hoc Network [2]

Wireless Mobile Ad Hoc Network, also known as wireless autonomous networks or wireless Ad Hoc networks, IEEE also call it MANET (mobile Ad Hoc network), It is a kind of wireless mobile network with no infrastructure, self-organization, self-adaptation and self-healing ability. In 1972 the United States Department of Defense Advanced Research Projects Agency (DARPA) launched the special projects of a packet radio network (Packet Radio Network, PRNET) for military applications, trying to ensure combat personnel and weapons in the complex battlefield environment to achieve the "communication in moving". 1993 DARPA launched a self-healing adaptive network (Survivable Adaptive Network, SURAN) project, then launched the global mobile information systems (Global Mobile Information Systems, GloMo) project, begun to conduct a comprehensive study on wireless autonomous networks for military requirement meet with the rapid deployment of wireless networks, high survivability and mobility.

From the point of comprehensive analysis, the concept of wireless mobile Ad Hoc network can include wireless sensor networks (Wireless Sensor Network, WSN) and mobile Ad Hoc networks (Mobile Ad Hoc, MANET) and the other fields.

The missile as a node of the network system of autonomous formation, its characteristics is fast moving, strong maneuverability, complex flight environment, high intensity of confrontation and so on, the communication network between missile has wireless self-organizing network (Ad Hoc) in common, but also has its particularity, especially the high requirements of survivability. For wireless Ad Hoc networks of MAF involved in the OSI model from the physical layer to many key technologies of application layer, the media access control (Media Access Control MAC) protocol is an important part of the Ad Hoc network protocol, which controls how the nodes access the wireless channel, packet is transmitted in the wireless channel on the direct control and reception.

If the channel access protocol can effectively use the limited bandwidth of wireless channel, it will play a decisive role in the performance of Ad Hoc network. The following is a brief introduction to the media access control protocol of Ad Hoc network, which has more important relationship with the content.

1.2.1.1   MANET Typical MAC Protocol

MACA (Multiple Access with Collision Avoidance) is a protocol for media access control protocol for single frequency network, which focuses on the solution to the hidden terminal in Ad Hoc network and the exposed terminal problem, but when the network load is relatively large, there is conflict for the MACA protocol in RTS/CTS during the interaction, and do not have the confirmation mechanism of the link layer. When a conflict occurs, the upper layer needs overtime retransmission, Therefore, the efficiency is reduced. MACAW (MACA for Wireless) protocol is an improvement of MACA protocol, it is better to solve the hidden and exposed terminal problems. The main drawback of MACAW is the number of interaction control information in the communication process too much; MACA-BI (MACA By Invitation) protocol is improved on the basis of MACA by the receiver driven MAC protocol. It does not use the RTS/CTS handshake signal, and using only the RTR (Ready To Receive) signal. For the CBR business, MACA-BI works very well, but for Burst traffic, its performance is not as good as the MACA protocol; DBTMA (Dual Busy Tone Multiple Access) dual busy tone multiple access protocol channel is divided into a control channel and a data channel, respectively transmit data information and control information. Compared with MACA and MACAW algorithm, DBTMA has greatly improved the efficiency and can also ensure that no conflict of data packets, but need for additional hardware equipment.

1.2.1.2   WSN Typical MAC Protocol

(1) MAC layer protocol based on Competition

Based on competitive random access MAC protocol uses on-demand channel mode, the typical protocol is CSMA (Carrier Sense Multiple Access, CSMA), which is representative of the improvement in the IEEE802.11 DCF TinyOS protocol, SMAC protocol and TMAC protocol.

(2) MAC protocol based on fixed allocation

The original Fixed allocation of MAC protocol has FDMA, TDMA and CDMA and so on, but in recent years, the following MAC protocol scheme based on fixed allocation was put forward aimed at the wireless sensor network. SMACS (Self-Organizing Medium Access Control for Sensor Networks) protocol is a distributed protocol, it does not require any global or local master node, and can discover neighbor nodes and establish the transmission/reception schedule with them; TRAMA (Traffic-Adaptive Medium Access) protocol uses two techniques to save energy: one is flow transmission schedule to avoid data packet conflict possibly occurred in the receiver, The second is when the node have no receiving the request, it can automatically enter the low energy consumption mode; DE-MAC (Distributed Energy-aware MAC) the core content is to let the nodes exchange information levels. It performs a local election process to choose the node with the lowest energy as the "winner", the "winner" has more sleep time than its neighbors, in order to balance the node energy, prolong the network life cycle; TDMA/FDMA is a hybrid scheme of time division multiplexing and frequency division multiplexing. The node maintains a special frame structure, similar to the TDMA time slot allocation table, node accordingly dispatch its communication with the adjacent nodes.

(3) MAC layer protocol based on power control

BASIC also called PARO (Power-Aware Routing Optimization protocol) protocol, is the MAC protocol for power control of the most simple application in Ad Hoc network, the hidden terminal problem will be more serious; PCM (Power Control

MAC) power control protocol, in order to solve the hidden terminal problem in the BASIC protocol; SMAC-CRPC the agreement is a multi-channel MAC protocol based on SMAC protocol; PCMA protocol is also a multi-channel MAC layer protocol, due to the use of multiple channels, and therefore SMAC-CRPC protocol has the same difficult to achieve in the sensor hardware problems.

1.2.2 Tactical Data Link

Tactical data link is a kind of system using radio communication equipment and data communication procedures, directly provide support and service for combat command and weapon control system with data communication and computer control combining closely. Referred to as TADIL (or TDL, NATO is often referred to as Link). At present, the main tactical data link have No. 4, No. 11, No. 16 and No. 22.

(1) No. 4 data link (Link4/TADIL-C) is a non-confidential data chain, that is used to provide wireless instructions to the fighter. Link4 data link has high reliability, and which is easy to maintain and operate, the link time is short, but it also has some problems, such as the confidentiality is not good enough, the anti-interference ability is not strong and so on.

(2) No. 11 data link (Link11/TADIL-A) is a data link for the exchange of tactical data, which exchange digital information among aircraft, land-based and Naval Tactical Data Systems by the way of using a network structure, polling call and node networking, The disadvantage of the Link11 data link is the use of fixed frequency technology, do not have the ability to resist interference; low transmission rate, system capacity is limited, do not have the function of speech transmission; must set up a network control platform in the network, so once hit will cause the breakdown of the network.

(3) No. 16 data link (Link16/TADIL-J) is a new type of tactical data link, which has the function of anti-interference and security, it adopts time division multiple access structure, scheduled time slot protocol and no center node networking mode, and use Time division multiple access (TDMA) and spread spectrum, frequency hopping, time hopping technology, the throughput of the system is not limited by the terminal throughput, and has a strong survivability.

(4) No. 22 data link (Link 22) is the next generation of data chain system jointly developed by NATO, is a kind of data link with over range of visibility and the anti-jamming ability which is developed in the remolding of the Link11 data chain, using TDMA or dynamic TDMA mode, increase the flexibility of the network. At the same time, its dynamic TDMA protocol can automatically locate capacity to units for large capacity transmission capacity in the network, simplified network and super network management, so it has good prospects for development.

1.2.3   Dynamic Route Planning

In the late 1980s, the United States start to put large manpower and material resources research on dynamic route planning system, The module of automatic route generation (Automatic Routing Module, ARM) developed by Control Arts Inc (SCT) for air-launched cruise missile and the mission planning software based on artificial intelligence developed by Boeing Aerospace Corporation (AI) as the representative have achieved some success, but the two route planning system still have many shortcomings in practice. NASA launched ANOE (Automatic Nap-of-the-Earth) research plan is to develop a dynamic route planning system designed to assist helicopter pilots to implement ground (Nap-of-the-Earth) flight. At the same time, there are many domestic scientific research institutions to carry out the relevant research in the field of route planning, mainly the off-line route planning method of the ground and the real time dynamic programming method of terrain avoidance and terrain following and so on, but it is less practical considerations, and compared with foreign research progress is relatively lagging behind.

The essence of missile dynamic route planning problem is to find an optimal and feasible flight path from the starting point to the target point in certain planning region and the given constraint. In practical applications, the dynamic route planning needs to take into account the missile maneuverability, the probability of collision and the flight time constraints. At the same time as the regional missile route planning is broad, forming a huge search space, the search algorithm is usually to obtain an optimal route convergence often requires long time and great memory space, which will cause great difficulties for the practical application.

In order to shorten the convergence time of route planning and reduce the search space, the usual method was only taken into account in the 2D plane search. these methods can't effectively avoid terrain avoidance and terrain avoidance due to the failure to use the terrain height information. Route planning is a NP problem, the direct solution often leads to the combination explosion, In order to improve the optimization and real-time performance of route planning, many domestic and foreign scholars have put forward a number of targeted algorithms. For example, skeleton or roadmap method, unit cell decomposition method, the artificial potential function, A^* mathematical algorithm, mathematical programming method, neural network method and evolutionary algorithm, etc. [3, 4].

1.2.4   Dynamic Task Planning/Dynamic Target Assignment

Dynamic task planning of missile formation refers to assign one or a set of tasks to each missile in the formation based on certain battlefield situation information and missile formation state according to the principle of maximum comprehensive efficiency of battlefield combat formation in the process of the missile formation to carry out the task. For example, specify the target of attack or a series of space-time cooperative nodes.

The dynamic target assignment of missile formation is that real-time assign one or a series of space-time cooperative nodes to each missile in the formation according to the principle of maximum comprehensive combat formation efficiency based on the state of the battlefield situation information and missile formation when the missile formation enter the area of guidance handover, and then the formation through the clustering formation into a single missile or small formation cooperative terminal guidance. Thus, the dynamic target assignment of missile formation is actually a dynamic task planning in the guidance handover area, is a part of the combat task planning guidance handover area. After nearly 10 years of development, several major solutions has formed in dynamic task planning field: enumeration method, branch and bound method and cutting plane method, task planning method based on modern optimization theory and the method of multi-agent system.

Dynamic target assignment is one of the most important part in the dynamic task planning of missile formation, when to carry out combat task for larger regional distribution target groups as the missile formation into the guidance handover area, real-time reasonable assign one target to each missile to ensure that the maximum damage probability and as far as possible to avoid repeated attacks and omissions on the basis of certain battlefield situation information and missile formation state, according to the principle of maximum comprehensive combat efficiency. Therefore, the modeling problem is the basic problem of dynamic target assignment method design need to be solved. methods commonly used are non-parametric method and parameter method on judging the target advantage is the premise of missile formation dynamic assignment targets. missile formation can implement dynamic assignment of target after obtaining the target advantage of judging by this method. Target assignment problem is a nondeterministic polynomial (NP) problem, which's solution is the Hungarian algorithm, the exhaustive method and the corresponding improvement method, and a large number of research methods based on modern optimization theory, such as expert system, artificial neural network, tabu search algorithm, genetic algorithm, information theory (such as rough set). When it comes to the problem of large scale variable, and the large computational cost is the difficulty for the NP problem, for most problems, the real-time calculation is difficult to guarantee, it is difficult to meet the goal of dynamic assignment of online real-time requirements. The essence for task planning/dynamic target assignment of MAF is a multi-constraint, strong coupled complicated multi-objective integer optimization problem, its obvious feature is the values of decision variables are discrete, If a multi-objective optimization problem has a non-inferior solution, There are often many, and the formation of non-inferior set, generally, the decision maker will choose the best one or a few non-optimal solutions as the final solution, so in general there is no single optimal solution for the multi-objective optimization problem, but is a Pareto optimal set, must be weighed against the optimal solution of Pareto optimal solution set, and obtain the optimal satisfactory solution [4].

In fact, there are a lot of similarities between the multi missile cooperative attack and multi UAV cooperative attack. Initially unmanned aerial vehicles are mainly used for reconnaissance missions, with the development of technology, the UAV gradually has the ability to fight and attack. As far as the missile is concerned, with the improvement of the capability of missile borne equipment detection, data processing and information transmission, the missile can't only accurately hit the target, but also provide information to other platforms to play the role of reconnaissance. But generally speaking, communication and data storage capacity of the UAV is more powerful than missiles, and can complete the task that many missiles unable to complete. So UAV cooperative combat tend to consider the battlefield situation, the enemy threat, its state, safe return, fire control and other kinds of factors, to accomplish more complex tasks than multiple missile cooperative attacks.

However, there are still many differences in the task, operational mode, operational environment and operational effectiveness evaluation of missile and unmanned aerial vehicle (UAV), as a result, there are many problems in the multi missile cooperative attack, which are different from the multi UAV cooperative attack. Multi missile salvo attack is a typical form of combat task, in this cooperative attack task, multiple missiles can be fired from one or more platforms and hit the target at the same time as possible, it can cause the enemy air defense and missile intercept system greater pressure in a short period of time, in order to achieve the high penetration probability, the comprehensive combat effectiveness of high damage ability and high cost-effective of missile attack. Obviously, the flight time of the missile salvo attack multiple constraints is a time-critical task. Although the missile launch platform, flight speed, model and so on are not the same, but through the information sharing, the use of a reasonable cooperative guidance law and distributed coordination algorithm can achieve multiple missile attack time consistency.

Compared with manned fighter aircraft and unmanned aerial vehicles, missile combat environment confrontation is stronger, the flight conditions are worse, the missile speed is faster, the maneuverability is greater, and the guidance time is shorter, This has raised a high demand for real-time and accuracy and on-board equipment volume, weight, power consumption, environmental adaptability and costs for CGC-of-MAF's algorithm of task planning/target assignment, formation decision-making and management, formation flight control, network communication, That's what this book is about the CGC technology of MAF.

1.2.5 Formation Configuration Control

At present, there are many research of the domestic and foreign control method about the robot formation and vehicle dense formation, the method for robot formation configuration control is rich, while the aircraft formation control intensive method is single. Related methods are Leader-follower formation control method and its improved method, and formation configuration control method based on behavior, virtual structure, graph theory, artificial potential field (Model Predictive Control, MPC) and model predictive control methods.

The traditional control method is mainly used to control with a clear model and deterministic environment, but in reality the environment is generally dynamic, and also has uncertainty. Model predictive control (MPC) takes into account the uncertainty of dynamic environment and process, replace the results of a one-time global optimization with a iterative finite optimization, the ideal combination of optimization and feedback can be realized in the process of predictive control and make full use of information, through online rolling optimization and combined with feedback correction of real-time information make the optimization of every moment are based on the actual process. Predictive model, rolling optimization and feedback correction make the predictive control satisfy many practical needs,-moreover, it has achieved great success in practical application. The MPC method has a strong theoretical foundation, but it has a large amount of computation, so it is necessary to do further research in the aspects of real-time computing, scalability and distributed implementation [5–26].

1.2.6 Network Control System

The network control system (NCS) refers to a control system, the part of the return circuit is connected through a network. Usually, the total delay between the signal emanating from transducing devices and the actuator receive signal can be divided into three parts, the delay τ_{sc} from sensor to controller, the controller calculates (including network protocol computing) delay τ_c and the delay τ_{ca} from controller to actuator. Depending on the network protocol and equipment used, the network delay may be fixed, bounded time-varying or random, they reduce the control performance of the system in different degrees, and even cause system instability. Especially when there are a plurality of control loop network, network delay which will make the coupling between loops, so that the network control system analysis and design more complex. Therefore, network delay is one of the key factors of the network control system analysis and design, must be reduced as much as possible Hours delay, reduce its uncertainty to overcome its adverse effects on the control system [27–31].

1. Research on Control

Due to the different network hardware and software, as well as the impact of network load, scheduling policy and node and network failures, network delay has different characteristics. The following will be constant delay, independent random delay and Markov chain random delay these 3 cases are introduced.

(1) The constant delay. It is a relatively simple delay model, which introduces the buffer way to simplify the original random delay into a constant. The constant delay is greater than the maximum delay of communication system, so the system can ignore the random variation of the influence caused by the delay. The mathematical expression is:

$$\begin{cases} \dot{x}(t) = Ax(t) + Bu(t - \tau) \\ \dot{y}(t) = Cy(t) + Du(t - \tau) \end{cases} \tag{1.1}$$

This method can effectively overcome the difficulties caused by the time delay stochastic model to modeling, greatly simplify the difficulty of analysis, but its biggest weakness is the system always work in the maximum delay, poor dynamic performance of the system, the conclusion is very conservative [32–35]. In the case of multiple delays, the [36] study the MIMO control system with multiple delays. In a discrete system model, considering the time delay between the sensor to the controller, the time delay between the controller and the actuator, and the sampling time jitter factors of different devices, therefore, the model can be used to design the network control system and optimize its global performance, both for the control system designer and the network system designer. Mostly using linear Matrix Inequation (LMI) method to study the Multiple delays, the controller is designed using robust or stochastic control methods.

(2) Independent random delay. If the entire network control system feedback loop delay can be reduced to $\tau(t) = \tau_{sc} + \tau_c + \tau_{ca}$, $\tau(t)$ A is a independent random time-varying, the network control system can be described as a continuous time delay system model

$$\begin{cases} \dot{x}(t) = Ax(t) + Bu(t - (\tau)) \\ u(t) = Kx(t - \tau(t)) \end{cases} \tag{1.2}$$

$$\begin{cases} x + y \\ z + s + x \end{cases}$$

Or discrete time delay system model

$$\begin{cases} x(k + 1) = Ax(k) + Bu(k) \\ u(k) = Kx(t - \tau(t)) \end{cases} \tag{1.3}$$

In the literature [37–46], the processing methods of random delay and the design method of the controller are given respectively for the system (1.2) and (1.3).

(3) Markov random delay. In the network control system, the size of the time delay in the network is often not independent of each other, but the network delay at the present time has a significant correlation with the previous or previous network delay. Therefore, many literatures [47–51] have modeled the network time delay as a random sequence of time delay probability distribution obeying Markov chain.

$$\begin{cases} \dot{x}(t) = Ax(t) + Bu(t - \tau(r(t))) \\ \dot{y}(t) = Cy(t) + Du(t - \tau(r(t))) \end{cases} \tag{1.4}$$

In formula, $r(t) \in [L, M, H]$, L indicates that the light delay, M represents a moderate delay, H said heavy delay. In Fig. 1.1, $q_{ij} = Pro\{r_{k+1} = j | r_k + i\}$, $i, j \in [L, M, H]$ represents the probability that the system will jump from one state to another.

2. Research on communication

This book introduces the current research status from two aspects, which are the MAC protocol and network scheduling algorithm

(1) Network structure and MAC protocol

Documents [52] under the assumption that without additional network clock synchronization and delay characteristics of off-line, estimate on line using the network protocol to the network delay, and design the optimal controller. Documents [53] send the Profibus tag to the protocol, and guarantee the quality of control (QoC) based on remote control strategy. Because of the network delay is affected by protocol parameters such as target cycle time and so on, so uses the genetic algorithm to select the target cycle time, thus ensures the QoC of the control information.

(2) Task scheduling algorithm

When the network bandwidth is limited, the system has to schedule the information to ensure the stability and performance of the control system with as little information as possible. Generally, there are two kinds of scheduling methods for the system information: static scheduling mode and dynamic scheduling mode.

Documents [50–52] study the static scheduling policies of TOD (Try-Once-Discard) and Token-Ring-Type for MIMO network control systems. The MTS (Mixed, Traffic, Scheduler) scheduling policy of documents [52–53] has high schedulability, and is easy to execute on CAN. Documents [49–54] uses MET-TOD strategy to process a networked control system with coupled dynamics, and dynamically allocates network bandwidth based on online acquired network transmission errors, the problem of node bandwidth allocation in SISO systems and MIMO systems is investigated respectively in documents [53] and [54]. The [55] proposed a dynamic bandwidth allocation strategy sharing time window, the system meets the real-time and stability at the same time, reduces the requirement for non emergency data buffer capacity of the system, has high utilization rate of cyber source.

Fig. 1.1 Schematic diagram of Markov jump

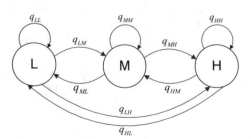

(3) Research on Cooperative Design of control and communication

In fact, a good agreement and the scheduling algorithm can't completely eliminate the network delay and conflicts, thus to improve the performance of the closed-loop system, not only network scheduling or single controller design, but the need for collaborative design from the two aspects. In document [56], the two stability problem of linear networked control systems with network scheduling protocol as TOD is investigated, and bilinear matrix inequalities are given sufficient conditions for the stability of the system in this design pattern and the maximum allowable interval of use. a method for designing exponential stability of networked control systems is studied in documents [57–62]. Firstly, a communication sequence that keeps the observability and reachability of the original system is designed, and then an observer based feedback controller is designed. a continuation of the stabilization problem of networked control systems It takes into account the delay problem, but only the delay is a constant delay. In document [63], a new delay state variable model is established for networked control systems with time-varying delays. The method of on-line estimation of time delay is given and a LQR fuzzy logic controller is designed.

1.2.7 Summary of Technique Application

Ever since the invention of controllable power aircraft, which has been tireless in the pursuit of how to overcome the limitations of space flight time, to further enhance the comprehensive application of vehicle performance, so the technology of formation flight of aerial refueling came into being. Flying from an early manned plane that relies entirely on artificial refueling to human in the loop of the semi-autonomous refueling formation for modern UAVs, spacecraft formation flying has been completely autonomous towards the future, is a typical example of aircraft formation cooperative guidance and control theory and technology by artificial stage to autonomous coordination stage.

Since the beginning of the second half of the last century, in order to effectively confront and attack the United States' powerful aircraft carrier combat groups, the former Soviet Union carried out relevant research on missile formation cooperative guidance and control theory and technology, and gradually equipped different cooperative guidance and control level weapons. Among them, the "carrier killer" called "granite" supersonic anti-ship missiles and "warship Nemesis" called "party - E" system and tactical cruise missile "ЗM-14Э/ТЭ", the most representative.

The Lockheed Martin Corporation "loitering attack missile (LAM)" and "small monitor attack cruise missile (SMACM)" also has a first grade the technical features of cooperative guidance and control level (CGCL) for MAF, it is an integral part of the U.S. Army's future combat system (FCS). And the FCS is a combat system which is lethal and easy to use that can connect ground combat vehicles, air planes and LAM missiles cooperative formation, is a major component of the U.S. military "cooperative combat capability" (CEC) system.

In view of the autonomous formation cooperative attack of medium and long range guided weapons, it has high defense capability and high military value target under complex strong electronic countermeasures, especially for the carrier battle group and wide distribution of major military facilities, with a prominent tactical and tactical deterrent capability. As well as the urgent need of the current and future military struggle, in recent years, the theory and technology of CGC-of-MAF have been paid great attention to in China.

1.3   Summary of the CGC-of-MAF [64]

The cooperative guidance and control system (CGCS) of MAF is a new type of Multidisciplinary and multi-disciplinary integrated integrated system related to wireless Ad Hoc network and missile guidance and control. Its bottom layer is based on wireless communication technology formation support network for information interconnection, interoperability until interoperability, through the upper autonomously complete formation of decision-making and management, make the formation have group consciousness ability, implement task planning and target assignment, cooperative route planning and cooperation navigation, ensure to complete complex formation cooperative combat task. In other words, the CGCS-of-MAF can complete penetration damage and damage assessment tasks through situational awareness, perception, decision-making, task planning and guidance and control based on the formation of supporting network. For cooperative operations of MAF, the battlefield information of the missile acquired autonomously and in real time relying on the formation support network, including missile members and formation status, target information and task environment, penetration damage, and battlefield damage assessment, is an important information input for the decision-making and management system of MAF, which directly affects the dynamic adjustment of missile autonomous formation combat effectiveness index, operational task planning, target dynamic assignment, cooperative route planning and other functions of the operation and dynamic quality, is an important part of analysis and design for the CGCS-of-MAF. Considering the functions and capabilities that the system should have, the CGCS-of-MAF should have seven functions based on the formation support network: situation awareness, group cognition, intelligent decision-making, Task planning, guidance and control, penetration, damage and damage assessment, and its functional structure as shown in Fig. 1.2.

If the CGCS-of-MAF should have seven functions based on the formation support network is focuses on the formation of the topology and network communication, then the whole system can be seen as a mobile Ad Hoc network system (MANET); If the system focus on the node of Sensor and network communication, then the whole system can be seen as a wireless sensor network (WSN); if the system focus on the control performance and network communication, then the whole system can be regarded as a network control system; If the system focuses on

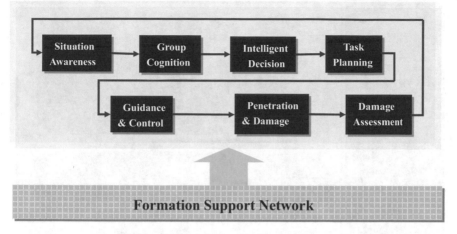

Fig. 1.2 Sketch map for functional structure of the CGCS-of-MAF

Networked autonomous formation overall comprehensive combat effectiveness, then this system is the book said the networked missile autonomous formation system, so that the system is the book of the networked missile autonomous formation system, namely missile autonomous formation system. The missile autonomous formation system emphasizes the consciousness of the whole formation, it focuses on the prominent problems on the formation of the overall behavior and overall behavior. The CGC-of-MAF is a cross fusion of the wireless self-organizing network and missile guidance and control. At present, the research work in this field is still in the stage of starting and accelerating.

To sum up, cooperative architecture of missile autonomous formation system mainly includes the following five parts (see Fig. 1.3):

(1) The system of formation support network, which is a kind of self-organizing network system can provide group cost information for co-ordinated action of members in the formation; except with information transmission and sharing of network communication function of missile formation

(2) The information acquisition system, which is the acquisition system includes the local node information of missile members, the network feature information of the formation and task environment information;

(3) The decision-making and management system, which is the central nervous system which can follow the basic principles of missile formation, can be responsible for weigh individual and group cost for the comprehensive combat effectiveness index, through the task planning/target assignment, route planning/cooperative guidance, mediation of the formation conflict, nodes from team management functions, and optimize the formation configuration and formation according to the requirements of guidance command, and ensure complete the task to follow the rules and laws of self-organization cooperative formation.

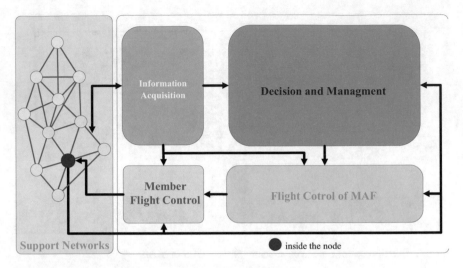

Fig. 1.3 System structure diagram of missile autonomous formation system

(4) The formation flight control system, according to the formation requirements and formation decision-making instruction, can real time optimize and form instruction of the formation control and maintain, ensure the system node implementation of collision avoidance maneuver control and high quality formation;

(5) The member flight control system, which is the guidance and control system for each missile member's flight posture and trajectory.

1.4 Major Contents

This book takes the basic principle of cooperative guidance and control system, design method and technical realization and verification as the main line by the method of combining theory with engineering practice. According to the seven function of the system and the five parts architecture of the system, systematically discusses the basic concept and definition of missiles autonomous formation, cooperative guidance control system the system structure and system integration method, information acquisition system, the decision-making and management system, formation flight control system, member's flight control system, formation support network and system simulation verification and other aspects of the main content.

    The book is divided into eight chapters. The first chapter is the introduction, first introduced the application background and the research status for CGC-of-MAF, then gives the seven function and five part of architecture of the cooperative guidance control system. The second chapter mainly introduces the relevant basic knowledge of the CGC-of-MAF, gives the basic concepts and definitions of MAF, deeply discusses the system structure of cooperative guidance control system and its five major components. The third chapter introduces the information acquisition system of CGCS-of-MAF, discusses the space-time registration of relative guidance and co ntrol, the relative navigation and the optimal allocation of task load, etc. The fourth chapter discusses the decision-making and management system of cooperative guidance control, mainly includes the formation principle and performance indicators, task planning and target assignment, route planning, cooperative guidance and guidance handover methods, etc. The fifth chapter introduces the formation configuration generation and guidance, system modeling and design method of formation flight control system, which focuses on the cooperative flight control problem under the loose formation and dense formation two formation framework. The sixth chapter introduces the modeling and system design method of member flight control system. The seventh chapter mainly introduces the basic concept of MAF support network, related contents and network support protocol, etc. In the eighth chapter introduces the digital simulation analysis of cooperative guidance control system, hardware in the loop (HIL) simulation test and test verification of embedded equivalent system, etc.

Chapter 2
Basis of Cooperative Guidance & Control (CGC)

2.1 Basic Concepts and Definitions

2.1.1 The Concept of Autonomy

The concept of Autonomous Operations (AO) was proposed by the Air Force Research Laboratory (AFRL) in 2000, it defines 10 Autonomous Control Level (ACL) [65, 66] of UAV, as listed in Table 2.1.

ACL1–ACL3 pursuits of the flight control capability of individual member, ACL4 begins to pursuit of operating capability of individual member, which represents the highest performance of individual member, while ACL1–ACL4 only focus on individual member's autonomy. However, ACL5–ACL10 pay more attention to the capability of group operations, from group harmony to group coordination, from tactical level of small-scale air fleet to strategic level of large-scale air fleet, until the groups have capability of full autonomy; ACL7–ACL10 is the US military's main direction of development among them. Figure 2.1 gives the ratings for the existing and future projects in UAV's roadmap, For more information, see Ref [66]. At present, the vast majority of UAV which was granted a number in US military all own a ACL level, for example, X-45 has already reached the level of ACL6 [67–68].

Since the edition of 2007–2032, US Defense Department has renamed "Unmanned Aircraft Systems Roadmap" as "Unmanned Systems Roadmap", which includes unmanned aerial vehicles, unmanned ground vehicles and unmanned ships etc. for further development of the concept of AO [69]. ACL gives the basic outline of autonomy: the variability of the adaption environment changes from certain to uncertain; the degree of human intervention changes from full-involved, half-intervened to totally not involved; the difficulty of the task to completed from a single simple to complex and difficult; the way to complete the task changes from acting alone to group cooperation, and from passive implement to active decision.

© National Defense Industry Press, Beijing and Springer Nature Singapore Pte Ltd. 2019
S. Wu, *Cooperative Guidance & Control of Missiles Autonomous Formation*,
https://doi.org/10.1007/978-981-13-0953-3_2

Table 2.1 ACL definitions of AFRL

ACL	Definition
1	Remotely guided
2	Real time health/diagnosis
3	Adapt to failure and flight conditions
4	Onboard route re-plan
5	Group coordination
6	Group tactical re-plan
7	Group tactical goals
8	Distributed control
9	Group strategic goals
10	Fully autonomous swarm

Fig. 2.1 Autonomous control level in UAV's roadmap

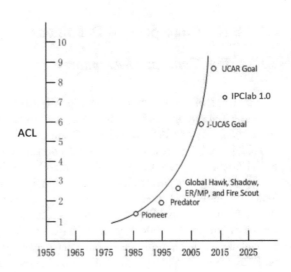

Autonomy Levels for Unmanned Systems (ALFUS) in National Institute of Standards and Technology (NIST) gave a definition of autonomy [70, 71]: "Unmanned systems have capabilities of observation, cognition, analysis, information exchange, planning, decision and management, while completing task assigned by operator through the man-machine interaction. Autonomy can rank the level of tasks according to the complexity of task, the difficulty of condition, the extent of human-computer interaction and etc. in order to complete different tasks."

The Department of High Altitude Long Endurance (DHALE) of Vehicle Systems Program (VSP) in National Aeronautics and Space Administration (NASA) proposed a more explicit definition of autonomy according to self-characteristics of High Altitude Long Endurance UAV. As listed in Table 2.2 [72].

Table 2.2 The NASA VSP DHALE s' definition of ACL

ACL	Name	Description	Characteristic
0	Remote control	Flight under remote control with man in the loop	Remote control aircraft
1	Simple automatic operation	Relying on automatic control auxiliary equipment, to carry out missions by the assistant of automatic control equipment with the operator's monitor	Autopilot
2	Remote operation	Operator pre-programmed to perform tasks	UAV integrated management, flight with preset waypoint
3	Highly automation	It can carry out complex missions automatically, and has some situational awareness and capability of routine decisions making	Automatic takeoff/landing can continue the mission after link interruption
4	Completely autonomy	It has extensive situational awareness (individual and environment), and has the ability and authority to do a comprehensive decision	Re-plan task automatically
5	Co-operation	UAVs can do teamwork together	Harmonious and Cooperative flight

In addition, according to the definition of ACL by AFRL, Indian Defense Ministry divided the ACL into 11 levels, each of which is subdivided into several sub-grade level and has a comprehensive and detailed exposition. Meanwhile, they pointed out the key role of communications and information played in realization of autonomous UAV [73].

IPCLab of Beihang University in China earlier proposed the conception of Cooperative Guidance & Control of UAVs Autonomous Formation [65]: It's about some principles, methods and techniques, which would make sure that the UAV can form certain scale formations through a dedicated information support networks, then enable the formations have situational awareness and cognitive ability, moreover, the formation can make decision and manage itself according to principle of maximum comprehensive operation effectiveness, and at last to guide and control the completion of combat task. Institute of UAV Design in Beihang University proposed a definition of ACL consists of 6 levels, which fit for UAV features. Shenyang Institute of Automation in Chinese Academy of Sciences also carried out relevant research for the methods of evaluation on unmanned systems autonomous. Aircraft Control Technology Integration Laboratory of Beihang University also made 9 levels for UAV technology's development [74].

2.1.2 Basic Principle of Cooperativity

It is a much common existence pattern in the nature that groups behave coopre-ratively. No matter in the top of ecological chain as predators, or in the bottom as prey, cooperative behavior in groups tend to be an efficient and stable way to maximum benefit during the process of surviving and evolution. For example, lions, ant colony, bees, wild geese, whales and other similar social animals all use some hierarchical, well organized, clear division of labor and flexibility tactical cooper-ative actions, in order to gain greater competitive advantage in occupying territo-ries, food capture, threat escape and thriving, etc. Although different types of harmonious groups have their own characteristics, they generally obey the fol-lowing 3 basic principles:

1. Principle of the necessary of group cooperative: Cooperative groups rather than action alone will get greater benefits, cooperation for each member of the group is necessary.
2. Principle of maximum comprehensive benefits: Comprehensive benefits embody the connection of coordinated balance between individual and groups, local and global, the recent and long-term benefit, group cooperative pursuit the maximum comprehensive benefits of the group.
3. Principle of returning patronage to every member of the group: Corporate action should fully take into account the differences between individual members, and patronage should be returned to all members of the group. It is a basic requirement of cooperative groups to be inclusive to all members of the integrity group.

According to definition about unmanned system's autonomy and control level by the NIST, AFRL and national relevant research institutions, the group consisting of unmanned system with higher level of control pays more attention to the capability of group operations. The development direction changes from group harmony to group cooperation, from tactical level of small-scale fleet to strategic level of large-scale, until transforming the groups with fully autonomy, which has become a basic consensus.

Missile, as an important member of the unmanned system's family, whose autonomy is in line with the US NIST s' definition of unmanned systems autonomy; cooperative formations composed of multi-missiles should follow the 3 basic principles for cooperative groups as well. Missiles shares common characteristics with unmanned systems, meanwhile, it also have some unique properties different from unmanned systems. In addition to the 3 basic principles mentioned above, cooperative formation which is composed of multi-missiles also has its unique properties in autonomy and cooperativity. The following will give basic concepts and definitions which are based on common characteristics of unmanned system and unique attributes of missile, thus suitable for analysis, design and evaluation of missiles autonomous formation.

2.1.3 Basic Concepts and Definitions of MAF

According to the basic principle of group cooperative given above, cooperative formations composed of multi-missiles should also obey the following 3 basic principles [64]:

1. Principle of the necessary of cooperative formation: cooperative formations composed of multi-missiles will obtain significantly improved comprehensive operation effectiveness compare with single missile or multi-missiles without cooperation when carry out a task. It's necessary to make the multi-missiles work as cooperative formation while carrying out a task to guarantee the maximum comprehensive combat-effectiveness.
2. Principle of maximum comprehensive benefits: Comprehensive combat-effectiveness embodies optimization balance relationship between missile individual and cooperative formations, the combat-effectiveness and cost. The goal of cooperative formation is to maximize the comprehensive combat-effectiveness of the missiles fleet.
3. Principle of returning patronage to every member of the formation: When the missiles formation carry out the task in a cooperative manner, we must fully take into account the different performance of missile individual, and all missile fleet members should be taken into consideration during the process of information acquisition and sharing, decision making and concerted action. It is a basic requirement of cooperative formation to be inclusive for all members of the integrity formation.

In other words, the 3 principles of cooperative formation is the basic premise for missiles formation to gain a significantly improved comprehensive combat-effectiveness and maximize it through the highest effectiveness of cooperation, highest quality of cooperation and highest satisfaction of cooperation.

Definition 2.1 In accordance with certain rules, at least two missiles carry out the task in the period $[t_0, t_f]$ based on required operation missions, and they form a cooperative formation, which is called Missiles Cooperative Formation, short by Missiles Formation (MF). Some formations are composed of mixed aerial vehicles, which are in accordance with certain rules of cooperative formation and take missiles as the mainstay, also belonging to the category of missiles formation according to the required combat-tasks.

Definition 2.2 The missile members in the missiles formation are also called nodes, represented by ε_i. Using $q_i(t)$ to represent the position of missile member ε_i in the inertial coordinate system, then the vector $q(t) = \text{col}\{q_1(t), q_2(t), \ldots, q_n(t)\}$ represent the formation formed by the missile members in the cooperative formation; $i \in I = \{1, 2, \ldots n(t)\}$ is the number of the missile members in the formation; $n(t)(\geq 2)$ is the scale of the missiles formation, which indicates the number of missiles of the formation in the period $t \in [t_0, t_f]$; $n_0 = n(t_0)$ is the initial scale of missiles formation, $n_f = n(t_f)$ is the final scale of missiles formation.

Definition 2.3 In the period $t \in [t_0, t_f]$, the distance between two missiles is called formation spacing i.e. $d_{ij}(t) = \|q_j - q_i\|$ $i \neq j, i, j \in I = \{1, 2, \ldots n(t)\}$, its mathematical expectation is $\mu_{ij} = E\{d_{ij}(t)\} = E\{\|q_j - q_i\|\}$, set $q_i(t) \sim N(q_{ei}, \sigma_i^2)$. $\|\bullet\|$ is Euclidean norm, $E\{\bullet\}$ is to solve mathematical expectation, $N(\bullet)$ is Gaussian distribution.

Definition 2.4 Missiles Autonomous Formation (MAF) is that missiles formation carry out the task autonomously in the period $[t_0, t_f]$ based on requirement of operation missions. That is, MAF is a missiles formation which has the capability of autonomous decision and autonomous management.

Definition 2.5 Cooperative Guidance & Control (CGC) is about some principles, methods and techniques, which would make sure that the missile members form a certain formation through formation support networks, then enable the formations obtain situational awareness and cognitive ability, moreover, the formation can make decision and manage itself according to principle of maximum comprehensive combat-effectiveness, and at last to guide and control the completion of operation missions. That is, CGC is about theories and technologies to research how to combine all the missile member's properties, to take advantages of the autonomy and cooperativity of missiles formation, then ensure the realization of the maximum comprehensive combat-effectiveness. CGC is one of the core research content in the field of missiles formation. Cooperative guidance & control system (CGCS) should have 7 functions based on formation support networks. The system should have capabilities as situational awareness, group acknowledge, intelligent decision, task planning, guidance and control, penetration and attack, battle damage assessment and so on. Its function structure showed as picture 1–13.

Definition 2.6 Formation support networks is a self-organizing network, which can transfer information with each other in the missiles formation and can sharing communications network as well. More importantly, it should also help the missile members be recognized of its position in the formation as well as relations with other members of the surrounding missile through some kind of protocols. It can also provide missile members with the costing information of the group when they were in the formation of cooperative action.

Formation support networks ensure missile member in the formation make an accurate assessment of the impact to the formation of the entire missiles caused by its own future behavior, which relies on information obtained from the local in the short time currently or information obtained in the past. It is one of characteristics that support networks is different from common communication network. It is also a sign of network that supports missiles formation with autonomous group intelligence.

Definition 2.7 Cooperative guidance & control level for missiles autonomous formation (CGCL), whose definition is listed as Table 2.3. According to the specific properties and the characteristics of tasks in the formation, every level of CGCL can be divided into several sub-levels for more detailed definition [64].

Table 2.3 Definitions of CGCL

CGCL	Main function	Autonomy	Cooperativity	Communication characteristics	Intelligence characteristics
1	Remote control formation flight	Semiautomatic	Man in the loop	Command and control link	Operator guide
2	Fly in default staging area and target zone	Automatic	Preprogrammed	Link among missiles	Acknowledge of individual characteristics
3	Cooperative planning for routine and management of leaving or entering into formation	Semiautonomous	Some of battlefield situational awareness and decision	Support networks	Part ability of decision and acknowledge of group characteristics
4	Dynamic plan tactical task and formation decision and management	Autonomous	Comprehensive battlefield situational awareness and decision	Tactical link and support networks	Total ability of decision and acknowledge of group characteristics
5	Operate task and dynamically planning tactical task	Autonomous and self-government	Comprehensive theater situational awareness and decision	Cyberspace, Tactical link and support networks	Ability of system cooperation and self-government

The capabilities of situational awareness, group acknowledge and intelligent decision are important characteristics of missiles autonomous formation whose level are higher than CGCL3. It's obvious that level CGCL3 is an iconic significance for autonomy and the degree of cooperativity of missiles formation. Missiles formation whose level is over CGCL3 is the main direction of future development.

CGC technology enables the missile members to form a suitable formation autonomously to cooperatively flight in accordance with the requirements of combat task. What's more, the formation can make decisions and be dynamically adjusted according to the task information, battlefield situation and the status of the formation. And it can make real-time analysis for combat-effectiveness, then re-planning the new combat task based on battlefield damage evaluation system. Missiles can be expected to achieve efficient coordinated operations in time, space, missions, launch platforms and some other aspects by CGC technology; it's expected to improve sophisticated electronic warfare capabilities and penetration capability; what's more, it's expected to improve the acquisition and tracking capability against targets and guidance performance. Thus achieve high comprehensive combat-effectiveness of maximum benefit-cost ratio with low-cost missiles.

The missiles formation are flight in a formation of flat and dense in the narrow safe passage, with the shield of terrain and physiognomy, which flight in the way of low altitude penetration. Then they carry out high density raid and saturation attack, which is the main form of operation. In this case, the probability of collision between missile and missile, missile and obstacle increases. In pursuit of the principle of high penetration probability, cooperative guidance and controller is required for a very high performance. There is almost no free space for collision avoidance maneuver in the vertical direction of the entire flat and dense formation, thus the whole formations collision avoidance maneuver can only carry out in the horizontal direction. That is to say, there is no free space for collision avoidance maneuver in the vertical direction because of the infinitely density, the formation cannot carry out collision avoidance maneuver by adjusting the height for flight. It can be seen that, the methods and conclusions of dense formation problem in two-dimensional plane can be easily extended to the problem of formation in three-dimensional space. Therefore, collision avoidance of high dynamic dense formation in the two-dimensional plane is the most complex and infrastructure problems. Based on the above analysis, the following study in the book will mainly discuss the issue of missiles autonomous formation within the two-dimensional plane.

Tightness degree of missiles formation (hereinafter referred to tightness) is a basic conception to describe the features of the distance between missile members of the formation. It has important implications for the design for CGCS of missiles autonomous formation. Density of missiles formation will be analysis in the following.

To describe the tightness of missiles formation, concepts and definitions of safe distance of missiles formation members will be given in the following. Usually, when it's suitable for missiles formation to carry out saturation attack, but the geography and meteorology is complex and changing, the missiles formation need to take advantage of the terrain and physiognomy and flight in the form of low altitude penetration in a narrow safe passage. If the distance between the missiles, missiles and the barriers is too close, the probability of collision will increase, thus threatening the flight safety of the entire missiles formation, and reduce overall combat-effectiveness of the missiles formation.

How to determine the appropriate safe distance between the missiles in the formation during the flight is the basic problem to design the CGCS of the missiles formation. A safe distance of a missile is usually determined by their actual situation, tasks and battlefield environment, and then it will be used to control the distance between missiles of the formation.

Definition 2.8 Safe distance $d_{si}(t)$ of the missile members can be described as that: when the distance $d_{ij}(t)$ between missile members ε_i and missile members ε_j of the formations or other threat obstacle is less than it, i.e. $d_{ij}(t) < d_{si}(t)$, the missile members must take appropriate measures to avoid collision; at that time, when $d_{ij}(t) = d_{si}(t)$, missile member ε_i should in the ready state to prepare to take appropriate measures to avoid collision. The average safe distance of the whole formation is $d_{sf}(t) = \frac{1}{n}\sum_{i=1}^{n} d_{si}(t)$.

There are four main influencing factors of safe distance of missile member ε_i in the formation as follows:

1. Sensor and control error component $d_{Ti}(t)$. Determined by the technical state of sensing and Guidance & Control systems, which reflects the performance about the sensor and Guidance & Control system for a missile member ε_i. If two missiles are closer than sensing and control error component, i.e. $d_{ij}(t) < d_{Ti}(t)$, it is considered that they will come into collision with each other.
2. Maneuvering capability component $d_{Mi}(t)$: it reflects the ability to avoid unexpected collision threat of missile member ε_i, the performance of maneuverability. Usually, when a single missile is in flight, if the spacing between it and an obstacle is less than the sum of sensing and control error component and maneuverability component, i.e. $d_{ij}(t) < d_{Ti}(t) + d_{Mi}(t)$, the missile should take measures to avoid collision.
3. Network induced component $d_{Ni}(t)$: it reflects the performance of the support networks for a missiles formation, which is a new restriction for safe distance of the missile member ε_i owing to the time delay and packet loss when information is transmitted from one to another.
4. Task and environment component $d_{Ei}(t)$: it takes the influences of complexity of the tasks, flight environment, tactical application etc. into account for formation flight security.

Usually, the safe distance of a missile member can be considered as the combined results of sensing and control error component, Maneuverability component, Network-induced component and task and environment component, i.e. $d_{si}(t) = d_{Ti}(t) + d_{Mi}(t) + d_{Ni}(t) + d_{Ei}(t)$. As shown in Fig. 2.2.

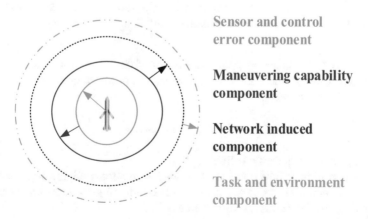

Fig. 2.2 Schematic of missile member's safe distance

Definition 2.9 Margin value of safe distance of missile member $\Delta d_{ij}(t)$ is to subtract the safe distance $d_{si}(t)$ from the actual distance of missiles $d_{si}(t)$, that is $\Delta d_{ij}(t) = d_{ij}(t) - d_{si}(t)$. Its mathematical expectation is $\Delta \mu_{ij}(t) = \mu_{ij}(t) - d_{si}(t)$.

In addition to the relationship of missiles distance and safe distance or margin value of safe distance, we should also study denseness degree of missiles formation (referred to denseness) to describe missiles formation's density. That is to say, the formation denseness is to consider the number of neighbor missiles in the surrounding who are influencing the free space used for the missile's maneuver to avoid collision. Because when the distance $d_{ij}(t)$ is between missile member ε_i and other missiles is less than safe distance $d_{si}(t)$, namely the margin value of safe distance $\Delta d_{ij}(t) < 0$, the missile member need to implement collision avoidance maneuvers, and this evasive maneuvers are often carried out in the surrounding free space. If there are no free space can be used for collision avoidance maneuver around the time, how can the missile member carry out collision avoidance maneuver? Thus, in addition to safe distance, we also need to introduce variable value to measure formation density. The value describe the free space where can be used for collision avoidance maneuver in the surrounding of the missile member, it's a variable value which is called adjacency, it is used to reflect how "dense" on earth the formation is, and then describe the formation density.

Next, we use the relevant conclusions of flocking to give out the concept and definition of MAFs' density.

Set $G = (\varepsilon, v, A)$ denoting an n order of the weighted directed graph. The set of node is $\varepsilon = \{\varepsilon_1, \ldots, \varepsilon_n\}$, the set of edge is $v \subseteq \varepsilon \times \varepsilon$, the weight of adjacency matrix is $A = [a_{ij}] \in R^{n \times n}$, $a_{ij} \geq 0$, $i, j \in I = \{1, 2, \ldots, n\}$ denote the number of node. Every edge of the graph is $e_{ij} = (\varepsilon_i, \varepsilon_j)$, where ε_j is the head of edge, ε_i is the tail of edge, it denotes that ε_i can obtain the information of ε_j. In the $e_{ij} = (\varepsilon_i, \varepsilon_j)$, ε_j is also called as the father node, while ε_i is called as the son node. In the undirected graph, the node of edge has no order, namely $e_{ij} = (\varepsilon_i, \varepsilon_j)$ denotes that ε_j and ε_i can obtain the information each other. The edges in the graph and the adjacency matrix elements in A is one-to-one correspondence, and if $e_{ij} \in v$, then $a_{ij} > 0$, otherwise $a_{ij} = 0$.

Definition 2.10 Adjacency $n_{ai}(t)$ is the amount of the neighbor nodes which are in the ball that refers to the node ε_i as the core, refers to adjacent distance $d_{imax} = k_{imax} \cdot d_{si} > 0$ as the radius:

$$n_{ai} = \{j \in \varepsilon : \mu_{ij} < d_{imax}\} \tag{2.1}$$

where, k_{imax} is the adjacent coefficient of ε_i.

For the two-dimensional problem of formation in horizontal, the adjacency is the amount of the neighbor nodes which are in the circular region that refers to adjacent distance d_{imax} as the radius, as shown in Fig. 2.3.

The average adjacency: $n_{av}(t) = \frac{1}{n(t)} \sum_{i=1}^{n(t)} n_{ai}(t)$

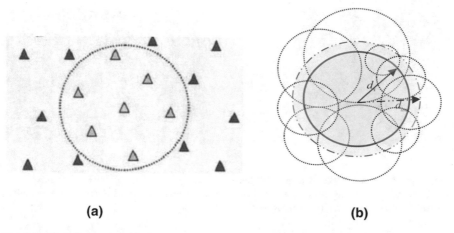

(a) **(b)**

Fig. 2.3 The adjacency $n_{ai}(t)$ and the adjacent distance of ε_i **a** the figure of the adjacency **b** the safe distance and the adjacent distance

By Definition 2.10, the average adjacency $n_{av}(t)$ of the formation is to describe the formation of the concept of average denseness, so the average adjacency $n_{av}(t)$ of the formation is also called as the density of the formation, which is used to measure the overall density of a formation.

By the definition of the adjacency $n_{ai}(t)$ and the average adjacency $n_{av}(t)$, the adjacency $n_{ai}(t)$ is an integer, while the average adjacency $n_{av}(t)$ is not necessarily an integer.

Definition 2.11 Adjacent groups $G(q) = (\varepsilon, v(q))$ is defined by ε and the node set below.

$$v(q) = \left\{ (i,j) \in \varepsilon \times \varepsilon : \mu_{ij} < d_{imax} , \ i \neq j \right\} \tag{2.2}$$

It totally depends on q. If the adjacency distances of the whole nodes are equal, then adjacent groups $G(q)$ becomes an undirected graph.

Some inspiration about value ranges of adjacent coefficient k_{imax} could be drawn from the characteristics of formation flying of migratory birds. After a long period of natural evolution, some kinds of birds, in order to reduce the threat of predators and the impact of energy consumption, usually perform formation flying using easy to observation escape and drag reduction formation. The commonly used formations of birds' formation flying include column and line abreast formation, wedge-shaped formation, diamond-shaped formation, shuttle-shaped formation and polygon formation; In order to reduce the influence of air resistance in formation flight, birds usually adopt highly tight formation flying. Figure 2.4 illustrates the relations between air resistance and wing tips spacing by experimental methods, of which b represents the length of the effective wingspan of birds and s denotes the wing-tip spacing. As can be seen from Fig. 2.4, in the case of $0.8 \leq R = \frac{b}{b+s} \leq 1$,

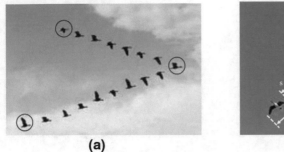 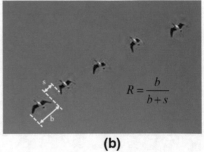

$$R = \frac{b}{b+s}$$

(a)   (b)

Fig. 2.4 Relationship between air resistance and wingtip spacing of birds' formation flying
a Birds formation flying with the spacing of 1/4 wingspan(Birds circled in red work the hardest)
b Relationship between air resistance and wingtip spacing

i.e. $0 \le s \le \frac{1}{4}b$, the drag reduction of birds' formation flying is significant. If the safe distance of birds is designed as $d_{si} = \frac{b}{2}$ and adjacent distance is $d_{imax} = \frac{b}{2} + s$, and then $k_{imax} = 1 + \frac{2s}{b}$ could be got after plugging them into the adjacent distance formula, the value range of adjacent coefficient can be obtained in highly tight formation flying, thereby validating the mechanism that the wing-tip spacing less than its 1/4 wingspan is commonly used in close formation flight during the long-distance migration of birds. According to this principle in birds' formation flying, the specific computing formula about adjacent coefficient k_{imax} of missile member ε_i is presented in the missiles formation with relatively high tightness.

Most commonly used missiles formation can be composed of four basic types, namely column and line abreast formation, wedge-shaped formation, diamond-shaped formation (Figs. 2.5, 2.6, and 2.7) No matter how large the formation scale is, it can be made up of these four basic types in principle.

Some commonly used, classic formations of aerial vehicles, such as shuttle-shaped formation and polygon formation, are composed of these four basic formations, which can be seen in Figs. 2.8 and 2.9.

With further improvement of complexity and requirements of tasks for missiles formation, the scale and density of formation would also increase. Therefore, the collision avoidance problem among individuals has become a basic issue needed to be tackled, which is directly related to the level of autonomy and cooperativity, as well as comprehensive combat effectiveness of the missiles formation. The following part will present the relationship between the probability in which the missile member, adjacent group and formation take collision avoidance maneuvering and tightness as well as density.

Suppose that the safe distance allowance $\Delta d_{ij}(t)$ of a missile member ε_i is an ergodic stochastic process ($\Delta d_{ij}(t)$ is abbreviated to Δd_{ij} in the following parts, similarly, all the variables related to time will also be abbreviated in this way), thus on the basis of the central limit theorem, with the comprehensive effect of random

Column   Line abreast

Fig. 2.5 Column and line abreast formation

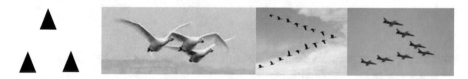

Fig. 2.6 Wedge-shaped formation

Fig. 2.7 Diamond-shaped formation

factors without notable causal relationship, it approximately obeys normal distribution:

$$\Delta d_{ij} \sim N(\Delta \mu_{ij}, \sigma_{ij}^2) \tag{2.3}$$

Fig. 2.8 Shuttle-shaped
formation

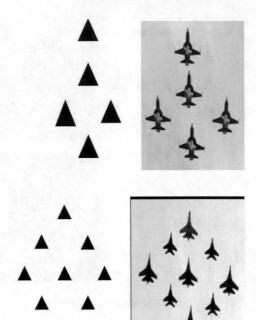

Fig. 2.9 Polygon formation

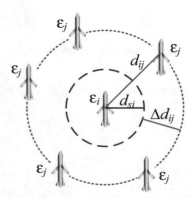

where $\Delta\mu_{ij}$ the expected safe distance allowance is set by the formation control
system and can be described as

$$\Delta\mu_{ij} = \mathrm{E}\{\Delta d_{ij}(t)\} = \mu_{ij} - d_{si} \tag{2.4}$$

where μ_{ij} the mathematic expectation of the distance between missiles is set by the
formation control system, and $\sigma_{ij}^2 = \sigma_i^2 + \sigma_j^2$ is the variance of the normal distribution. If $\Delta d_{ij} \leq 0$, i.e. $d_{ij} \leq d_{si}$, ε_i should prepare to take steps to prevent collision,
as is shown in Fig. 2.10.

Fig. 2.10 The relative
position between ε_i and ε_j

The probability of collision avoidance measures between ε_i and ε_j should be taken can be described as

$$p_m(i,j) = P\{\Delta d_{ij} \leq 0\} = \int_{-\infty}^{-\frac{\Delta\mu_{ij}}{\sigma_{ij}}} \frac{1}{\sqrt{2\pi}} \cdot e^{-\frac{\tau^2}{2}} d\tau \tag{2.5}$$

$$= \Phi\left(-\frac{\Delta\mu_{ij}}{\sigma_{ij}}\right) = \Phi(-\lambda_{ij})$$

where

$$\Phi(x) = \int_{-\infty}^{x} \frac{1}{\sqrt{2\pi}} \cdot e^{-\frac{\tau^2}{2}} d\tau \tag{2.6}$$

is the normal probability integral and

$$\lambda_{ij} = \frac{\Delta\mu_{ij}}{\sigma_{ij}} \tag{2.7}$$

is the safe factor between ε_i and ε_j, which reflects the relationship between $\Delta\mu_{ij}$ and σ_{ij} (Fig. 2.11).

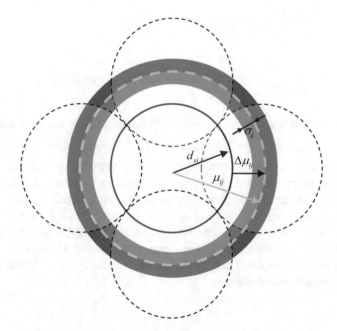

Fig. 2.11 The relationship among μ_{ij}, d_{si}, $\Delta\mu_{ij}$ and σ_{ij}

If $\Delta d_{ij} > 0$, i.e. $d_{ij} > d_{si}$, there is no need for ε_i to take steps to avoid collision, the probability of it is

$$q_m(i,j) = P\{\Delta d_{ij} > 0\} = 1 - p_m(i,j) = 1 - \Phi(-\lambda_{ij}) \tag{2.8}$$

From (2.5) and (2.8), one can see that $p_m(i,j)$ only depends on λ_{ij}. Figure 2.12 shows how the probability varies with the change of λ_{ij}. If $\lambda_{ij} \geq 0$, then $p_m(i,j) \leq 0.5$, and it decreases with the increasing of λ_{ij}. Contrarily, if $\lambda_{ij} < 0$, then $p_m(i,j) > 0.5$, and it increases with the decreasing of λ_{ij}. Furthermore, the smaller $p_m(i,j)$ is (i.e. the bigger λ_{ij} is), the less likely ε_i and ε_j come into collision.

However, the probability formula for collision avoidance maneuvering derived in (2.5) and (2.8) just considers the impact caused by tightness. Since the missile individual ε_i is likely to collide with its neighbors in adjacent groups in any direction, it is more reasonable to take influences on this probability imposed by density into account as well. Based on this, the probability for the member ε_i to perform collision avoidance maneuvering can be expressed as

$$
\begin{aligned}
p_g(i) &= \sum_{j=1}^{n_{ai}} p_m(i,j)[q_m(i,j)]^{n_{ai}-1} + \sum_{j=1}^{C_{n_{ai}}^2} [p_m(i,j)]^2[q_m(i,j)]^{n_{ai}-2} \\
&\quad + \cdots + \sum_{j=1}^{C_{n_{ai}}^k} [p_m(i,j)]^k[q_m(i,j)]^{n_{ai}-k} + \cdots + \prod_{j=1}^{n_{ai}} p_m(i,j) \\
&= \sum_{k=1}^{n_{ai}} \sum_{j=1}^{C_{n_{ai}}^k} [p_m(i,j)]^k[q_m(i,j)]^{n_{ai}-k} \\
&= \sum_{k=1}^{n_{ai}} \sum_{j=1}^{C_{n_{ai}}^k} [\Phi(-\lambda_{ij})]^k[1 - \Phi(-\lambda_{ij})]^{n_{ai}-k}, \quad (n_{ai} \geq 2)
\end{aligned}
\tag{2.9}
$$

where n_{ai} expresses the neighbor number of the individual ε_i, i.e. adjacency; $C_{n_{ai}}^k = \frac{n_{ai}!}{k!(n_{ai}-k)!}$ denotes the number of combinations. In the case of $n_{ai} = 1$, the probability of collision avoidance maneuvering $p_g(i)$ can be calculated using (2.5).

Especially, when the safe factor $\lambda_{ij} = 0$ (i.e. the safe distance of missile members is regarded as the desired distance between them), the probability $p_g(i) = 0.75$, in the case of $n_{ai} = 2$; but in the case of $n_{ai} = 6$, $p_g(i) = 0.984375$. In other words, if safe factor $\lambda_{ij} = 0$, then the individual, which is in a formation with high tightness and density, almost certainly takes maneuvers for collision avoidance.

For a formation composed of n members, the probability for the entire formation to take avoidance maneuvering is

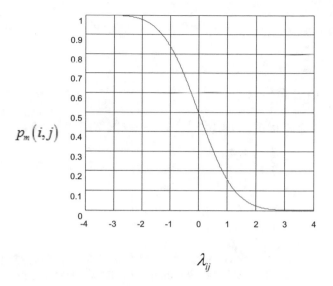

Fig. 2.12 The relationship between $p_m(i,j)$ and λ_{ij}

$$p_f = \sum_{i=1}^{n} p_g(i)[1 - p_N(i)]^{n-1} + \sum_{i=1}^{C_n^2} [p_g(i)]^2 [1 - p_g(i)]^{n-2}$$

$$+ \cdots + \sum_{i=1}^{C_n^k} [p_g(i)]^k [1 - p_g(i)]^{n-k} + \cdots + \prod_{i=1}^{n} p_g(i)$$

$$= \sum_{k=1}^{n} \sum_{i=1}^{C_n^k} [p_g(i)]^k [1 - p_g(i)]^{n-k}$$

$$= \sum_{k=1}^{n} \sum_{i=1}^{C_n^k} [p_g(i)]^k [q_g(i)]^{n-k} \qquad (2.10)$$

where, $q_g(i) = 1 - p_g(i)$ represents the probability in which the individual ε_i need not to take collision avoidance maneuvers.

According to (2.5)–(2.10), in the case that safe factor is given, the probability for the entire formation taking collision avoidance maneuvering p_f is increasing with the growth of adjacency n_{ai} and formation scale n.

Especially, when $p_m(i,j) = \alpha (0 \leq \alpha \leq 1)$ is constant, (2.9) and (2.10) can be transformed

$$p_g(i) = n_{ai}\alpha\beta^{n_{ai}-1} + C_{n_{ai}}^2\alpha^2\beta^{n_{ai}-2} + \cdots + C_{n_{ai}}^k\alpha^k\beta^{n_{ai}-k} + \cdots + \alpha^{n_{ai}}$$

$$= \sum_{k=1}^{n_{ai}} C_{n_{ai}}^k\alpha^k\beta^{n_{ai}-k} \tag{2.11}$$

$$= 1 - \beta^{n_{ai}} = \xi$$

$$P_f = n\xi\eta^{n-1} + C_n^2\xi^2\eta^{n-2} + \cdots + C_n^k\xi^k\eta^{n-k} + \cdots + \xi^n$$

$$= \sum_{k=1}^{n} C_n^k\xi^k\eta^{n-k} \tag{2.12}$$

$$= 1 - \eta^n$$

$$= 1 - \beta^{n \cdot n_{ai}}$$

where $\beta = 1 - \alpha$, $\eta = 1 - \xi$.

If the collision probability of the missile members, adjacent groups and formation needs to be calculated, then the margin of safe distance of missile members, Δd_{ij} follows normal distribution, i.e. $\Delta d_{ij} \sim N(0, \sigma_{ij}^2)$. When the distance between missile members ε_i in the formation is not greater than its sensor and control error component, that is $d_{ij} \leq d_{Ti}$, the individual ε_i would collide with the individual ε_j. As described in Fig. 2.11, if the safe factor in (2.5) is transformed as $\lambda_{ij} = \frac{\Delta\mu_{ij}}{\sigma_{ij}} = \frac{d_{Mi} + d_{Ni}}{\sigma_{ij}}$, then the collision probability of corresponding missile members, adjacent groups and formation could be calculated using (2.5)–(2.12).

The other important variables used to describe the characteristics of the entire formation are the average mean square error of the formation $\sigma_f = \frac{1}{\bar{n}}\sum_i^{n-1}\sum_{j=i+1}\sigma_{ij}$, the average distance mathematical expectation of the formation $\mu_f = \frac{1}{\bar{n}}\sum_i^{n-1}\sum_{j=i+1}\mu_{ij}$, the average safe factor of the formation $\lambda_f = \frac{1}{\bar{n}}\sum_i^{n-1}\sum_{j=i+1}\lambda_{ij}$, where $\bar{n} = \frac{n(n-1)}{2}$. These variables are to be employed in the establishment of comprehensive combat-effectiveness assessment model of the missiles formation.

The following part will take practical examples to illustrate respectively influences on the probability of missile members, adjacent groups and formation taking collision avoidance maneuvers, caused by using distinct safe factor λ_{ij}, adjacency n_{ai} and formation scale n.

Example 2.1 When safe factor is assigned the specific value, as $\lambda_{ij} = 1, 2, 3$ (i.e. 1σ, 2σ, 3σ), the missiles numbers range from 3 to 64, which formed the basic formation of wedge-shaped, diamond-shaped, shuttle-shaped and polygon formations respectively. (Namely adjacency is about 2–7, $n_{ai} = 2 \sim 7$) By using Eqs. (2.11) and (2.12), it's easy to calculate the probability of taking collision avoidance maneuvering while forming into columns; the calculation of probability is divided into single missile member and the whole formation respectively, as shown in Fig. 2.13 ($p_g(i)$) and Fig. 2.14 ($p_f(i)$).

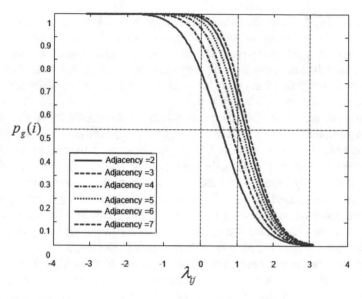

Fig. 2.13 The relation between probability of taking collision avoidance maneuvering ($p_g(i)$) and safe factor (λ_{ij})

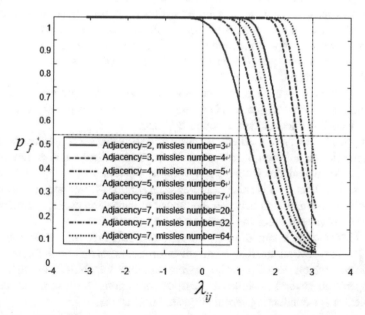

Fig. 2.14 The relation between probability of taking collision avoidance maneuvering ($p_f(i)$) and safe factor (λ_{ij})

In Fig. 2.14, it shows that when safe factor is equal or greater than 3 ($\lambda_{ij} \geq 3$, namely 3σ), at the aspect of whole formation, no matter what kind of basic formation type is used, the probability of taking collision avoidance maneuvering while forming into columns is small; only when the number of missiles is more than 72, the probability of taking collision avoidance maneuvering while forming into columns is greater than 50%. ($p_f > 0.5$).

In consideration of tactical feature of missiles autonomous formation, when the missiles number exceeds 72, the common way is the method of cluster formation based on enemy defense capability, through which the formation of large-scale is sub-divided into several small-scale formation. When safe factor is less than one ($\lambda_{ij} < 1$, namely 1σ), especially when value of adjacency or missile number is great, regardless of what kind of basic formation type is, probability of taking collision avoidance maneuverings (both $p_f(i)$) and $p_g(i)$) is great. Thus it can be seen that setting value to safe factor has great influence on formation type, mobility and combat-effectiveness, etc., for autonomous missiles formation.

Therefore, under the case safe factor is set equal to 3 ($\lambda_{ij} = 3$, namely 3σ), probability value of taking collision avoidance maneuvering can be used as a benchmark, by which size of missiles formation can be divided into the following five types:

1. $2 \leq n \leq 9$ (n is number of missiles) this type is called adjacency-scale formation. This is because the number of whole missiles does not exceed the maximum of member's adjacency, as a result type of formation is mainly restricted to the several basic type mentioned before(wedge-shaped, diamond-shaped, shuttle-shaped and polygon formations); about which we will give a more precise interpretation afterwards.
2. $9 < n \leq 18$ This type is called small-scale formation. The number of missiles in formation is between the number ranges of a salvo commonly used by The Advanced Precision-guided Weapon System.
3. $18 < n \leq 36$ This type is called medium-scale formation. The number of missiles in formation is between the ranges of Basic Ammunition Load commonly used by The Advanced Precision-guided Weapon System.
4. $36 < n \leq 72$ This type is called a larger scale missiles formation. The number of missiles in formation is between the ranges of saturation attack commonly used by The Advanced Precision-guided Weapon System.
5. $n > 72$ This type is called massive formation. This type of formation is usually divided into several small-scale missiles formation loaded by different launcher platform, which based on the enemy defense capability; through tactical co-operation several small-scale missiles formations are assembled as a new formation for conducting repeatedly saturation attack.

Next we use grid structure to describe geometry of the desired spatial structure formation. And there is an assumption for the formation with the ideal spatial structure: in adjacent groups, the distance between each node and other nodes is equal to safe distance (d_{si}) with probability of 100%, namely

$$P\{d_{ij} = d_{si}\} = 1, \quad \forall j \in n_{ai}(q) \tag{2.13}$$

For conveniently describing, this structure is defined as follows:

Definition 2.12 σ-adjacent formation refers to formation q satisfying the restrain of Eq. (2.13).

Definition 2.13 σ-adjacent formation whose safe distance is equal to d ($d_{si} = d$), which is called α-adjacent formation.

In the equation (2.13), d_{si} is also known as mesh density; Adjacent coefficient of node ε_i is $k_{imax} = d_{imax}/d_{si}$, is also called ratio of mesh density.

As can be seen from the Definitions 2.12 and 2.13, both σ-adjacent formation and α-adjacent formation are ideal spatial structure, whose character is that distance between any nodes is equal to safe distance (d_{si}) with probability of 100%. Equation (2.5) shows that, both σ-adjacent formation and α-adjacent formation are high-tightness formation with zero safe factor ($\lambda_{ij} = 0$). The results from Figs. 2.13 and 2.14 confirm that such formation has a high probability of taking collision avoidance maneuvering; which requires formation has the control precision of zero error. So that the cooperative guidance and control of multi-missile system need very high performance, which is a much-rigorous limitation for actual engineering of the missiles formation system. As a result, it's difficult to achieve autonomous missiles formation with structure of σ-adjacent formation and α-adjacent formation in the actual project.

In order to facilitate actual engineering, by relaxing the constraints in Eq. (2.13), we can get a more general quasi-ideal spatial structure as following: safe distance margin (Δd_{ij}) is placed within the closed internal $[-\sigma_{si}, \ \sigma_{si}]$ with probability 100%, that is,

$$P\{-\sigma_{si} \le \Delta d_{ij} \le \sigma_{si}\} = 1 - \zeta, \quad \forall (i,j) \in v(q) \tag{2.14}$$

In the formula: $\sigma_{si} = \rho d_{si}, \ 0 \le \rho, \zeta \ll 1$
Or written as follows:

$$P\{(1 - \rho)d_{si} \le d_{ij} \le (1 + \rho)d_{si}\} = 1 - \zeta, \quad \forall (i,j) \in v(q) \tag{2.15}$$

Definition 2.14 Quasi-σ-adjacent formation refers to formation q satisfied the Eqs. (2.14) or (2.15).

When $\sigma_{si} = \sigma = \rho d$, Eq. (2.15) turns to

$$P\{(1 - \rho)d \le d_{ij} \le (1 + \rho)d\} = P\{-\sigma \le \Delta d_{ij} \le \sigma\} = 1 - \zeta, \quad \forall (i,j) \in v(q)$$

By Definition 2.12–2.14, we can get the following characteristics about σ-adjacent formation, α-adjacent formation and quasi-σ-adjacent formation:

1. The mesh density of α-adjacent formation is uniform, namely $d_{si} = d$, while the mesh density of σ-adjacent formation and quasi-σ-adjacent formation are non-uniform.
2. α-adjacent formation is a particular case of σ-adjacent formation and quasi-σ-adjacent formation, σ-adjacent formation is a particular case of quasi-σ-adjacent formation.
3. All of the three types are formation with zero safe factor (namely $\lambda_{ij} = 0$).

This proves, with the above definition of α-adjacent formation, σ-adjacent formation and quasi-σ-adjacent formation, it is easy to describe accurately the property of several common basic formation: column and line abreast formation, wedge-shaped, diamond-shaped, shuttle-shaped and polygon formations.

Suppose safe distance margin of missile member is $\Delta d_{ij} \sim N\left(\Delta\mu_{ij}, \sigma_{ij}^2\right)$, and set $t = \frac{\Delta d_{ij} - \Delta\mu_{ij}}{\sigma_{ij}}$, therefore

$$
P\{-3\sigma_{ij} \leq \Delta d_{ij} \leq 3\sigma_{ij}\} = P\left\{\frac{-\Delta\mu_{ij} - 3\sigma_{ij}}{\sigma_{ij}} \leq t \leq \frac{-\Delta\mu_{ij} + 3\sigma_{ij}}{\sigma_{ij}}\right\}
$$

$$
= \Phi\left(\frac{-\Delta\mu_{ij} + 3\sigma_{ij}}{\sigma_{ij}}\right) - \Phi\left(\frac{-\Delta\mu_{ij} - 3\sigma_{ij}}{\sigma_{ij}}\right)
$$

$$
= \Phi\left(-\lambda_{ij} + 3\right) - \Phi\left(-\lambda_{ij} - 3\right)
$$

When the safe factor $\lambda_{ij} = 0$, the above formula turns to be

$$
P\{-3\sigma_{ij} \leq \Delta d_{ij} \leq 3\sigma_{ij}\} = \Phi(3) - \Phi(-3) \approx 0.9973, \quad \forall(i,j) \in v(q) \qquad (2.16)
$$

By Definition 2.14, formation q which satisfies the Eq. (2.16) is a quasi-σ-adjacent formation, and $\sigma_{si} = 3\sigma_{ij}$, $\zeta \approx 0.0027$. Equation (2.16) shows that the quasi-σ-adjacent formation is a quasi-ideal spatial structure for actual engineering; The quality of formation depends entirely on the precision σ_i of guidance and control system of every single missile members ε_i.

Next, the relationship between safe factor λ_{ij} and adjacency coefficient k_{imax} (ratio of mesh density) of every missile member is given below.

Equation (2.1) shows that $\mu_{ij} \leq k_{imax} \cdot d_{si}$, considering definition equation of safe factor λ_{ij}, which turns to $\Delta\mu_{ij} = \lambda_{ij} \cdot \sigma_{ij} \leq (k_{imax} - 1)d_{si}$; as $d_{si} > 0$, the adjacency coefficient formula for missile member is

$$
k_{imax} \geq 1 + \frac{\lambda_{ij} \cdot \sigma_{ij}}{d_{si}} \qquad (2.17)
$$

In α-adjacent formation, σ-adjacent formation and quasi-σ-adjacent formation, when safe factor is equal to zero ($\lambda_{ij} = 0$), the adjacency coefficient of missile members satisfy the following constraints:

$$k_{imax} \geq 1 \tag{2.18}$$

And we can know that the minimum of adjacency coefficient is 1.

For α-adjacent formation, $k_{imax} = \kappa$ (ratio of mesh density) can be set a more stringent constraints as:

$$1 < \kappa < \sqrt{m} \tag{2.19}$$

Here, m is the dimension of space.

By the Eq. (2.19), we can give a rigorous interpretation about value interval of adjacency coefficient ($1 < k_{imax} < 1.5$) resulted from bird cluster flying in tight formation, which showed in Fig. 2.4.

Figure 2.15 shows that adjacency of node ε_i in α-adjacent formation is equal to:

$$n_{ai} = n_{av} = n_a = \frac{2\pi d_{imax}}{d_{sj}} = \frac{2\pi d_{max}}{d} = [2\pi\kappa], \quad \forall j \in n_{ai}(q) \tag{2.20}$$

Here, $[\cdot]$ is rounding to an integer.

Substituting Eq. (2.19) into Eq. (2.20), we can get the range of adjacency value to α-adjacent formation in 2-D space (m = 2):

$$6 = [2\pi] \leq n_{ai} = n_a \leq \left[2\pi\sqrt{m}\right] = 8 \tag{2.21}$$

The range of adjacency value to α-adjacent formation in 3-D space (m = 3):

$$6 = [2\pi] \leq n_{ai} = n_a \leq \left[2\pi\sqrt{m}\right] = 10 \tag{2.22}$$

As a result, that is the reason why the type of missiles number ranges from 2 to 9 ($2 \leq n \leq 9$) is called adjacency-scale formation, which aforementioned in formation classification based on missiles number.

Fig. 2.15 Adjacency and adjacent distance of node ε_i

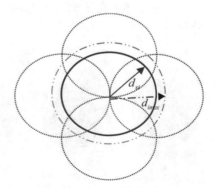

The range of adjacency value to a node ε_i in σ-adjacent formation and quasi-σ-adjacent formation is

$$\frac{2\pi d_{imax}}{d_{smax}} \le n_{ai} \le \frac{2\pi d_{imax}}{d_{smin}}, \quad \forall j \in n_{ai}(q) \tag{2.23}$$

When $d_{smax} = d_{smin}$, it shows $n_{ai} = 6$. Here, d_{smax} and d_{smin} are called maximum safe distance and minimum safe distance of adjacent group respectively.

About range of adjacency value to α-adjacent formation in 2-D space (m = 2), which given by—Eq. (2.21), it can be corroborated by a classics question—"Circle packing in a circle". A circle packing is an arrangement of circles inside a given unit circle such that no two overlap and some (or all) of them are mutually tangent; and solve the maximum of circles' radius, as showed in Fig. 2.16.

All 1–13 and 19 circles have been strictly proved optimal in theory. But the rest cases merely can be solved by heuristic algorithm, which cannot be proved as the only optimal solution. In the figure, r is the ratio of unit circle diameter and inside circle diameter.

The conclusion above applied to missiles formations will turn to be: the maximum number of small circles of a given radius can be placed in a unit circle. The study confirmed that average number (after rounding to integer) of neighbors of any inside circles except the one tangent with boundary of unit circle, is between 6 and 8, which is adjacency range of 2-D space (m = 2) showed in Eq. (2.21).

In order to interpret the difference between σ-adjacent formation and arbitrary formation q, bias function can be defined as below:

$$P_E(q) = \frac{1}{|v(q)|+1} \sum_{i=1}^{n} \sum_{j \in n_{ai}} \psi(d_{ij} - d_{si}), \quad \forall (i,j) \in v(q) \tag{2.24}$$

Here, $\psi(z) = [E\{z\}]^2$ is the bias function, which can be regard as a non-smooth potential energy function, while $E\{\cdot\}$ is mathematical expectation. It is obvious that the potential energy function has global minimum which is equal to zero when q is σ-adjacent formation.

The bias function of quasi-σ-adjacent formation and σ-adjacent formation satisfies:

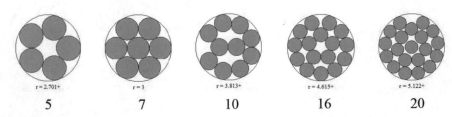

r=2.701+	r=3	r=3.813+	r=4.615+	r=5.122+
5	7	10	16	20

Fig. 2.16 Some cases of "Circle packing in a circle"

$$P_E(q) = \frac{1}{|v(q)|+1} \sum_{i=1}^{n} \sum_{j \in n_{ai}} \psi(d_{ij} - d_{si})$$
$$\leq \frac{|v(q)|}{|v(q)|+1} \sigma_{max}^2 \leq \sigma_{max}^2$$

(2.25)

Here, $\sigma_{max}^2 = \max_{ij} \sigma_{ij}^2$

Thus it can be seen, the mean square $\sigma_i(t), t \in [t_0, t_f]$ of each missile member in the quasi-σ-adjacent formation determines the gap between its formation and expectant σ-adjacent formation; and maximum of this gap is equal to maximum variance σ_{max}^2 of formation. In other words, if the accuracy of guidance and control in each missile member is higher (mean square $\sigma_i(t)$ is smaller), the potential energy function of quasi-σ-adjacent formation is smaller, while the formation is more close to the ideal quasi-σ-adjacent formation structure and the quality of its formation is higher.

Bias function between quasi σ-adjacent formation and σ-adjacent formation meets the following formula:

$$P_E(q) = \frac{1}{|v(q)|+1} \sum_{i=1}^{n} \sum_{j \in n_{ai}} \psi(d_{sj} + \sigma_{ij} - d_{si})$$
$$\leq \frac{|v(q)|}{|v(q)|+1} (\sigma_{\Delta max}^2 + \sigma_{max}^2) \leq (\sigma_{\Delta max}^2 + \sigma_{max}^2)$$

(2.26)

In this formula, $\sigma_{\Delta max}^2 = \max_{i,j} [E\{d_{sj} - d_{si}\}]^2$.

As it is showed in relationship between safe factor and collision avoidance maneuver probability in Figs. 2.13, and 2.14, we can know that value range of safe factor λ_{ij} and adjacency n_{ai} decide the tightness and the density of the missiles formation. The control and guidance accuracy of member ε_i directly effects on the potential energy of the formation—the quality of the formation.

It is showed above that the collision avoidance maneuver probability has an important influence on the dynamic stability of missiles formation. Let us discuss the collision avoidance maneuver probability and the dynamic stability and classify formations by the safe factor λ_{ij} and the adjacency n_{ai}. This "type" above refers to the classification of missiles formations maintaining the coordinated flight within a certain period of time. Missiles autonomous formations could be classified into 3 types according to the tightness, the density and the potential energy of the formation: Loose Formation, Tight Formation and Dense Formation.

Definition 2.15 Loose Formation refers to the formation in which all members in the adjacent group have a safe factor not less than 3 ($\lambda_{ij} \geq 3$, $\forall(i,j) \in v(q)$). That is to say, expectation $\Delta\mu_{ij}$ of safe distance margin in all loose formation missile members is not less than $3\sigma_{ij}$. There is less probability of collision among missile members when they are in a loose formation. Tightness and density are stay in low while potential energy $P_E(q)$ is high. The requirement for the coordinate guidance and controller is low.

Definition 2.16 Tight Formation refers to the formation in which all members in the adjacent group have a safe factor between 1 and 3 ($1 \leq \lambda_{ij} < 3$, $\forall(i,j) \in v(q)$). That is to say, expectation $\Delta\mu_{ij}$ of safe distance margin in all tight formation missile members is between σ_{ij} and $3\sigma_{ij}$. Being different from loose formation, safe factor λ_{ij} is smaller. Therefore, the distance between missiles d_{ij} is further close to the safe distance d_{si}. Formation members tend to adopt collision avoidance maneuver frequently. There is high probability to have collisions among missile members. As to the density of the formation, conclusions below could be got from formula (2.17): missile member ε_i has a middle adjacency level n_{ai}, there is free space for missile members to adopt collision avoidance maneuver and the size of the free space is in a middle level for missile members. There is a higher requirement for the cooperative guidance and controller of the missiles formation than the loose formation.

Definition 2.17 Dense Formation refers to the formation in which all members in the adjacent group have a safe factor between 0 and 1 ($0 \leq \lambda_{ij} < 1$, $\forall(i,j) \in v(q)$). That is to say, expectation $\Delta\mu_{ij}$ of safe distance margin in all loose formation missile members is between 0 and σ_{ij}. Especially, the formation becomes a normal distribution quasi σ-adjacent formation with a mathematical expectation equals to safe distance d_{si} when the safe factor $\lambda_{ij} = 0$. With respect to the tightness level, when safe factor $\lambda_{ij} = 0$, d_{ij}—distance between formation members will get close to d_{si}—the safe distance with high probability. Therefore, there is much probability for missile members in dense formation to adopt collision avoidance maneuver than those in tight formation. In addition, it is showed in formula (2.17) that missile members ε_i in dense formation have an adjacency in a high level. Thus is, the formation has a high density level so that missile members do not have enough free space to adopt collision avoidance maneuver and there is a high probability for collisions among missile members. Therefore, there is a higher requirement for the cooperative guidance and controller of the missiles formation than the tight formation.

The potential energy $P_E(q)$ in the three kinds of missiles formations (loose formation, tight formation and dense formation) is showed in the following formula:

$$P_E(q) = \frac{1}{|v(q)|+1}\sum_{i=1}^{n}\sum_{j\in n_{ai}}\psi(d_{ij}-d_{si}) = \frac{1}{|v(q)|+1}\sum_{i=1}^{n}\sum_{j\in n_{ai}}\left(1+\lambda_{ij}^2\right)\sigma_{ij}^2 \quad (2.27)$$

Formula (2.27) could be used as a performance evaluation formula for the Missiles Formation Cooperative guidance and controller.

Range of potential energy in loose formation:

$$P_{E\_LF}(q) \geq \frac{10}{|v(q)|+1} \sum_{i=1}^{n} \sum_{j \in n_{ai}} \sigma_{ij}^2 \geq \frac{10|v(q)|}{|v(q)|+1} \cdot \sigma_{min}^2 = 10 \cdot \bar{\sigma}_{min}^2$$

$$P_{E\_LF}(q) \geq 10 \cdot \bar{\sigma}_{min}^2 \tag{2.28}$$

Range of potential energy in tight formation:

$$P_{E\_TF}(q) \geq \frac{2}{|v(q)|+1} \sum_{i=1}^{n} \sum_{j \in n_{ai}} \sigma_{ij}^2 \geq \frac{2|v(q)|}{|v(q)|+1} \cdot \sigma_{min}^2 = 2 \cdot \bar{\sigma}_{min}^2$$

$$2 \cdot \bar{\sigma}_{min}^2 \leq P_{E\_TF}(q) < 10 \cdot \bar{\sigma}_{min}^2 \tag{2.29}$$

Range of potential energy in dense formation:

$$P_{E\_DF}(q) \geq \frac{1}{|v(q)|+1} \sum_{i=1}^{n} \sum_{j \in n_{ai}} \sigma_{ij}^2 \geq \frac{|v(q)|}{|v(q)|+1} \cdot \sigma_{min}^2 = \bar{\sigma}_{min}^2$$

$$\bar{\sigma}_{min}^2 \leq P_{E\_DF}(q) < 2 \cdot \bar{\sigma}_{min}^2 \tag{2.30}$$

Minimal potential energy in loose formation is 5 times as big as it in tight formation, 10 times as big as it in dense formation. Minimal potential energy in tight formation is 2 times as big as it in dense formation.

The classification and characteristics of each kind were showed in Table 2.4:

Example 2.2 Assume that a certain kind of missile has the safe distance $d_{si} = 200\,\text{m}$, mean square error $\sigma_{ij} = 10\,\text{m}$, given formation distance $\mu_{ij} = 230\,\text{m}$, where $i,j = 1,2,\ldots,7$. The safe factor $\lambda_{ij} = \frac{\Delta\mu_{ij}}{\sigma_{ij}} = \frac{\mu_{ij}-d_{si}}{\sigma_{ij}} = \frac{230-200}{10} = 3$, according to the missile member safe distance formula $\lambda_{ij} = \frac{\Delta\mu_{ij}}{\sigma_{ij}}$. This is a loose formation according to Definition 2.15. The probability that collision avoidance maneuver was adopted between missile members ε_i and ε_j: $p_b(i,j) = \alpha = \Phi(-3) = 0.00135$.

Table 2.4 Kinds and characteristics of formations

Formation	Safe factor λ_{ij}	Tightness level	Density level	Adjacency	Potential energy $P_E(q)$
Loose formation	$\lambda_{ij} \geq 3$	Low	Low	Low (many)	$[10 \cdot \bar{\sigma}_{min}^2, +\infty)$
Tight formation	$1 \leq \lambda_{ij} < 3$	Middle	Middle	Middle (middle)	$[2 \cdot \bar{\sigma}_{min}^2, 10 \cdot \bar{\sigma}_{min}^2)$
Dense formation	$0 \leq \lambda_{ij} < 1$	High	High	High (few)	$[\bar{\sigma}_{min}^2, 2 \cdot \bar{\sigma}_{min}^2)$

If 7 members ($n = 7$, $n_{av} = 6$) form into a formation whose basic shape is a polygon, according to formula (2.9) and (2.10), the probability that collision avoidance maneuver was adopted among missile members: $p_g(i) = 0.008073$ $<1\%$. The probability that the formation adopt collision avoidance maneuver: $p_f = 0.05516 < 6\%$. Assume that the given formation distance gets more close to the safe distance with other condition invariant, $\mu_{ij} = 210\,\text{m}$ $\lambda_{ij} = 1$. According to the Definition 2.17, the formation belongs to a tight formation, $p_m(i,j) = \alpha = \Phi(-1) = 0.1587$. In this condition, The probability that collision avoidance maneuver was adopted among missile members: $p_g(i) = 0.64543 > 60\%$. The probability that the formation adopt collision avoidance maneuver: $p_f = 0.999295$ $\approx 100\%$. That is to say, the probability that the formation needs to adopt collision avoidance maneuver increases from less than 6% to nearly 100% when the safe factor decreases from 3 to 1 (the safe distance margin decreases from 3σ to σ). Loose formation changes into tight formation in this condition. The formation will almost certainly adopt collision avoidance maneuver because the probability is close to 1. It is showed that the safe factor λ_{ij} is not only the mark parameter to classify the formation class but also the parameter to affect the dynamic stability, the index to measure the tightness level and density level of a formation and the parameter to evaluate the performance of the formation cooperative guidance and controller.

Nevertheless, comparing potential energy in loose formation, tight formation, dense formation (formula (2.28)–(2.30)) and relationship between safe factor λ_{ij}, probability of collision avoidance maneuver p_f in Fig. 2.14, a conclusion could be got: As to the formation flight, there is one thing in common—missile members are likely to adopt collision avoidance maneuver at any moment. The distinction is in the different size of the free space for collision avoidance maneuver. Therefore, tight formation could be combined into the frame of dense formation because the requirements in distance-keep and collision avoidance maneuver strategy of the tight formation are same to the dense formation.

The main task for missiles formation in formation flight region is to follow the TF/TA^2 off-line route planning using the cover of terrain. The formation should execute the on-line route planning to keep the safety and integrality of the formation when emergency occurs. Various kinds of threats and obstacles could occur during flight. Generally, terrain and surface features are treated as non-cooperative real threats which could be divided into two kinds: off-line prior known threats (e.g. mapping data loaded before launching) and on-line sudden threats (threats need real time detection, e.g. high-voltage cable and chimneys). The on-line sudden threats could be fatal threats if evasions are not adopted although they are fixed to the land. Generally, the threatening sensitive region (e.g. radar detection region and air defense scope of positional defense) are treated as fixed non-cooperative probabilistic threats while the effective combat range of enemy mobile combat units are treated as mobile non-cooperative probabilistic threats (mobile units are real threats themselves). Formation should evade from probabilistic threats and adopted defense penetration with the lowest costs. The difference between fixed and mobile non-cooperative probabilistic threats is that formation has to calculate the trace of the mobile non-cooperative probabilistic threats during evasion. The two kinds of threats above could also divided into two kinds-on-line and offline. Formation members will treat neighboring members as mobile cooperative probabilistic threats to avoid collision because there is measuring error, external disturbance, quality of service reduction of maneuver flight and communication when the distance between members is smaller than the safe distance [74].

It is showed that collision avoidance problem among formation members is a kind of cooperative problem in tight formation frame. This is a special point of collision avoidance problem in tight formation frame. According to the characteristics of different types of missiles formations described above, later problems of missiles formation flight will be divided into two frames—loose formation frame and tight formation frame.

As showed in the three formation principles, formation cooperative quality is mainly determined by cooperative efficiency, cooperative quality and cooperative overall benefit level. Cooperative effectiveness is mainly effected by formation scale, information update rate, member response time adjacency level, etc.; Cooperative quality is mainly evaluated by formation scale, information exchange quality (e.g. QoS), member safe factor and adjacency level; Cooperative overall benefit level is mainly determined by cooperative participation level, cost-effectiveness, etc.

The relationship between cooperative guidance and control level (CGCL) of missiles autonomous formation and some factors of formation cooperative quality, formation type of formation cooperative quality is in the Table 2.5. It is showed that the missiles formations in level CGCL1–CGCL2 only fit relative small scale formations with low tightness level and density level. Only missiles formations over level CGCL3 are able to adopt large scale tight formation and dense formation with high level of tightness and density.

2.1.4 Typical Flight Area of Winged Missiles Autonomous Formation

2.1.4.1 Typical Trajectory of Winged Missile

1. Launching mode of winged missile

   (1) Classified by launching base.
       Classified into: land-based launching; seabased launching; onboard launching; submarine launching.

Table 2.5 Relationship between CGCL and formation cooperative quality, formation type

CGCL	Formation type	Autonomy	Cooperativity	Communication characteristics	Intelligence characteristics
1	Loose	Semiautomatic	Precision and time of cooperative response decided by operator (poor and slow), small scale, low level of tightness and density	Command and control link	Operator guide
2		Automatic		Link among missiles	Acknowledge of individual characteristics
3	Loose (tight)	Semiautonomous	Precision and time of cooperative response (nice and fast), middle scale, middle level of tightness and density	Support networks	Part ability of decision and acknowledge of group characteristics
4	Tight Dense	Autonomous	Precision and time of cooperative response (nice and fast), large scale, high level of tightness and density	Tactical link and Support networks	Total ability of decision and acknowledge of group characteristics
5	Dense Large	Autonomous and self-government	Precision and time of cooperative response (nice and fast), middle scale, middle level of tightness and density	Cyberspace, Tactical link and Support networks	Ability of system cooperation and self-government

(2) Classified by stationary base or moving base:

① Launching on stationary base: Launch platform do not move when launching;

② Launching on moving base: Launch platform move when launching. For example, vehicle and land-based launching, sea-based launching, onboard launching and submarine launching belong to launching on moving base.

(3) Classified by the initial launching posture

① Oblique launching: generally divided into rail launching and railless launching (zero-length launching);
② Vertical launching: generally divided into launching powered by engine in the missile and launching powered by engine outside.

2. Typical trajectory of winged missile

Typical trajectory of winged missile was divided into 4 courses: initial course, cruise flight course (midcourse guidance course), guidance handover course and terminal guidance course.

1. Initial course

The initial course is also called launch course. Missile translates from launching state to cruise flight state during this course. In order to make the missile to the expected speed and altitude as soon as possible, besides the main engine for cruise missile, there is also a big thrust booster working in short time installed. The booster will be abandoned after working. Generally this course lasts only 1–5 min. The initial course is same to the climbing course during land-based launching (sea-based launching). The missile accelerates to climb using the power of the booster after launching. The booster separates from the missile and main engine relay to accelerate after expected speed and altitude were reached. At the same time missile get into cruise flight course smoothly subjected to the flight control system. As to the air-based launching, main engine starts to work after the missile glides down without power from launching altitude to a certain altitude (sometimes with power). The missile gets into cruise flight course smoothly after reaching the expected altitude.

2. Cruise flight course

Cruise flight is the main course of cruise missiles. The main work in trajectory design is in this course. Generally, missile flies with a constant high speed and height in one or more altitudes. Sometimes the missile take some maneuvers in orientation and altitude according to the given navigation point. Cruise flight course is also called midcourse guidance course because seeker usually does not work and missile fly in the guidance of project trajectory and relay order rely on the inertial guidance devices, GPS guidance devices assisted by altimeter and terrain contour matching.

3. guidance handover course

Guidance handover course is a key period in the transition from the midcourse guidance course to the terminal guidance course for cruise missiles. In other words, the handover region where missile transmits from the midcourse guidance state to the terminal guidance state is a connection section.

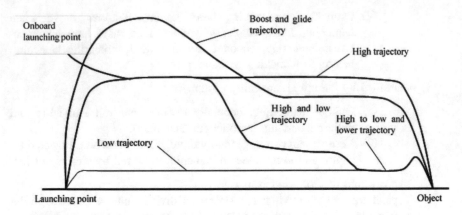

Fig. 2.17 Schematic diagram of cruise missile typical trajectory

4. Terminal guidance course

    After the object caught by the seeker, the missile flies to the object according to the guidance law and then hit the object with the given incidence angle. Typical trajectory of a cruise missile is showed in Fig. 2.17.

2.1.4.2   Typical Flight Area of Autonomous Formation

Typical trajectory of the missiles and the task requirements of the formation are the main basis to define typical flight area of winged missiles autonomous formation. Usually the typical flight area of winged missiles autonomous formation is divided into the following four parts:

1. Formation Staging Area

    This area mainly includes original segment and cruising flight segment. The missiles launched by different platforms from different launching points using different launching modes complete the formation buildup in accordance with the initial binding or according to the requirement for combat task, and prepare for the next step of formation coordinated action.
2. Formation Flight Area

    This area mainly covers the cruising flight segment, and is also named the midcourse guidance of the missiles. According to the comprehensive combat effectiveness maximization principle, missiles formation complete autonomously the decision and management and formation flight for the tasks such as general electronic countermeasures and defense penetration.
3. Formation Handover Guidance Area

    This area mainly includes handing over midcourse to terminal segment, and it's the connection point for the cruising flight segment and the terminal guidance segment, and is also the key stage for the missiles to change over to terminal guidance from midcourse guidance, and in this area, missiles formation complete the decision and management for the tasks such as task planning, target acquisition, dynamic weapon-target assignment and handover guidance, according to the comprehensive combat effectiveness maximization principle.

4. Terminal Guidance Area

   This area mainly includes the terminal guidance segment, and in this area, the missiles acquire target information by its own seeker or the seekers on the adjacent missiles in the same small teams. In order to insure the comprehensive combat effectiveness maximization, the missiles formation will usually be divided into some small-scale formations even one-missile group, when the formation complete the handover guidance. Then the child formations will be guided to their targets by the seekers and terminal guidance law, and at last hit the target in a given incident angle or feedback the battle damage assessment information. The typical flight area of winged missile is as showed in Fig. 2.18.

In order to improve penetration probability, missiles formation usually uses the earth's curvature to implement a super-low-altitude penetration flight, so missiles formation will use a flat formation as much as possible in the two-dimensional plane, and in this way, the center of mass movement of each missile is mainly controlled with the ballistic angle and speed. As described in [2, 4, 75], in the process of formation flight, the whole missile dynamic presents a double time scale characteristics, and the flight control system of MAF can be designed respectively as the double loops structure (that is, the inner loop and outer loop). The inner loop is the traditional member flight control system mainly used for the control of attitude Angle, while the outer loop is based on the inner loop and is mainly used to control the trajectory in order to keep the desired distance between the missiles. Therefore, the missiles formation flight control system design scheme in [1, 75, 76] can be used, so firstly the inner loop flight control system will be designed and the

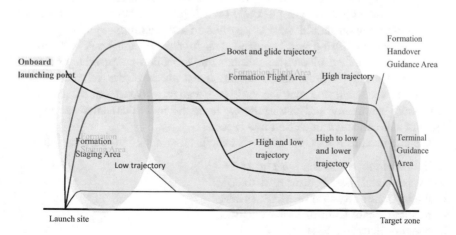

Fig. 2.18 Schematic diagram for the typical flight area of winged missiles autonomous formation

time constant of the ballistic angle and speed will be obtained, and secondly the formation spacing control and maintain system can be designed using one order inertia of speed and ballistic angle. The design method for member flight control system can be seen in [1, 75].

2.2   Cooperative Guidance & Control System (CGCS)

2.2.1   Architecture of CGCS

As can be seen from the classification of autonomy and cooperativity, missiles formations should be on a high level of the capability of autonomy and cooperativity. Individual automation cannot be key point to reflect the autonomy and cooperativity, while the ability of cooperation and coordination among the members shows the great significance, what's more, the NASA put the cooperation and coordination on the highest level. To achieve an advanced degree of autonomy such as cooperation tasks, cooperative flight and cluster autonomy, in the first place, the members must have the ability of group cognizance, which means the ability to evaluate the characteristics of the group, the status of itself in the group, and the influence of individual behavior over the group, only with this community consciousness, the members can realize high-level autonomy, can weigh the individual interests and group interests and can discipline themselves, and their autonomy accords with the autonomy law of intelligent creatures to a certain extent.

This book's research focus on the cooperative guidance & control of missiles autonomous formation, which involves in the whole process of the missiles autonomous formation carrying out the tasks, which include the missiles launching, midcourse guidance, handing over midcourse to terminal and terminal guidance. Under the condition of the large-scale dense formation flight together, the threat of collision among members of the missiles formation is particularly prominent when the dense formation flies synergistically on a large-scale. With the need of penetration, the formation should as far as possible do a low-altitude flight, so promoting altitude to avoid collision will be abandoned. As a result, this book will discusses the theory and method of missiles autonomous formation cooperative guidance & control technology, around the core problems which is the missiles collision avoidance when carrying out the task cooperatively, based on the concept of community autonomy and cooperativity, from the individual interests and local interests and collective interests three aspects.

The architecture for CGCS-of-MAF includes five important parts as follows.

1. Information Acquisition System, IAS
2. Decision and Management System, DMS
3. Flight Control System of MAF, FCSM
4. Member Flight Control System, MFCS
5. Support Networks System, SNS

In the system architecture showed in Fig. 1.3, the members in the formation are named node in the support networks. The nodes acquire information from support networks and the sensor networks, and the information includes the information of local (or all, depending on the scale and ability of the network) nodes, the characteristics information of the network or the task environment. Then the nodes analyze the information by decision and management system, weigh the individual and group costs of the members, and do kinds of tasks such as task planning/ weapon-target assignment, cooperative route planning/formation optimization, generating formation configuration guiding law and guidance law, producing formation guidance trajectory and optimization indexes, and at last, complete formation flight control according to the formation configuration guiding instruction and formation requires, using formation flight control system and member flight control system. A cooperative guidance & control system architecture for missiles autonomous formation is given in Fig. 2.19, which is more detailed than that in Fig. 1.3, and the related system architecture is given in Fig. 2.20.

As shown in Figs. 2.19, and 2.20, the SNS which is the tie between each member and the virtual part attached to every member plays a vital role in the coordination and cooperation of the formation. In a sense, the SNS of MAF reflects the swarm intelligence in a networked way. It can enable every member to assess the influence of its future behavior to the whole formation based only on the present and past information acquired from local interaction in a short time. In formation flight, the maneuver of collision avoidance of local nodes could lead collision in other parts of the formation. The SNS should cognize the characteristic of the formation and assess the influence of chain effect to the formation. To accomplish this assessment, the SNS should enable a single node to cognize its status in the

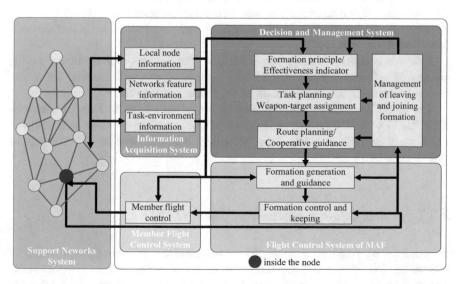

Fig. 2.19 Architecture for CGCS-of-MAF

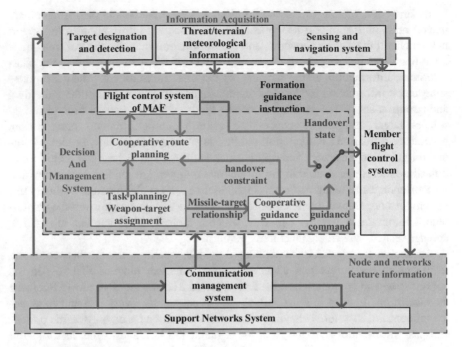

Fig. 2.20 The related architecture for CGCS-of-MAF

networks and its relationship between other nodes and offer collective cost information to every node in autonomous flight formation control. This is one of the characteristics the SNS different from normal communication networks. The main idea of CGC of autonomous formation in this monograph is according to this architecture and introducing local and global cost to accomplish the autonomous formation flight with CGCL.

It is worth noting that the autonomous of the formation also reflects on the abilities of self-organization of the SNS. The monograph will research the autonomous of the SNS and autonomous of the formation respectively.

The seven functions of CGCS-of-MAF are based on the five major compositions. Tied by the SNS, every composition not only performs its own functions but also contact each other closely. Now we will introduce the five major compositions of CGCS-of-MAF in detail.

2.2.2 Information Acquisition System (IAS)

Information Acquisition System (IAS) provides the following three types of information in real time and accurately: the local node information, network feature information and task environmental information. Among them, the local node information is contained in the target information that is obtained by the missile detector members and the neighbor node information that is measured by the relative navigation device, as well as the perception information of their own sensor, including the flight attitude, speed, overload, location, endurance and other motion information of the member of missiles respectively, as well as the seeker, warhead, fuze, electronic countermeasures such as task load information. Network characteristic information is obtained through the fleet support networks system, including fleet size, fleet formation, node status, neighbor nodes, and connectivity, packet loss, delay and update rate of network and other network health information. Task environment information is mainly binding of the fire control system, target designation and damage assessment system before launch, and through the formation of a support networks system to obtain the online dynamic combat task and operations environment information. Mainly includes operations category, tactical plan, target information (type, characteristic, quantity number, location and battle damage state, etc.), threat information (type, characteristic, quantity, location and threat level status, etc.), flight environment (climate, topography, a no-fly zone, etc.) as well as the optimal allocation of task load information.

2.2.3 Decision and Management System (DMS)

The decision and management system (DMS) is the core system which is the reflection of autonomous degree of decision and management level of self-organization. It follows the basic principles to balance the cost between individual and group, mediate the conflict and manage the leaving and joining formation. It carries out the decision and management in the process of task planning, weapon-target assignment, cooperative route planning and cooperative guidance. It optimizes the shape of formation and the instruction of formation guidance. The DMS is the center system of formation to guarantee the rules and laws of self-organization and cooperation in the accomplishment of tasks.

In order to guarantee a reliable and smooth handing over midcourse to terminal guidance, task planning and dynamic weapon-target assignment play important roles in the connection of route planning and cooperative terminal guidance. Task planning and dynamic weapon-target assignment are based on the targets information and task load and damage property of every missile member to plan the tasks of different property or same property of missile members. It can give full play to every member's property and guarantee the propose of maximum the integrative combat effectiveness of MAF. After task planning and dynamic weapon-target assignment, the missile members hand over midcourse to terminal guidance respectively, and the MAF also accomplish the handing over midcourse to terminal guidance.

After that, the MAF will divide into smaller sub-formation or single member to attack their targets cooperatively, reach the propose of maximum the integrative combat effectiveness of MAF.

In the course of MAF cooperative attack, in order to promote the missile penetration probability and the effectiveness-cost ratio, the technology of dynamic weapon-target assignment is adopted to select and sort the missiles of different properties according to the integrated factors of targets properties, defensive weapons properties around targets, available attack direction, angle of sight restrictions, missiles warheads types, terminal guidance modes, guidance accuracy, missile weapon costs, and etc.

Multiple missiles can be launched from single or multiple platforms simultaneously. The missiles adopt two-tier cooperative guidance architecture. The upper cooperative strategy is distributed weighted average consensus algorithm which calculates the expected guidance time of MAF according to the expected guidance time provided by each missile. The lower cooperative strategy executes the guidance and control of missiles according to the expected guidance time of MAF to realize the saturation attack at the same time promoting the penetration probability and damage effectiveness.

In addition, based on the missiles and targets dynamics and kinematics model and navigation, guidance and control property of MAF, the formation midcourse spreading probability areas, targets spreading probability areas and handing over midcourse to terminal guidance areas can be modeled to achieve the MAF targets cooperative capture and cooperative terminal guidance promoting the autonomous capture probability and attack accuracy of targets.

The propose of cooperative route planning is to guarantee every member in MAF reach the specified target areas according to the spatio-temporal distribution by combat task. Based on the principles of maximum the integrative combat effectiveness, the cooperation of assigned spatio-temporal of MAF and the cooperation of the tasks execution is guaranteed.

Normally, the status of each member in MAF is different, e.g. in the common "Lead-Follow" formation pattern, the member can be simply divided into lead missile and follow missile of two types. The lead missile is being dominant in the shape formation, guidance route navigation, targets confirmation and etc. The follow missile is dominated by lead missile to accomplish the formation keeping and reconstruction through its control system. The role of lead and follow could probably change in some circumstance in the autonomous formation process. Usually, the lead missile is charged for route guidance and the follow missile also need the abilities of obstacle avoidance and local real-time planning. The route guided by lead missile should guarantee that MAF is easy to keep and avoid the high acceleration of turning as far as possible. Also the route towards targets area should be short as far as possible and avoid the enemy's radar or the probability exposure to the enemy's radar is lowest to maximum the penetration probability. It is necessary to research the proper TF/TA^2 technology for MAF to solve the trade-off between route planning accuracy and real-time in calculation. Combined the global off-line planning and the local real-time planning, we apply the tree topology of dynamic programming and route section slope/curvature design approach to solve the optimal route planning problem in low altitude penetration of MAF.

2.2.4 Flight Control System of MAF (FCSM)

Flight control system of MAF (FCSM) is that according to the formation of decision and management system to generate the formation optimization indexes and requirements, real-time optimization and generated the instructions of formation control and holding to ensure the realization of nodes collision avoidance maneuvers control and high-quality formation flight system. Missiles can achieve security and stability of autonomous formation flight and complete combat task as required, it is highly dependent on the level of interconnection and intercommunication and interoperability between the missiles to information obtained by information acquisition system, such as navigation and positioning systems, sensors and detectors and ect. Formation of missile members not only need timely acquire local information node information and task characteristics, but also to obtain timely and accurate information on network characteristics, so that the characteristics cognition of group, and realize the formation of autonomous decision and management, and the information exchange and processing operations are mainly by means of missile wing support networks. However, the vast majority of the research on formation flight hypothesis support networks has very good health and accurate wireless communication, there is no packet loss and information delay and so on. However, due to the inherent characteristics of the network itself in the practical system, it tends to affect the stability of the flight control system. Therefore, when designing the formation and the members of the flight control system, it is extremely necessary to analyze the support networks protocol itself.

When formations and members through the support networks of autonomous formation flight control to constitute the network control system, the network delay and packet loss problem of the control system will have a huge impact on the performance of the flight control system. Time delay and packet loss will cause the instability of the controlled system and oscillation, the performance indicators of overshoot, rising time, steady-state error and adjustment time and ect. will become larger and longer, the change of time delay will affect the performance and stability of the control system, it can cause or exacerbate oscillation of the control system.

According to the above problem, this book will combine network control system theory and method of stochastic robustness analysis and design (SRAD) [1, 75, 76] for flight control system, and give full consideration to the stability of the closed-loop system and the requirements of dynamic performance index, which included adjusting time, response time, peak time, delay time and overshoot and so on, to establish a random robust design index of each design point and channel, and apply the modern optimization algorithm to design robust flight controller which could measure system stability and dynamic performance index.

The flight control function of MAF is that following route planning and coordination guidance commands under the dispatch from the fleet management module, in the process of flight control and maintain fleet formation stable flight in accordance with the requirements, the functional structure as shown in Fig. 2.21.

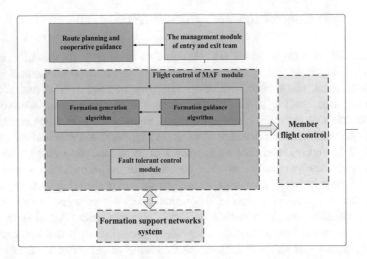

Fig. 2.21 Flight control function of MAF

According to the combat task for the requirement of high dynamic missiles formation, when designing formation configuration, it is based on the different members of the missile motion ability, considering the missile aerodynamic influence among members, task requirements and the connectivity of formation support networks, to determine the parameters of the formation configuration and generate formation guidance instruction, then avoid the behavior of the conflict between the formation and the members of missile when formation maneuver flight, and the collision among the members of missile. Therefore, it need to study the guidance instruction of formation configuration, threat avoidance, autonomous coordination between formation and members of missile, modeling and design methods of formation and members of the flight control system.

In order to avoid the destruction of autonomous formation caused by being intercepted or communications failure, it need to use formation flight tolerance control technology, the scale of formation and formation parameters can be adjusted quickly after the change, and achieved the desired ideal formation, to improve the robustness of autonomous formation system.

Figure 2.21 shows that flight control of MAF module under the support of formation support networks system and under the dispatching of the management module of entry and exit team, the route instructions or guidance commands of route planning and coordination guidance module as input, it's responsible for completing to generate formation, guide formation, tolerant control of formation, and with the route planning form a closed, it can guarantee the missiles formation in a suitable formation flight and penetration, and provide the maximum acquisition probability of shift conditions.

2.2.5 Member Flight Control System (MFCS)

Members of the flight control system(MFCS) is the guidance and control system for flight attitude of every missile member and the trajectory, the system performance is one of the main factors that determine the guidance and control precision of missile member.

Formation support networks is the foundation to achieve synergy guidance and control, and through missiles formation support networks transmit the information of respective sensors and detectors between members. Formation support networks requires dynamic independent networking capability and good robust stability to guarantee the communication quality in the process of the formation flight. At the same time, high dynamic wireless communication inherent in the network latency and packet loss and other issues have brought new challenges to design missiles autonomous formation cooperative guidance and control system, how to guarantee better autonomous formation quality is a problem of coordinated guidance and control system design to be solved in poor high dynamic wireless communication quality.

2.2.6 Support Networks System (SNS)

The support networks system (SNS) of MAF is not only the communication networks which transmit the target information captured by seeker and the information acquired by sensors of missile member itself, but also a reflection of networked autonomous swarm intelligence which is more important in a sense. The SNS should support the cognition of the feature of the group and enable the member in the group to assess the chain effect by its action in the future precisely based on the information acquired from local in a short time at present and past to avoid the collision among other members in the local collision avoidance maneuvering. The SNS enable the member to cognize its status in the networks and the relationship between others through some protocols. The protocols can provide a single node for the group cost in autonomous formation flight control. This is one of features that the SNS different from normal communication networks. Based on the characteristic of networked autonomous formation, we proposed the SNS and networks support protocols which are the extension and supplement for the traditional communication networks and networks communication protocols. The effect of the cooperative combat is depended on the information exchange ability among missile weapon system in a great degree. The SNS should accomplish the large capacity and high update rate of information sharing under highly dynamic circumstances. Also, the SNS need to have the ability of dynamic autonomous networking in order to quickly establish the dynamic network. Because of the dynamic of MAF and the uncertainty of formation size, network structure changes constantly. The protocol should adapt to the change so as to quickly join the network and guarantee insensitive to network structure at the same time. The SNS should alleviate the leaving formation phenomenon because of the network local break off by interference of distance, environment and other factors. In addition, the SNS should support multiple nodes merge according to communication protocols to integrate a larger network with a larger network (entry question) while guarantee the system stable and reliable.

2.2.7   Combat Effectiveness Assessment (CEA) of MAF

According to three basic principles to be followed in missiles formation can obtain the better integrated combat effectiveness include penetration probability, the probability of damage and cost-effectiveness ratio and other indicators, it is one of the necessary conditions for missile to use autonomous formation cooperative combat methods. Therefore, the so-called missiles autonomous formation combat effectiveness assessment is that according to the missiles formation following three basic principles to evaluate quantitatively the integrated combat effectiveness of missiles autonomous formation system.

Currently a lot of methods were used in missile combat effectiveness assessment, such as ADC guidelines, SEA criterion, information entropy evaluation method, etc. However, these methods were not taken into account the interaction among missile members when missiles formation combatting and the missiles formation cooperative guidance and control technology brought about the enhancement of comprehensive combat effectiveness. Therefore, if the above evaluation method applied to missiles autonomous formation combat system will need to do the corresponding modification and extension.

This book will be based on the ADC model which puts forward by American industry Weapon System Effectiveness Advisory Committee (WSEIAC), it take into account the impact of cooperative guidance and control technology for comprehensive combat effectiveness and introduce the missiles autonomous formation cooperative guidance and control level (CGCL) coefficient to establish the Combat effectiveness assessment criteria which is applicable to the missiles autonomous formation cooperative guidance and control system.

In addition, the battle damage assessment system is based on the information acquisition system of formation to assess the battle damage state of the target, in order to determine whether to continue to fight against this target, whether to need to re-plan some members of missile combat task. Therefore, the result of the battle damage assessment is one of the main input information of task planning and target dynamic assignment system.

The cooperative ideology of maneuverable missiles autonomous formation mainly reflects in the high integration of the formation support networks, information acquisition system, formation decision and management system, flight control system of MAF, member flight control system, etc. When overall designing the coordinated guidance and control system, it considers the main function of each subsystems and the mutual coordination relationship from the following aspects.

1. From the different launch platform, different types of hitting targets, the complexity situation of task environment and the movement characteristics of maneuverable missile to determine whether conform to the basic principles of missiles autonomous formation, then analyzed on formation system general requirements in aerodynamic, structure, dynamics and guidance and control for the members of missile, and come up with the specific technical requirements of formation ability for the member of missile: the capabilities of independent networking and communication, target detection and guidance, route planning and correction, the necessary flexibility and speed control, and the necessary information acquisition of formation, etc.

2. Study the autonomous formation decision and management system. It refers to the specific combat task, through targeted information, topographical information, threat information, missiles information and effectiveness assessment information to judge the intelligent threat and estimate situation. According to the change of combat environment and combat task to complete the judgment of formation basic principles and generate the combat effectiveness indicator, to implement the choice of combat task, re-planning and target assignment, timely mediate conflict formation and the management of entry and exit formation, and modify partial track or real time route planning. According to the attack requirement of time-sensitive targets to carry out air patrols and standby attacks, and ultimately achieve the purpose of improving the comprehensive combat effectiveness index.

3. Study the key technology of flight control system of MAF. Including formation generation and guidance technology of flight control system of MAF, formation control and maintain technology, the flight control technology of missile members. In particular, it is important to solve the chain effect and the control of collision avoidance maneuver and other issues in the close formation conditions.

4. Study the support networks technology of applying to missiles autonomous formation. In addition to meet to transmit target information obtained by the various onboard detector and communication requirements of perception information from its own sensor, more importantly, the missile member can through the support networks to make an accurate evaluation, it can only rely on the current information acquired from local in a short period of time or the information obtained from the past to cause the effect of their behavior in the future and the whole formation. Therefore, the safe, reliable and robust support networks system is one of the key aspects of coordinated guidance and control system overall design, wherein the support networks protocol design is particularly important.

5. Through the modeling, analysis, synthesis design for cooperative guidance and control system of missiles autonomous formation to build the corresponding digital simulation analysis system, the hardware in the loop (HIL) simulation test systems and flight test systems, to evaluate the combat effectiveness of MAF.
6. Based on the design concept of standard components, to achieve comprehensive integration of key technologies, to form the technical specifications and standards of the modularization, standardization and generalization of CGCS.

Chapter 3
Information Acquisition System (IAS) of MAF

3.1 Compositions of IAS

According to the architecture of CGCS-of-MAF given in Fig. 2.19, the information acquisition system should ensure that the following three types of information are provided in real time: local node information, characteristic information of SNS and task & battlefield information, as shown in Fig. 3.1.

3.1.1 Node Local Information

The node local information is the information obtained by the detectors and missile-borne sensors.

(1) The target information obtained by the seeker such as the type, characteristic, size, position and war damage state of the targets, and the neighboring node information measured by the relative navigation measuring device.
(2) The motion information of MAF members perceived by the sensors such as altimeter, accelerometer and inertial navigation system, as well as the task load information, including the flight attitude, speed, overload, position, endurance and other movement information of MAF members, and the seeker, warhead, fuze, electronic warfare and other task load information.

3.1.2 Characteristic Information of SNS

Characteristic information of SNS includes the formation scale, node status, neighbor node provided by SNS, as well as network connectivity, packet loss rate, delay and update rate and other network health status information.

© National Defense Industry Press, Beijing and Springer Nature Singapore Pte Ltd. 2019
S. Wu, *Cooperative Guidance & Control of Missiles Autonomous Formation*,
https://doi.org/10.1007/978-981-13-0953-3_3

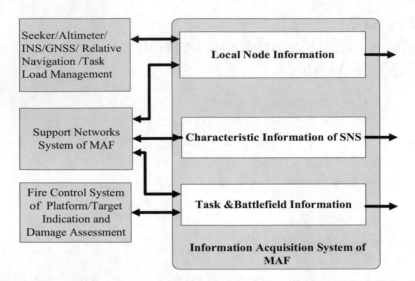

Fig. 3.1 Information acquisition system of MAF

3.1.3   Task and Battlefield Information

Task & battlefield information mainly includes combat task category, tactical program, target information (type, characteristic, quantity, position, war damage state, etc.), threat information (type, characteristic, quantity, position, threat level etc.), flight environment (weather, terrain, no-fly zone, etc.) and its task load optimization configuration and other information. Part of the task & battlefield information is initialized by the fire control system, the target indicator and battle damage assessment system. The other part is given through SNS in real-time.

In this chapter, we focus on time-space registration to ensure the normal operation of IAS and relative navigation system which is the flight basis of MAF. Characteristic information of SNS is introduced through SNS in Chap. 7, including the acquisition ways and methods of the formation scale, formation geometric configuration, node status, neighbor nodes, and network connectivity, packet loss rate, delay and update rate and other network health status. Task & battlefield information is introduced by information requirements of the decision and management of MAF in Chap. 4.

3.2   Time-Space Registration for IAS

Time-space registration is the prerequisite for effective work of IAS. Its task is to solve the multi-sensor and multi-detector data fusion failure of MAF, owing to the time asynchronous and inconsistent coordinate systems problem when measuring the same target. The task of time registration is to synchronize the same target's measurement information of different sensors or detectors that are not synchronized to the same reference time. The task of space registration is the transformation of the coordinate system and its orientation.

3.2.1   Task of Time-Space Registration

The task of time-space registration of MAF is mainly to solve the time synchronization and inconsistent coordinate system when the missile-borne multi-sensors (the inertial navigation system, seeker, etc.) measure the same target, which leads to data fusion failure. The task of time registration is to synchronize the same target's measurement information of different sensors or detectors that are not synchronized to the same reference time. The task of space registration is the transformation of the coordinate system and its orientation. It can be seen that the main part of time-space registration of MAF is the deviation estimation and compensation. The relationship of time-space registration of MAF is shown in Fig. 3.2.

3.2.2   Deviation Estimation and Compensation

The deviation estimation and compensation are realized by the multi-sensor measurements of the common target. The main sources of multi-sensor registration error includes: (1) The deviation of the sensor itself. (2) The azimuth, pitch and distance deviation measured in the reference coordinate system of each sensor. (3) Sensor position error and timing error relative to the common coordinate system. The main purpose of time-space registration is to estimate and compensate for the target state detected by the sensor and the systematic deviation of the sensors.

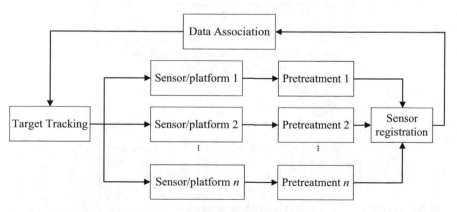

Fig. 3.2 Time-space registration relationship of MAF

3.2.3   Space Registration Methods

Space registration of MAF can be divided into two categories, platform-level and system-level. Platform-level space registration problem is about space registration issues of the same missile, such as the multi-mode composite seeker's space registration problem. The problem of system-level space registration is the issue of sensor space registration for multiple missile members of the entire missile formation. The relationship between platform-level and system-level space registrations is shown in Fig. 3.3.

1. **Platform-level multi-sensor space registration**

The main task of platform-level space registration is to convert sensor measurements into a specified common coordinate system. This registration method can obtain better registration effect under certain conditions. However, this registration method is only applicable when zero-position error, small-deviation and time-invariable assumptions of distances between sensors (for example, the sensor is located in the same platform). So application of this method is limited. The algorithm for platform-level space registration is shown in Fig. 3.4.

2. **System-level multi-sensor space registration**

Select the common coordinate system $S_0 - x_0 y_0 z_0$ for space registration. Assuming that sensor A is located at the origin of the common coordinate system, sensor B is at the point (x_B, y_B, z_B) of the common coordinate system. The positional relationship between sensor A, sensor B and the same target in the common coordinate system $S_0 - x_0 y_0 z_0$ is shown in Fig. 3.5. The related definition of reference coordination can be seen in Ref. [1].

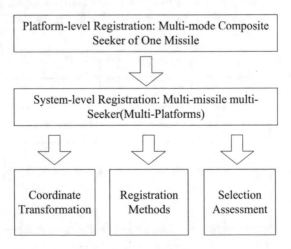

Fig. 3.3 Platform-level and system-level space registration

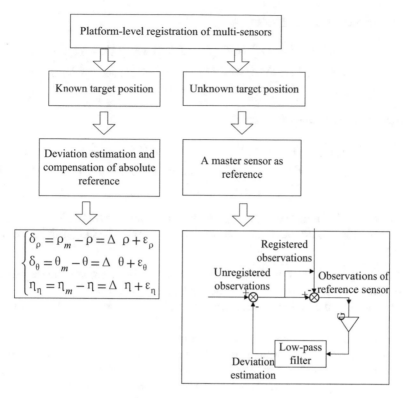

Fig. 3.4 Schematic diagram of platform-level spatial registration

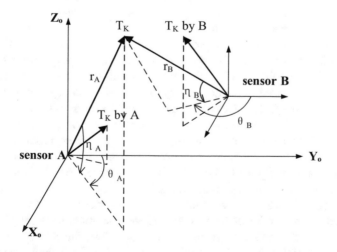

Fig. 3.5 Schematic diagram of the target and sensor in the common coordinate system

Measurement results of sensor A:

$$\left. \begin{array}{l} x'_{Al} = [r_A(k) - \Delta r_A] \sin[\theta_A(k) - \Delta\theta_A] \cos[\eta_A(k) - \Delta\eta_A] \\ y'_{Al} = [r_A(k) - \Delta r_A] \cos[\theta_A(k) - \Delta\theta_A] \cos[\eta_A(k) - \Delta\eta_A] \\ z'_{Al} = [r_A(k) - \Delta r_A] \sin[\eta_A(k) - \Delta\eta_A] \end{array} \right\} \quad (3.1)$$

Measurement results of sensor B:

$$\left. \begin{array}{l} x'_{Bl} = [r_B(k) - \Delta r_B] \sin[\theta_B(k) - \Delta\theta_B] \cos[\eta_B(k) - \Delta\eta_B] + x_B \\ y'_{Bl} = [r_B(k) - \Delta r_B] \cos[\theta_B(k) - \Delta\theta_B] \cos[\eta_B(k) - \Delta\eta_B] + y_B \\ z'_{Bl} = [r_B(k) - \Delta r_B] \sin[\eta_B(k) - \Delta\eta_B] + z_B \end{array} \right\} \quad (3.2)$$

Results of different methods applied to the system-level multi-sensor space registration are also different; analyses of five commonly used algorithms are listed as follows:

(1) Real-time quality control program method is to calculate the registration deviation by averaging the measurement of each sensor.
(2) Least square method is the least squares problem of general or no weight space registrations.
(3) Generalized least squares method is an extension of the least squares method, which determines the weight of each measure from the measured variance.
(4) Three-dimensional Exact Maximum Likelihood (EML) is to use the characteristics of the space distributions between platforms to obtain registration model between the sensors and to add the appropriate measurement noise in the obtained registration model. On the basis of this, the model is simplified and transformed to obtain the standard registration equation. The registration estimation is obtained by obtaining the maximum likelihood solution for the standard registration equation.
(5) Three-dimensional geometric projection method is the commonly used sensor registration method. The advantage of the stereoscopic projection method is to reduce the complexity of registration algorithms. The disadvantages are, (1) errors are introduced into the regional sensor and regional system measurement by the three-dimensional geometric projection method; (2) data distortion due to three-dimensional geometric projection method. For multi-platform multi-sensor systems, this defect is more prominent, because the impact of the Earth's curvature on the three-dimensional geometric projection cannot be ignored when the distance between the platforms increases. The exact maximum likelihood method based on the geocentric coordinate system (ECEF) takes into account the effects of measurement noise. The registration estimation is obtained by finding the maximum likelihood function of the geocentric coordinate system. Figure 3.6 shows the relationship between the five commonly used algorithms for system-level multi-sensor space registration.

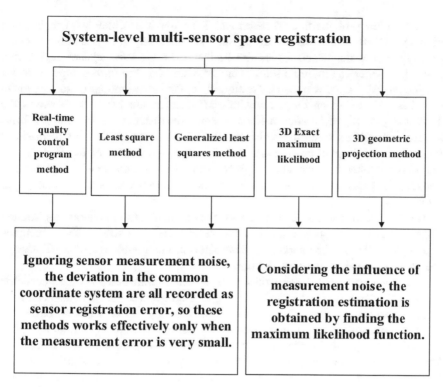

Fig. 3.6 System-level space registration algorithms

3.2.4   Assessment and Selection of Space Registration Methods

The platform-level space registration algorithm is only suitable if the sensor position error does not exist and the small distance assumption between the sensors (for example, the sensor is in the same platform). Under the above conditions, its registration effect is better than other registration methods. However, as the distance between the sensors becomes larger and deviation angles increases, the registration effect deteriorates or even does not work effectively. So this registration algorithm can be applied to single-platform multi-sensor registration only. In the simulation analysis, we found that when the distance between the sensors is more than 100 m, the algorithm simulation program will not converge.

To sum up, the above-mentioned platform-level registration methods are clearly not applicable in the CGC-of-MAF composed of several missiles only equipped with a single-mode seeker. However, if there exist some missile members equipped with multi-mode seeker in the formation, system-level registration can be carried out by the above method, and then system-level registration is achieved to the integration center.

The accuracy of the three-dimensional exact maximum likelihood registration algorithm (EML) for multi-sensor registration processing is high and the application range is wide, which solves the limitation of the least squares registration algorithm. When the multiple sensors are far apart, the algorithm is superior to the traditional least squares registration algorithm. First, the amount of measurement data has little effect on the amount of EML registration bias, which shows the consistency of this registration method. Second, the distance between the platforms usually has little effect on the registration results. This means that this method can be used for multi-platform registration. However, the impact of the Earth's curvature is not taken into account by this registration algorithm, so when the distance between the platforms is large, the registration method becomes inapplicable, even failure.

It can be seen that the three-dimensional exact maximum likelihood registration algorithm can reduce requirement of update rate and data size of SNS. Accuracy and applicability can be guaranteed. This algorithm is suitable when the distance of the missile formation is not very large.

The exact maximum likelihood registration algorithm based on the geocentric coordinate system not only has better registration accuracy for medium and long distance targets. As the distance between the sensor and the target increases, the algorithm can still maintain good angular registration accuracy (azimuth and pitch angle deviation), only the distance deviation from the long distance target increases. But it is still much smaller than the distance from the sensor to the target. In particular, we can transform the measurement of different measurement coordinate system into a unified geodetic coordinate system through this algorithm, to get a unified measurement value for further processing. The shortcomings of the algorithm are that the data processing capacity is larger than the EML, and the consistency is not as good as the former.

It can be seen that the data processing capacity of the exact maximum likelihood registration algorithm based on the geocentric coordinate system is large, considering factors such as the curvature of the earth. Appropriate space registration algorithm is selected considering the performance of the missile-borne computer, the distance of the formation and the performance of SNS.

3.2.5   Multi-sensors Time Registration

Figures 3.7 and 3.8 show the overall framework and multi-sensor time techniques for solving multi-sensor time registration problems, respectively.

There are three kinds of system time synchronization in the multi-sensor timing mode with regard to the system timing mode: (1) In peacetime, the working clock of information source takes the national standard time (e.g., Beijing time); (2) The other is the combat state, the clock in command center is used as the time base; (3) The third is in the comprehensive processing of intelligence, a processing cycle at different times in the measurement of the radar track points unified to the same time. In this

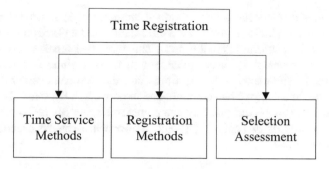

Fig. 3.7 Overall framework of multi-sensor time registration

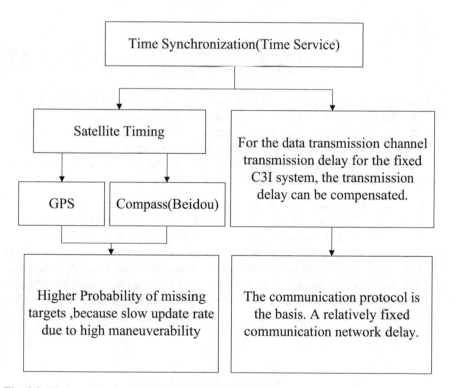

Fig. 3.8 Timing technology of multi-sensors

chapter, the latter two are discussed how to solve the fixed communication delay in the case of two issues:

(1) to ensure the time base of each station sensor are the same, that is system time-checking problem.
(2) the sensor sampling and measurement cycle is inconsistent; the data sampling time is not the same time.

The command center sends information at the moment T_1 in accordance with its own time. The sub-platform received the information at the platform internal time t_1 and replied at the platform internal time t_2. The command center received the reply message at the moment T_2 in accordance with its own time. Assuming that the difference of the sub-platform display time t and the command center display time T is dt and the time delay of two communication network is the same, Δt. Therefore, time setting between the sub-platform and command center can be completed by the calculation results of dt and Δt. The above procedure can be described by the following equations:

$$\begin{cases} T_1 + dt + \Delta T = t_1 \\ t_2 - dt + \Delta T = T_2 \end{cases} \tag{3.3}$$

So,

$$\begin{cases} dt = \frac{1}{2}(t_1 + t_2 - T_1 - \Delta T) \\ \Delta T = \frac{1}{2}(t_1 - t_2 - T_1 + T_2) \end{cases} \tag{3.4}$$

Therefore, the command center compensates network time delay, to unify time between the command center and sub-platform, as shown in Fig. 3.9.

The commonly used time registration methods are the least squares time registration method, the interpolation/extrapolation time registration method and the multi-platform multi-sensor time registration method based on the maximum entropy inference machine, as shown in Fig. 3.10.

(1) Least squares method. Assuming sampling periods of sensor A and sensor B are T1 and T2, and the ratio is n. Sensor B has n measurements between two consecutive target status updates of sensor A. Therefore, the least squares method can be used to fuse the measured value of sensor B into a virtual measurement as the measured value of the sensor B at time k and then converge with the measured value of the sensor A. Then eliminate the synchronization of the target state measurements due to time deviation, thus eliminating the effect of time deviation on multi-sensor data fusion.

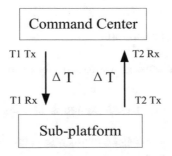

Fig. 3.9 Timing diagram between command center and sub-platform

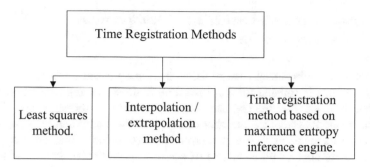

Fig. 3.10   Time registration methods

(2) Interpolation/extrapolation time registration method. This method assumes that time registration is to interpolate and extrapolate the target observation data collected by each sensor at the same time, and achieves time synchronization of two sensors by calculating the data of high precision observation time to the low-precision observation time point.

(3) Multi-platform multi-sensor time registration method based on maximum entropy inference engine. For all probability distributions that satisfy certain conditional constraints, the probability distribution with the largest entropy should be selected, that is, the solution containing most stochastic input satisfying the known constraints. The original high rate signal is estimated by the minimum mean square estimation of the low rate signal by the power spectral density estimation of the input high rate signal through the maximum entropy criterion.

With the improvement of sensor technology and the improvement of computing power, each sensor platform has higher sampling rate and transmission rate. In order to achieve high-quality integration of global sensor information, it is necessary to carry out high-speed communication between platforms. However, data blocking occurs due to the increasingly busy communications of modern multi-platform systems. It requires a reduction in data communication rates and the ability to achieve high-quality global integration of sensor information at lower communication rates. One of the most straightforward ways to reduce the rate of data communication is to reduce the amount of data transmitted by reducing the sampling rate. However, reducing the sampling rate means a reduction in the amount of information, which can be compensated by a number of different low rate sensors. However, in this case the least squares method and the interpolation/extrapolation time registration method have limitations, and the maximum entropy reasoning machine can still be used to solve such problems.

3.2.6   Assessment and Selection of Multi-sensors Time Registration

Multi-platform multi-sensor time registration method based on maximum entropy inference engine is suitable for multi-platform multi-sensor registration at different rates, but it depends on the premise of unified coordinate measurement. This registration method overcomes the shortcomings of the traditional time registration method (least squares method and interpolation extrapolation method), can directly deal with a number of different rate of sensors. In particular, it can use low-rate sensors with different filtering characteristics to handle high-speed varying signals, which cannot be done by other methods.

Nevertheless, there are many unsolved problems with this approach: (1) the influence of different types of sensor transfer functions on the registration effect; and (2) the influence of different sampling rates on the registration effect.

Considering the actual application of MAF, such as high-maneuverability, high Mach-number, fast changing data signals, higher requirements of SNS's update rate and data amount, the time-space registration scheme and performance requirements of multi-platform seeker and other sensors should be designed by analyzing the task requirements of time-space registration and the applicability of various registration algorithms. In time registration, the multi-platform multi-sensor time registration algorithm based on the maximum entropy inference engine is suitable. The time synchronization precision of the medium-sized MAF should be in nanoseconds. The network communication delay compensation can be used as the timing mode. In terms of space registration, three-dimensional exact maximum likelihood (EML) can be used.

3.3   Relative Navigation of MAF

3.3.1   Task of Relative Navigation

Relative navigation means that the MAF member obtains motion information such as flight attitude, velocity and position relative to other missile members through the measuring device and the formation SNS. It requires a sufficiently high accuracy and update rate. To achieve high-quality autonomous formation flight, it is necessary to control and maintain relative distance, azimuth, speed and other movement variables of MAF members accurately in real-time. Therefore, it is necessary to use various measurement devices to measure flight status, relative distance, relative velocity, line of sight and other parameters of the missile members themselves, and obtain relative motion information of high-maneuverability missile members by the nonlinear filtering algorithms and measurements through formation SNS.

Relative navigation system is based on the different requirements of the missile formation, by a variety of navigation sensors/systems in accordance with a certain configuration. Relative distances, relative speeds and line of sight of high-maneuver ability missile members can be obtained through information sharing through SNS and efficient non-linear information fusion algorithms, mainly including the sensor/ system and the appropriate relative navigation algorithm. Measuring sensors/ systems commonly used for relative navigation of moving objects are as follows:

$$
\text{Measurement Sensors/Systems}
\begin{cases}
\text{Inertial Navigation Systems} \\
\text{Global Navigation Satellite System} \\
\text{Vision Navigation}
\begin{cases}
\text{Lidar, Sonar (Active Vision)} \\
\text{CCD Camera (Passive Vision)}
\end{cases} \\
\text{Doppler Radar} \\
\text{Data Link} \\
\text{Radio Navigation}
\begin{cases}
\text{DME} \\
\text{VOR/NDB}
\end{cases}
\end{cases}
$$

Because of the non-linearity between the measurement data of the sensor/system and the fusion state, information fusion system has strong non-linear characteristics. Therefore, it requires an efficient nonlinear information fusion algorithm, such as Gauss-Hermite filtering algorithm [1, 65, 77]. It can overcome the problem of system convergence due to the use of extended Kalman filter (EKF).

3.3.2 Technical Requirements of Relative Navigation

In the aspect of ranging, ranging methods of formation SNS based on time synchronization and TOA measurement can be adopted. Its key technologies include time synchronization, geometric precision factor (GDOP) selection and formation support network performance. In the aspect of angle measurements, angle measurement method based on beacon transmission/reception can be adopted, which needs to consider how to meet the requirements such as accuracy and measurement range and missile-borne adaptability. Through the highly efficient nonlinear Gauss-Hermite filter algorithm to optimize the fusion, from the inertial navigation system/satellite navigation system/beacon/formation SNS data is one of the mature methods considering the navigation accuracy and real-time requirements. For medium-sized missile formation, relative navigation accuracy should be in meters, and time synchronization accuracy should be in nanosecond. For more information, see Refs. [1, 65, 77].

In summary, the relative navigation system of MAF is configured of all kinds of navigation sensors/systems in accordance with different combat task requirements and combat environment constraints. The design of the appropriate relative navigation algorithm is a crucial issue, as well as selection of the measurement sensor and its configuration options. Aiming at a kind of relative navigation algorithm based on global satellite positioning system, the following section briefly introduces the relative navigation algorithm design and analysis for MAF based on a low latency, high data rate differential GPS relative positioning system and carrier phase time difference correlation studies [78–80].

3.3.3 Relative Navigation Method Based on GPS

1. Relative position algorithm based on carrier phase

(1) Single Difference

The single difference is the difference between the two receivers and a common satellite. The satellite clock difference δt_{SV} can be removed while reducing atmospheric disturbance errors (T and I), as shown in Fig. 3.11.

The carrier phase difference between the two receivers A, B and a common satellite k is defined as:

$$\Delta\phi_{AB}^k = \phi_A^k - \phi_B^k \tag{3.5}$$

ϕ_A^k is the phase measurement between the receiver A and the satellite k. It represents the phase value of the carrier propagating on the star-station path at the moment the user receives instantaneous relative satellite firing. The observation equation is as follows:

$$\phi = \lambda^{-1}(r + c(\delta t_r - \delta t_{SV}) + T - I + m_\phi + v_\phi) + N \tag{3.6}$$

$\varnothing$ is the carrier phase measurement (cycles) of the satellite positioning system, λ is the carrier wavelength, $m_\varnothing$ is the multipath error (m) of the carrier phase measurement, $v_\varnothing$ is the noise (m) of the carrier phase measurement. N represents the integer ambiguity (cycles), it is an unknown constant in absence of signal interruption or cycle-slip.

Substituting Eq. (3.6) into Eq. (3.5), we have:

$$\Delta\phi_{AB}^k = \lambda^{-1}\left(\Delta r_{AB}^k + c \cdot \Delta\delta t_{r_{AB}} + \Delta T_{AB}^k - \Delta I_{AB}^k + \Delta m_{\phi_A\phi_B}^k + \Delta v_{\phi_A\phi_B}^k\right) + \Delta N_{AB}^k \tag{3.7}$$

Fig. 3.11 Single difference
(one satellite, two receivers)

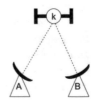

It can be seen that the satellite clock difference δt_{SV} can be removed while reducing the atmospheric disturbance error (T and I); ΔN_{AB}^k is the single-difference integer ambiguity of carrier phase.

(2) Double Difference

Double difference is the difference between the two single difference measurements, so the satellite clock and receiver clock error are eliminated, as shown in Fig. 3.12.

The double difference is defined as $\nabla\Delta\phi_{AB}^{jk} = \Delta\phi_{AB}^j - \Delta\phi_{AB}^k$. Substituting Eq. (3.7), we have:

$$\nabla\Delta\phi_{AB}^{jk} = \lambda^{-1}\left(\Delta\nabla r_{AB}^{jk} + \Delta\nabla T_{AB}^{jk} - \Delta\nabla I_{AB}^{jk} + \Delta\nabla m_{AB}^{jk} + \Delta\nabla v_{AB}^{jk}\right) + \Delta\nabla N_{AB}^{jk} \quad (3.8)$$

It eliminates the receiver clock error δt_r, $\Delta\nabla N_{AB}^{jk}$ is the carrier phase double differential integer ambiguity.

(3) Modeling

Relative position vector in ECEF, floating point values of the carrier phase double differential integer ambiguities are selected as state vectors. Relative speed and relative acceleration can also be selected as the state vector according to actual situations.

Here, only the simple position pattern is discussed, and the state vector is defined as follows:

$$x_P = \begin{bmatrix} X & Y & Z & \Delta\nabla N^{1,2} & \dots & \Delta\nabla N^{1,n} \end{bmatrix} \quad (3.9)$$

$x_1 = X = X_A - X_B, x_2 = Y = Y_A - Y_B, x_3 = Z = Z_A - Z_B$. X, Y, and Z denote the relative positions (m) of the ECEF coordinate system in the three-axis directions, A and B represent the following missile and the leading missile, respectively. Satellite 1 is selected as the base satellite, $\Delta\nabla N$ represents the carrier phase difference between the reference satellite and other non-reference satellites.

The measurement equation should have the following form, establishing the relationship between the measurement vector and the state vector.

$$z = Hx + v \quad (3.10)$$

Fig. 3.12 Double difference (two satellites, two receivers)

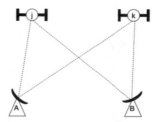

If four satellites are selected and satellite 1 is the base satellite, z can be defined as:

$$z = \begin{bmatrix} \Delta\nabla\rho_{A,B}^{1,2} & \Delta\nabla\rho_{A,B}^{1,3} & \Delta\nabla\rho_{A,B}^{1,4} & \Delta\nabla\phi_{A,B}^{1,2} & \Delta\nabla\phi_{A,B}^{1,3} & \Delta\nabla\phi_{A,B}^{1,4} \end{bmatrix}^T$$

To facilitate the derivation of the observation array H, the relationship diagram of the missile formation and satellite is shown in Fig. 3.13.

Since the distance between the receiver and the satellite is much larger than that of the short baseline missile formation, it is reasonable to assume that all formation members have the same line of sight vector for arbitrary given satellites. e^j represents the unit of sight vector for the j-th satellite. According to the geometric relationship, we have:

$$r_A^j - r_B^j = \Delta\mathbf{x} \cdot \mathbf{e}^j \tag{3.11}$$

This associates the distance between the j-th satellite and each receiver with the state vector and the unit line of sight vector.

The double difference pseudorange measurements are defined as follows:

$$\Delta\nabla\rho_{A,B}^{jk} = \Delta\rho_{A,B}^j - \Delta\rho_{A,B}^k \tag{3.12}$$

Pseudorange equation:

$$\rho = r + c(\delta t_r - \delta t_{SV}) + T + I + m_\rho + v_\rho \tag{3.13}$$

Using Eq. (3.13) to expand Eq. (3.12), and ignoring the secondary item:

$$\Delta\nabla\rho_{A,B}^{jk} = (r_A^j - r_B^j) - (r_A^k - r_B^k) \tag{3.14}$$

Fig. 3.13 Position relationship of formation and satellite

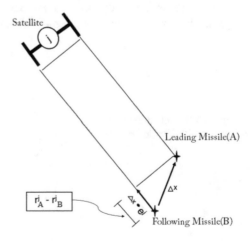

Substituting Eq. (3.11) into the above equation

$$\Delta\nabla\rho_{A,B}^{jk} = \left(\mathbf{c}^j - \mathbf{c}^k\right)\cdot\Delta\mathbf{x} \tag{3.15}$$

Similarly, the following equation can be obtained based on the analysis of the integer ambiguities of double differential carrier phase:

$$\Delta\nabla\phi_{AB}^{jk} = \lambda^{-1}\left(\mathbf{e}^j - \mathbf{e}^k\right)\cdot\Delta\mathbf{x} + \Delta\nabla N_{AB}^{jk} \tag{3.16}$$

If noise is omitted, a complete matrix of H can be established according to Eqs. (3.15) and (3.16):

$$
\begin{bmatrix}
\Delta\nabla\rho_{A,B}^{1,2} \\
\Delta\nabla\rho_{A,B}^{1,3} \\
\Delta\nabla\rho_{A,B}^{1,4} \\
\Delta\nabla\phi_{A,B}^{1,2} \\
\Delta\nabla\phi_{A,B}^{1,3} \\
\Delta\nabla\phi_{A,B}^{1,4}
\end{bmatrix}
=
\begin{bmatrix}
(\mathbf{e}^1 - \mathbf{e}^2) & 0 & 0 & 0 \\
(\mathbf{e}^1 - \mathbf{e}^3) & 0 & 0 & 0 \\
(\mathbf{e}^1 - \mathbf{e}^4) & 0 & 0 & 0 \\
\lambda^{-1}(\mathbf{e}^1 - \mathbf{e}^2) & 1 & 0 & 0 \\
\lambda^{-1}(\mathbf{e}^1 - \mathbf{e}^3) & 0 & 1 & 0 \\
\lambda^{-1}(\mathbf{e}^1 - \mathbf{e}^4) & 0 & 0 & 1
\end{bmatrix}
\begin{bmatrix}
\Delta x \\
\Delta y \\
\Delta z \\
\Delta\nabla N_{A,B}^{1,2} \\
\Delta\nabla N_{A,B}^{1,3} \\
\Delta\nabla N_{A,B}^{1,4}
\end{bmatrix}
\tag{3.17}
$$

(3) Calculation of relative position

If the code value of the Eq. (3.17) is omitted, relative position value is separated from the other elements. We have:

$$
\begin{bmatrix}
\Delta\nabla\phi_{A,B}^{1,2} \\
\Delta\nabla\phi_{A,B}^{1,3} \\
\Delta\nabla\phi_{A,B}^{1,4}
\end{bmatrix}
=
\begin{bmatrix}
\lambda^{-1}(\mathbf{e}^1 - \mathbf{e}^2) \\
\lambda^{-1}(\mathbf{e}^1 - \mathbf{e}^3) \\
\lambda^{-1}(\mathbf{e}^1 - \mathbf{e}^4)
\end{bmatrix}
\begin{bmatrix}
\Delta x \\
\Delta y \\
\Delta z
\end{bmatrix}
+
\begin{bmatrix}
1 & 0 & 0 \\
0 & 1 & 0 \\
0 & 0 & 1
\end{bmatrix}
\begin{bmatrix}
\Delta\nabla N_{A,B}^{1,2} \\
\Delta\nabla N_{A,B}^{1,3} \\
\Delta\nabla N_{A,B}^{1,4}
\end{bmatrix}
\tag{3.18}
$$

$$
\begin{bmatrix}
\lambda^{-1}(\mathbf{e}^1 - \mathbf{e}^2) \\
\lambda^{-1}(\mathbf{e}^1 - \mathbf{e}^3) \\
\lambda^{-1}(\mathbf{e}^1 - \mathbf{e}^4)
\end{bmatrix}
\begin{bmatrix}
\Delta x \\
\Delta y \\
\Delta z
\end{bmatrix}
=
\begin{bmatrix}
\Delta\nabla\phi_{A,B}^{1,2} \\
\Delta\nabla\phi_{A,B}^{1,3} \\
\Delta\nabla\phi_{A,B}^{1,4}
\end{bmatrix}
-
\begin{bmatrix}
1 & 0 & 0 \\
0 & 1 & 0 \\
0 & 0 & 1
\end{bmatrix}
\begin{bmatrix}
\Delta\nabla N_{A,B}^{1,2} \\
\Delta\nabla N_{A,B}^{1,3} \\
\Delta\nabla N_{A,B}^{1,4}
\end{bmatrix}
\tag{3.19}
$$

Let $x \equiv [\Delta x \ \ \Delta y \ \ \Delta z]^T$, $H \equiv [\lambda^{-1}(\mathbf{e}^1 - \mathbf{e}^2) \ \ \lambda^{-1}(\mathbf{e}^1 - \mathbf{e}^3) \ \ \lambda^{-1}(\mathbf{e}^1 - \mathbf{e}^4)]^T$, we have:

$$
z \equiv
\begin{bmatrix}
\Delta\nabla\phi_{A,B}^{1,2} \\
\Delta\nabla\phi_{A,B}^{1,3} \\
\Delta\nabla\phi_{A,B}^{1,4}
\end{bmatrix}
-
\begin{bmatrix}
1 & 0 & 0 \\
0 & 1 & 0 \\
0 & 0 & 1
\end{bmatrix}
\begin{bmatrix}
\Delta\nabla N_{A,B}^{1,2} \\
\Delta\nabla N_{A,B}^{1,3} \\
\Delta\nabla N_{A,B}^{1,4}
\end{bmatrix}
\tag{3.20}
$$

So, there exists

$$\mathbf{Hx} = \mathbf{z}$$

Depending on the number of selected satellites, the matrix H is not always a square matrix. So the relative position vector x cannot be simply multiplied by matrix H^{-1}. Therefore, pseudo-inverse can be used:

$$\mathbf{x} = (\mathbf{H}^T\mathbf{H})^{-1}\mathbf{H}^T\mathbf{z} \qquad (3.21)$$

(4) Simulations

(1) Single station positioning method. The relative position is obtained from the position difference between the leading missile and the following missile. Considering multipath interference of carrier and pseudorange and receiver noise, simulation results are showing in Fig. 3.14. It can be seen that the error term has a great influence on the accuracy of single station positioning method.

(2) Double differential pseudorange method. From Eq. (3.17), we have
$\begin{bmatrix} \Delta\nabla\rho_{A,B}^{1,2} \\ \Delta\nabla\rho_{A,B}^{1,3} \\ \Delta\nabla\rho_{A,B}^{1,4} \end{bmatrix} = \begin{bmatrix} (\mathbf{e}^1 - \mathbf{e}^2) \\ (\mathbf{e}^1 - \mathbf{e}^3) \\ (\mathbf{e}^1 - \mathbf{e}^4) \end{bmatrix} \begin{bmatrix} \Delta x \\ \Delta y \\ \Delta z \end{bmatrix}$. Considering multipath interference of

carrier and pseudorange and receiver noise, simulation results are showing in Fig. 3.15. It can be seen that the relative position obtained by the carrier phase double differential pseudorange method is larger and only suitable for the loose formation flight.

(3) Double differential carrier phase method. Using the carrier phase relative position calculation method described above, simulation results are showing in Fig. 3.16. From the figure, it can be seen that the carrier phase double differential phase method has a good effect; error of the interference term on the calculation results is small.

Fig. 3.14 Simulation results of single station positioning method

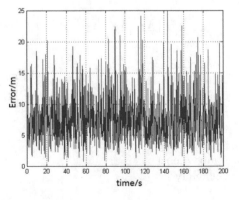

Fig. 3.15 Simulation results of double differential pseudorange method

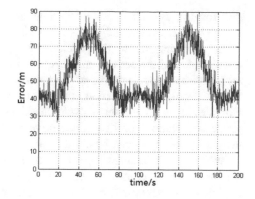

Fig. 3.16 Simulation results of double differential carrier phase method

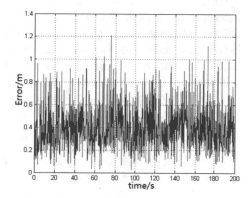

3.4 Relative Velocity Algorithm Based on Carrier Phase Time Difference Observation

Compared with the pseudorange method, the carrier phase of satellite positioning system has a high resolution advantage [81, 82]. From the carrier phase observation Eq. (3.6), there exists an integer ambiguity N in the carrier phase observation. The integer ambiguity N has two basic properties: first, it is an unknown integer that is randomly generated by the receiver's carrier tracking loop after capturing the carrier signal; second, if no cycle slip occurs, it will remain as a constant.

Since the integer ambiguity N is a random integer, it is fundamentally different from other error terms in the carrier phase observation equation and cannot be compensated or corrected by model or other means. Therefore, the carrier phase cannot be applied in real-time navigation unless one of the following two conditions is satisfied: (1) Real-time solution of the integer cycle ambiguity N; (2) Eliminate the integer ambiguity N by a certain algorithm.

According to the previous discussion, it can be seen that in the absence of a base station (i.e., a precision single point positioning), the calculation of the integer cycle ambiguity generally needs about 30 min. In the help of a base station (i.e., RTK positioning), the calculation of the integer cycle ambiguity usually takes 30–60 s. So the integer cycle ambiguity cannot be calculated in real-time.

In addition, when the cycle slip occurs, integer ambiguity should be re-calculated. If the convergence cycle of integer cycle ambiguity calculation is too long, real-time requirements of navigation applications cannot be achieved. It can be seen from the observation equation of carrier phase that the integer ambiguity will remain as a constant if cycle slip does not happen. Therefore, if the carrier phase observations of two consecutive measurement periods are subtracted, the integer ambiguity N can be eliminated.

While the differential operation of successive carrier phase observations is performed, the error terms contained in the carrier phase observations are subjected to differential operation. First, the satellite clock error, satellite orbit error, iono-spheric error and tropospheric error are a strong time-correlation errors, all belonging to errors of constant deviation. Thus, the differential operations between adjacent observations substantially eliminate these four errors. Secondly, the mul-tipath error and the receiver noise are random errors. Differential operation is one of additive operations. According to the error propagation law, the additive operation will amplify the random error. Thus, it can be concluded that the differential operation between successive carrier phases eliminates the deviation of the observed error, but increases the random noise of the observed value.

(1) Carrier phase time difference observation equation

In order to apply the carrier phase time difference to navigation and integrated navigation, observation equation should be derived firstly. From Eq. (3.6), we can get carrier phase expression between missile A and the j-th satellite

$$\phi_A^j(t) = \lambda^{-1}\left\{ r_A^j(t) + c\left[\delta t_{r_A}(t) - \delta t_{SV}^j(t) \right] + T_A^j(t) - I_A^j(t) + m_{\phi A}^j(t) + v_{\phi A}^j(t) \right\} + N_A^j$$

$$(3.22)$$

Assuming that t_1 and t_2 are two consecutive measurement time points, the carrier phase difference at these two time points can be expressed as:

$$\phi_A^j(\Delta t) = \phi_A^j(t_2) - \phi_A^j(t_1)$$
$$= \lambda^{-1}\left\{ r_A^j(\Delta t) + c\left[\delta t_{r_A}(\Delta t) - \delta t_{SV}^j(\Delta t) \right] + T_A^j(\Delta t) - I_A^j(\Delta t) + m_{\phi A}^j(\Delta t) + v_{\phi A}^j(\Delta t) \right\}$$

$$(3.23)$$

In the above equation, the integer ambiguity N is a constant and is eliminated; $r_A^j(\Delta t)$ represents the change of the geometric distance from the receiver to the satellite during time interval $[t_1, t_2]$.

Ionospheric error, tropospheric error, satellite clock error and satellite orbit error remain the same level of carrier phase measurements noise of the receiver within one second. It can be considered that these four errors have been eliminated by the difference between the carrier phases at time t_1 and t_2.

Define a new variable denotes that $\phi_A^j(\Delta t)$ corrects the deviation from the ionospheric error and tropospheric error, while combining the multipath error with the receiver noise.

$$\tilde{\phi}_A^j(\Delta t) = \lambda^{-1}\left\{r_A^j(\Delta t) + c\left[\delta t_{r_A}(\Delta t) - \delta t_{SV}^j(\Delta t)\right] + \xi_A^j(\Delta t)\right\} \qquad (3.24)$$

Similarly, for missile B also have:

$$\tilde{\phi}_B^j(\Delta t) = \lambda^{-1}\left\{r_B^j(\Delta t) + c\left[\delta t_{r_B}(\Delta t) - \delta t_{SV}^j(\Delta t)\right] + \xi_B^j(\Delta t)\right\} \qquad (3.25)$$

Subtracting Eqs. (3.24) and (3.25) to get a new variable:

$$\Delta\tilde{\phi}_{AB}^j(\Delta t) = \tilde{\phi}_A^j(\Delta t) - \tilde{\phi}_B^j(\Delta t) = \lambda^{-1}\left\{\Delta r_{AB}^j(\Delta t) + c\cdot\Delta\delta t_{r_{AB}}(\Delta t) + \Delta\xi_{AB}^j(\Delta t)\right\}$$

$$(3.26)$$

In Eq. (3.26), $\Delta r_{AB}^j(\Delta t)$ will be expanded below to calculate the concrete expression of $\Delta\tilde{\phi}_{AB}^j(\Delta t)$. Figure 3.17 shows the positional relationship between satellites and formation in two successive measurements.

Positions of the j-th satellite at time t_1 and t_2 are expressed as $SV^j(t_1), SV^j(t_2)$, respectively. $\overrightarrow{R^j(t)}$ and $\overrightarrow{b_A(t)}$ denote the position of the j-th satellite and missile A in the ECEF coordinate system at time t, respectively. $\overrightarrow{r_A^j(t)}$ represents the distance

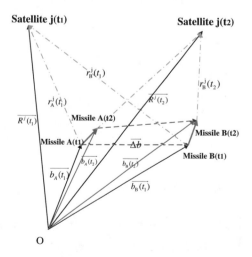

Fig. 3.17 Positional relationship between satellites and formation in two successive measurements

vector from missile A to the satellite at time t. $\overrightarrow{b_A(\Delta t)} = \overrightarrow{b_A(t_2)} - \overrightarrow{b_A(t_1)}$ represents the position increment from t_1 to t_2. Expression of $\Delta r_{AB}^j(\Delta t)$ can be written as:

$$
\begin{aligned}
\Delta r_{AB}^j(\Delta t) = r_A^j(\Delta t) - r_B^j(\Delta t) &= \left[r_A^j(t_2) - r_A^j(t_1) \right] - \left[r_B^j(t_2) - r_B^j(t_1) \right] \\
&= \left\{ \left[\overrightarrow{R^j(t_2)} - \overrightarrow{b_A(t_2)} \right] \cdot \mathbf{e}_A^j(t_2) - \left[\overrightarrow{R^j(t_1)} - \overrightarrow{b_A(t_1)} \right] \cdot \mathbf{e}_A^j(t_1) \right\} \\
&\quad - \left\{ \left[\overrightarrow{R^j(t_2)} - \overrightarrow{b_B(t_2)} \right] \cdot \mathbf{e}_B^j(t_2) - \left[\overrightarrow{R^j(t_1)} - \overrightarrow{b_B(t_1)} \right] \cdot \mathbf{e}_B^j(t_1) \right\}
\end{aligned}
\tag{3.27}
$$

The first term of the right brace in Eq. (3.27) is:

$$
\begin{aligned}
&\left\{ \overrightarrow{R^j(t_2)} \cdot \mathbf{e}_A^j(t_2) - \overrightarrow{R^j(t_1)} \cdot \mathbf{e}_A^j(t_1) \right\} - \left\{ \overrightarrow{b_A(t_2)} \cdot \mathbf{e}_A^j(t_2) - \overrightarrow{b_A(t_1)} \cdot \mathbf{e}_A^j(t_1) \right\} \\
&= \left\{ \overrightarrow{R^j(t_2)} \cdot \mathbf{e}_A^j(t_2) - \overrightarrow{R^j(t_1)} \cdot \mathbf{e}_A^j(t_1) \right\} - \left\{ \overrightarrow{b_A(t_1)} \cdot \mathbf{e}_A^j(t_2) - \overrightarrow{b_A(t_1)} \cdot \mathbf{e}_A^j(t_1) \right\} - \overrightarrow{b_A(\Delta t)} \cdot \mathbf{e}_A^j(t_2)
\end{aligned}
\tag{3.28}
$$

It is not difficult to find that the first term of Eq. (3.28) represents the Doppler effects caused by satellite motion, which is proportional to the average Doppler shift generated by the motion of the satellite along the line of sight (LOS). The value is only related to the motion of the satellite, but not with the receiver along the LOS motion. The second term of Eq. (3.28) represents the relative relationship between the satellite and the receiver. Let:

$$
\begin{aligned}
Dop_A^j &= \left\{ \overrightarrow{R^j(t_2)} \cdot \mathbf{e}_A^j(t_2) - \overrightarrow{R^j(t_1)} \cdot \mathbf{e}_A^j(t_1) \right\} \\
Geo_A^j &= \left\{ \overrightarrow{b_A(t_1)} \cdot \mathbf{e}_A^j(t_2) - \overrightarrow{b_A(t_1)} \cdot \mathbf{e}_A^j(t_1) \right\}
\end{aligned}
\tag{3.29}
$$

Substituting Eq. (3.27),

$$
\begin{aligned}
\Delta r_{AB}^j(\Delta t) &= \left\{ Dop_A^j - Geo_A^j - \overrightarrow{b_A(\Delta t)} \cdot \mathbf{e}_A^j(t_2) \right\} - \left\{ Dop_B^j - Geo_B^j - \overrightarrow{b_B(\Delta t)} \cdot \mathbf{e}_B^j(t_2) \right\} \\
&= \Delta Dop_{AB}^j - \Delta Geo_{AB}^j - \left[\overrightarrow{b_A(\Delta t)} \cdot \mathbf{e}_A^j(t_2) - \overrightarrow{b_B(\Delta t)} \cdot \mathbf{e}_B^j(t_2) \right]
\end{aligned}
\tag{3.30}
$$

Substituting Eq. (3.30) into Eq. (3.26):

$$
\Delta \tilde{\phi}_{AB}^j(\Delta t) = \lambda^{-1} \left\{ \begin{array}{l} \Delta Dop_{AB}^j - \Delta Geo_{AB}^j - \left[\overrightarrow{b_A(\Delta t)} \cdot \mathbf{e}_A^j(t_2) - \overrightarrow{b_B(\Delta t)} \cdot \mathbf{e}_B^j(t_2) \right] \\ + c \cdot \Delta \delta t_{r_{AB}}(\Delta t) + \Delta \xi_{AB}^j(\Delta t) \end{array} \right\}
\tag{3.31}
$$

In the Doppler term, the satellite orbit data is used to obtain $\overrightarrow{R^j(t)}$, using the satellite orbit data and a single station receiver to calculate the LOS vector. In the Geo item, $\overrightarrow{b(t)}$ is obtained through the single station. Define a new variable:

$$\Delta \tilde{\tilde{\phi}}^j_{AB}(\Delta t) = \Delta \dot{\tilde{\phi}}^j_{AB}(\Delta t) - \lambda^{-1} \left\{ \Delta Dop^j_{AB} - \Delta Geo^j_{AB} \right\}$$
$$= \lambda^{-1} \left\{ - \left[\overrightarrow{b_A(\Delta t)} \cdot \mathbf{e}^j_A(t_2) - \overrightarrow{b_B(\Delta t)} \cdot \mathbf{e}^j_B(t_2) \right] + c \cdot \Delta \delta t_{r_{AB}}(\Delta t) + \Delta \xi^j_{AB}(\Delta t) \right\}$$

(3.32)

Subtracting $\Delta \tilde{\tilde{\phi}}^j_{AB}(\Delta t)$ between the different satellites eliminates the receiver's clock error term. $\Delta \nabla \tilde{\tilde{\phi}}^{jk}_{AB}(\Delta t) = \Delta \tilde{\tilde{\phi}}^j_{AB}(\Delta t) - \Delta \tilde{\tilde{\phi}}^k_{AB}(\Delta t) = \lambda^{-1} \left\{ - \left[\overrightarrow{b_A(\Delta t)} \cdot \Delta \right. \right.$
$\mathbf{e}^{jk}_A(t_2) - \overrightarrow{b_B(\Delta t)} \cdot \Delta \mathbf{e}^{jk}_B(t_2)] + \Delta \nabla \xi^{jk}_{AB}(\Delta t) \}$
Since the distance between the receiver and the satellite is much larger than the short-range missile formation, it is assumed that for all given satellites, all formation members have the same line of sight vector.

$$\Delta \nabla \tilde{\tilde{\phi}}^{jk}_{AB}(\Delta t) = \lambda^{-1} \left\{ - \left[\overrightarrow{b_A(\Delta t)} - \overrightarrow{b_B(\Delta t)} \right] \Delta \mathbf{e}^{jk}_A(t_2) + \Delta \nabla \xi^{jk}_{AB}(\Delta t) \right\} \quad (3.33)$$

The above equation is the observation equation for the phase difference of the double missile carrier phase. The left side of the equation is a known quantity, which can be deduced from the carrier phase observation step by step. The right side of the equation is the unknown quantity to be measured.

(2) Relative velocity algorithm based on carrier phase time difference

From the observation equation of the carrier phase time difference, it can be seen that the carrier phase time difference is a non-fuzzy high-precision observation of satellite positioning system. The relative velocity algorithm based on the carrier phase time difference is given below.

From Eq. (3.33), the carrier phase time difference contains the position increment of the two missiles between successive measurements of satellite positioning system. If this position increment can be calculated from the observation equation of the carrier phase time difference, the relative velocity is obtained by dividing the position by the interval between the two measurements. If this position increment can be calculated from the observation equation of the carrier phase time difference, the relative velocity is obtained by dividing the position by the interval between the two measurements.

Position increments of receiver A, B between two consecutive measurements are derived, and then the velocity is calculated. For the convenience of derivation, define a new variable Y_{jk}.

$$Y_{jk} \triangleq \Delta\nabla\tilde{\phi}_{AB}^{jk}(\Delta t) = \lambda^{-1}\left\{-\left[\overline{b_A(\Delta t)} - \overline{b_B(\Delta t)}\right]\Delta e_A^{jk}(t_2) + \Delta\nabla\xi_{AB}^{jk}(\Delta t)\right\} \quad (3.34)$$

Assuming a total of n satellite observations, choose the first satellite as the reference. According to Eq. (3.34), the total observed equation formed by the carrier phase time difference of all the satellites can be expressed as:

$$\begin{bmatrix} Y_{1,2} \\ Y_{1,3} \\ \vdots \\ Y_{1,n} \end{bmatrix} = -\begin{bmatrix} \Delta e_A^{1,2}(t_2)^T \\ \Delta e_A^{1,3}(t_2)^T \\ \vdots \\ \Delta e_A^{1,n}(t_2)^T \end{bmatrix}\overrightarrow{b_{AB}(\Delta t)} + \begin{bmatrix} \Delta\nabla\xi_{AB}^{1,2}(\Delta t) \\ \Delta\nabla\xi_{AB}^{1,3}(\Delta t) \\ \vdots \\ \Delta\nabla\xi_{AB}^{1,n}(\Delta t) \end{bmatrix} \quad (3.35)$$

To simplify (3.35), define the following variables:

$$\vec{Y} = \begin{bmatrix} Y_{1,2} \\ Y_{1,3} \\ \vdots \\ Y_{1,n} \end{bmatrix}, G = -\begin{bmatrix} \Delta e_A^{1,2}(t_2)^T \\ \Delta e_A^{1,3}(t_2)^T \\ \vdots \\ \Delta e_A^{1,n}(t_2)^T \end{bmatrix}, \overrightarrow{\Delta\xi} = \begin{bmatrix} \Delta\nabla\xi_{AB}^{1,2}(\Delta t) \\ \Delta\nabla\xi_{AB}^{1,3}(\Delta t) \\ \vdots \\ \Delta\nabla\xi_{AB}^{1,n}(\Delta t) \end{bmatrix} \quad (3.36)$$

So,

$$\vec{Y} = G\overrightarrow{b_{AB}(\Delta t)} + \overrightarrow{\Delta\xi} \quad (3.37)$$

In (3.37), $\overrightarrow{b_{AB}(\Delta t)}$ is a three-dimensional unknown quantity. When the observation $\vec{Y}$ is not less than three dimensions, that is, the number of observation satellites is not less than four, $\overrightarrow{b_{AB}(\Delta t)}$ can be obtained according to the least squares principle as follows:

$$\overrightarrow{b_{AB}(\Delta t)} = (G^T G)^{-1} G^T \vec{Y} \quad (3.38)$$

Assuming that the time interval for two consecutive measurements is Δt, the relative velocity calculated by the carrier phase time difference observation can be expressed as:

$$\vec{V_r} = \frac{\overrightarrow{b_{AB}(\Delta t)}}{\Delta t} \quad (3.39)$$

Simulation results of receiver noise and multipath interference of the carrier and pseudorange are shown in Fig. 3.18.

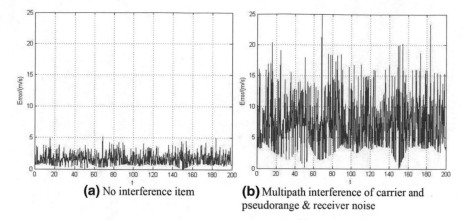

(a) No interference item   **(b)** Multipath interference of carrier and pseudorange & receiver noise

Fig. 3.18 Simulation results of relative velocity algorithm

It can be seen from the above figure, that the relative velocity algorithm based on the carrier phase time difference is not particularly ideal. For the relative navigation, more effective technical approaches should be researched to achieve higher relative navigation accuracy requirements of the intensive formation flight.

Chapter 4
Decision and Management System (DMS) of MAF

4.1 Compositions of DMS

The structure of MAF cooperation guidance and control system is showed in Fig. 2.19. Formation decision and management system is to follow the basic principles of cooperativity. It is the central system bearing the following works: weighing individual costs and group costs and giving operational effectiveness indicators; adjusting the formation conflict; nodes management of leaving and joining from the team; implementation of task planning/target assignment; decision and management of cooperative route planning/cooperation guidance and other processes; optimizing the formation and formation guidance instructions; Ensuring that the required tasks to complete the required tasks to follow the self-organizing rules and rules. The system performance determines the degree of autonomy of the formation and the level of self-organization and management of the formation. This is showed in Fig. 4.1.

In order to ensure a smooth and successful completion of the mid-final guidance, task planning/target dynamic assignment plays an important role in connecting route planning with cooperation ordination. In consideration of the target information, the missile members of their own task load and damage characteristics, task planning/target dynamic assignment is to conduct task planning for different performance or the same performance of the missile members, so that each missile members can give full play to their respective characteristics, to achieve a comprehensive combat effectiveness of the missile formation. Missile members turn into the final guidance stage from the middle of the guidance stage after task planning/target dynamic assignment. Therefore, missile formation complete the mid-final guidance duty shift. Subsequently, the missile formation will be divided according to the target assignment results in a smaller formations or in a single missile member independent operations on their own targets to carry out coordinated attack, autonomous missile formation to achieve the goal of maximizing the integrated operational effectiveness.

© National Defense Industry Press, Beijing and Springer Nature Singapore Pte Ltd. 2019
S. Wu, *Cooperative Guidance & Control of Missiles Autonomous Formation*,
https://doi.org/10.1007/978-981-13-0953-3_4

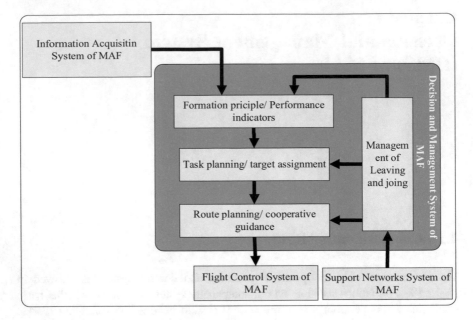

Fig. 4.1 Decision and management system of MAF

4.2   Basic Principle of MAF

The three basic principles of the missile formation given in Sect. 2.1.2 are the basis to ensure that the formation of multiple missile synergies combat operations and get the comprehensive combat effectiveness to maximize. As can be seen from Fig. 4.1, formation decision and management system as the center of the MAF, should be based on the information provided by the information acquisition system, as well as conducting the three basic principles testing before the launch and during the task. Formation decision and management system form a comprehensive operational effectiveness indicators in order to provide the basis to the following task planning, route planning and cooperation guidance control.

4.3   Combat Effectiveness Indication of MAF

4.3.1   Combat Effectiveness Indication of Missile Weapon System

The ADC model proposed by Weapons System Effectiveness Industry Advisory Committee (WSEIAC) takes into account the effectiveness, reliability and combat capability of the combat system, synthesizing the performance evaluation of the system and the specific combat tasks, and can reflect the completion of the established task. The structure of missile weapon system is showed in Fig. 4.2 [1].

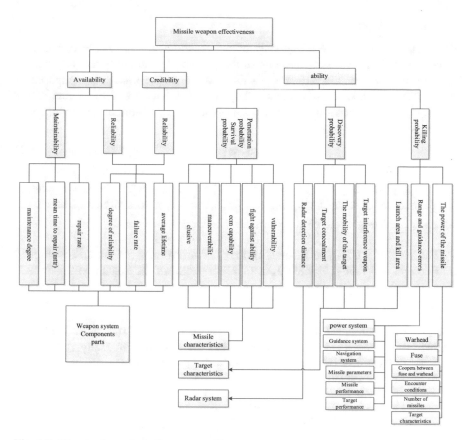

Fig. 4.2 The structure of missile weapon efficacy

The state of the system at the beginning of the task, the state of the task in progress, and the degree of completion of the given task, the three together constitute the system performance. The state of the system at the beginning of the task is described by the "availability" of the system. The state of the system during the execution of the task is described by the "credibility" of the system. The degree to which a system completes a given task is described by the "capability" of the system. Thus, the "availability", "credibility" and "capability" of the system constitute the system's performance, called the three elements of system performance. These three parts are represented by the probability. The expression was propose by WSEIAC:

$$E = A \cdot [D] \cdot C \tag{4.1}$$

E—performance vector; A—availability vector; $[D]$—credibility vector; C—capacity vector

1. Availability vector

Availability is a measure of the state of the system at the beginning of the task. Thus, the available vector is made up of the probability that the system could be in all possible states at the beginning of the task. Since the system is composed of many different links or subsystems, they may not have the same status at the beginning of the task. In this way, the general expression of the availability vector is:

$$A = (a_1, a_2, \ldots, a_n) \tag{4.2}$$

a_i—the probability that the system is in the first state when the task is started; n—the number of states the system may be in; n kinds of possibilities constitute the sample space. Therefore $\sum_{i=1}^{n} a_i = 1$.

The state of the missile weapon system at the beginning of the task is divided into normal and faulty. Therefore, availability vector of the system is:

$$A = (a_1, a_2) \tag{4.3}$$

a_1—the probability that the system is in a normal working state when the task is started (availability); a_2—the probability that the system is in faulty state at the beginning of the task (Unavailability). Obviously, $a_1 + a_2 = 1$.

According to the reliability theory:

$$a_1 = \frac{MTBF_s}{MTBF_s + MTTR_s} = \frac{1/\lambda_s}{1/\lambda_s + 1/\mu_s} \tag{4.4}$$

$$a_2 = \frac{MTTR_s}{MTBF_s + MTTR_s} = \frac{1/\mu_s}{1/\lambda_s + 1/\mu_s} \tag{4.5}$$

$MTBF_s$—the average time between failures of the system; $MTTR_s$—the average repair time of the system; λ_s—system failure rate; μ_s—system repair rate.

2. Credibility matrix

Credibility is a measure of the state in which a system performs a task. Every state at the beginning of the implementation of the task, in the course of the implementation of the task through the system performance changes and maintenance of their own, may be converted to other states. The credibility matrix is composed of the probability that the various states are transformed into other states.

Every state at the beginning of the task could maintain the original state or change into the other $n - 1$ possible states. Therefore, every state at the beginning of the task corresponds to n kinds of possible conversion states. If the system has

n kinds of possible states at the beginning of the task, there are $n \times n$ possible transition states during the execution of the task. Therefore, credibility matrix is a $n \times n$ order square matrix:

$$[D] = \begin{pmatrix} \alpha_{11} & \alpha_{12} & \cdots & \alpha_{1n} \\ \alpha_{21} & \alpha_{22} & \cdots & \alpha_{2n} \\ \vdots & \vdots & & \vdots \\ \alpha_{n1} & \alpha_{n2} & \cdots & \alpha_{nn} \end{pmatrix} \qquad (4.6)$$

α_{ij}—it is known that the system is in the i-th state when the task begins while in the j-th state in the middle of the task.

For every state of the system when the task begins the n possible transformation states during the execution of the task constitute the sample space for this state transformation. Therefore, the sum of the elements of each line in the matrix $[D]$ equals to one:

$$\sum_{j=1}^{n} \alpha_{ij} = 1 \quad i = 1, 2, \cdots, n \qquad (4.7)$$

Every element α_{ij} in the confidence matrix $[D]$ is related to both the original state i and the state j in which the task is executed. When the system is in the same state j during the execution of the task, α_{ij} differs with the change of the state of the system in the middle of the task. α_{ij} could be calculated based on the state of the system and its elements at the time of the execution of the task and the state of the task, the reliability and maintainability of the system and its elements, and the principle of probability theory.

The missile weapon system is a non-repairable system during mission execution. Assuming that the element α_{ij} in $[D]$ is sorted by the number of the lower right corner of the element, the greater the fault is more or more serious. Therefore, $i > j$ is impossible (because of non-repairable system). Thus, $\alpha_{ij} = 0$. Therefore, the system's credibility matrix becomes a triangular matrix:

$$[D] = \begin{pmatrix} \alpha_{11} & \alpha_{12} & \alpha_{13} & \cdots & \alpha_{1n} \\ 0 & \alpha_{22} & \alpha_{23} & \cdots & \alpha_{2n} \\ 0 & 0 & \alpha_{33} & \cdots & \alpha_{3n} \\ \vdots & \vdots & \vdots & & \vdots \\ 0 & 0 & 0 & \cdots & \alpha_{nn} \end{pmatrix} \qquad (4.8)$$

3. **Capacity vector**

The capacity of the system is the degree to which the system finally completes a particular task, generally expressed by the probability of completing a particular task. This probability is closely related to the state in which the system is performing the task. The same system, due to the different state, the probability of completion of specific tasks are different.

When the missile weapon system contains only one quality factor, the system's capability is expressed as a capability vector:

$$C = \begin{pmatrix} c_1 \\ c_2 \\ \vdots \\ c_n \end{pmatrix} \tag{4.9}$$

c_i—The capacity value of the system in the state i during executing the task.

Some systems can perform specific tasks as long as they are working properly. For weapons systems, the system may not be able to complete a specific task even if it works properly. Working properly is a necessary condition for accomplishing a particular task, but not a sufficient condition. For the winged missile (WM) weapon system, the capability vector C mainly includes the missile's penetration probability P_1, duty shift success rate P_2, hit probability P_3, damage probability P_4, etc.

$$P_1 = f(Ma, h, n_a, RCS_{wm}, \lambda, \Xi, \Lambda) \tag{4.10}$$

Ma—missile flight mach number; h—missile flight height; n_a—missile overload available; RCS_{wm}—radar cross section of missile; λ—current density of missile; Ξ—missiles autonomous formation **CGCL** level; Λ—The performance of the enemy defense system.

Usually the missile in the medium guidance section flies to guidance duty shift section mainly rely on inertial navigation system, satellite navigation system, terrain/geomagnetic matching and other navigation equipment, with the help of altimeter and autopilot. Winged missiles rely on their own seeker or the formation support network system to capture the specified target and turn into the final guidance phase. The success rate of the shift is to measure the probability that the flight missile is successfully transferred to the final guidance state by the mid-guidance state, which can be described by the following:

$$P_2 = f(D, \dot{D}, \Psi, \dot{\Psi}, \Delta_D, \Delta_\Psi, \Delta_M, RCS_T, \Xi, \Lambda) \tag{4.11}$$

D—distance from missile to target; $\dot{D}$—closing speed between missile and target; Ψ—relative angle between missile and target (azimuth angle and pitch angle); $\dot{\Psi}$—change rate of relative angle between missile and target (change rate of azimuth angle and pitch angle); Δ_D—seeker detection distance of missile; Δ_Ψ—seeker field of view of missile; Δ_M—working mode and dynamic characteristics of missile's seeker; RCS_T—radar cross section of target.

The hit probability P_3 is mainly related to the guidance accuracy of the winged missile, the movement characteristics of the target and the target dimensions; which could be described by the following equation:

$$P_3 = f(\varepsilon_G, n_T, \Omega_T) \tag{4.12}$$

ε_G—guidance precision of missile; n_T—maneuvering overload of target; Ω_T—dimensions of target.

The probability of damage is mainly related to the damage capability of the winged missile and the vulnerability of the target. The specific expression is:

$$P_4 = f(\varepsilon_W, \zeta_{FW}, \Psi_T, \varsigma_T) \tag{4.13}$$

ε_W—warhead effectiveness of missile; ζ_{FW}—the degree of cooperation between the fuze and the warhead; Ψ_{wm}—missile state when reaching the target; ς_T—vulnerability of target;

To sum up, the basic capability indicators of winged missile (WM) can be expressed as:

$$P_0 = P_1 \cdot P_2 \cdot P_3 \cdot P_4 \tag{4.14}$$

4.3.2   Combat Effectiveness Indication of Winged Missile

Taking into account the impact factors of missile formation, combat effectiveness indication of winged missile was proposed based on the WSEIAC system performance expression:

$$E = \frac{A \cdot [D] \cdot C}{\Sigma} \tag{4.15}$$

$\Sigma = \sum\limits_{i}^{n} \frac{b_i}{B_i}$—the sum of the cost ratios of missile formation. Among them, b_i and B_i—respectively, the actual cost of the i-th warfare missiles and their corresponding standard allocation of the standard cost.

1. **Availability vector**

The availability vector A has $n + 1$ dimensions:

$$A = (A_0, A_1, A_2, \ldots, A_k, \ldots, A_n) \tag{4.16}$$

A_0—Every missile is in normal state, and the system is in the available state, $A_0 = a_{wm}^n$. A_k—there are k missiles in faulty state, and the system is in the available state, $A_k = C_n^k a_{wm}^{n-k} \cdot (1 - a_{wm})^k$. A_n—n missiles are in faulty state, system is in the unavailable state, $A_n = 1 - \sum_{i=0}^{n-1} A_i$. Among them, a_{wm}—reliability of winged missile (WM).

2. **Credibility**

$$[D] = [\alpha_{ij}]_{(n+1)\cdot(n+1)}, \quad i = j = 1, 2, \ldots, n \qquad (4.17)$$

When $i > j$, $\alpha_{ij} = 0$; When $i = j$, $\alpha_{ij} = \exp\left[-(n-i+1) \cdot \frac{T}{MTBF_s}\right]$;

When $i < j$, $\alpha_{ij} = C_{n-i+1}^{j-i} \cdot \exp\left[-(n-j+1) \cdot \frac{T}{MTBF_s}\right] \cdot \left[1 - \exp - \left(\frac{T}{MTBF_s}\right)\right]^{j-i}$.

T—Task working time of the system.

3. **Capacity value**

With the optimal configuration of the task load and the formation decision and management system, the winged missile formation could use different types and specifications of the seeker to carry out high and low match. It is possible to effectively improve the basic capacity indicators $P_0 = P_1 \cdot P_2 \cdot P_3 \cdot P_4$ of the winged missile (WM) and reduce the total cost ratio of the missile formation $\sum = \sum_i^n \frac{b_i}{B_i}$ by exchanging the target information through the formation support network.

The capacity value corresponding to the system in the i-th state during the execution of the task is calculated by the following:

$$c_i = \sum_{j=1}^{M} w_{ij} c_{ij} \qquad (4.18)$$

M—amount of targets, w_{ij}—weight value in the state i to attack target j, c_{ij}—capacity value in the state i to attack target j.

$$c_{ij} = 1 - \left(1 - \frac{1}{1 - \Xi_j(1 - P_0(i))} \cdot \frac{P_0(i)}{T_j}\right)^{N_j} \qquad (4.19)$$

$P_0(i)$—basic capacity indicator for each missile when the system is in the state i. T_j—average missile amount to hit to destroy the j-th target; N_j—missile amount to attack target j; Ξ_j—CGCL level of the missile formation to attack target j ($\Xi_j \leq 1$), It can be determined by simulation test that the initial value can be set to $\Xi_{j0} = \frac{CGCL_j}{5}$, $CGCL_j$ is the CGCL level of the missile formation to attack the target j.

4.4   Task Planning and Dynamic Weapon-Target Assignment

4.4.1   Summary

As the relationship between task planning and target dynamic assignment of MAF, and the characteristics of multi-constrained, strongly coupled complex multi-objective integer optimization problems with task planning and dynamic target assignment, discussed in Chap. 1, this chapter will first introduce the static task planning strategy for missile formation in a defined environment, and designs a hybrid optimization strategy based on modern optimization algorithm to accelerate the convergence velocity and local search ability of the algorithm to complete the task contingency planning task, aiming at the large-scale complex multi-objective integer optimization problem involved in task planning. After completing the task contingency planning in the determined environment, the strategic task planning strategy and cooperative mechanism of missile formation in uncertain environment are given, and the feasibility of verifying the dynamic task planning strategy in the dynamic process is designed, overcoming the shortcomings of multiple rounds of frequent communication [4].

MAF should implement a rational assignment of targets in the theater within the specified time and realize real-time reconstruction of battlefield situation and real-time task planning when attacking target groups. It is asked that the battlefield situation reconstruction time of a single missile members should be best in the millisecond level, and the battlefield situation reconstruction time of a small-scale missile formation should be best in the second level. The principle of assignment is to maximize the probability of destroying the target group and to avoid repeated attacks and omissions. The target assignment problem is a NP problem. The solution method includes Hungarian algorithm, exhaustive method and improved exhaustive method, and a large number of research algorithms based on modern optimization theory, such as simulated annealing, genetic algorithm, taboo search algorithm, neural network, etc. When the problem is large, the large computational cost is the difficulty of the NP problem. Real-time is difficult to guarantee for most problems and it is difficult to meet the requirements of online real-time dynamic assignment.

In the following, we first give the target dynamic assignment model and the effective assignment algorithm of formation co-guidance. The initial solution is obtained by using an auction, and then the initial solution is corrected based on the improved algorithm of taboo search. In this way, the sub-optimal target assignment scheme of the missile members can be obtained within the specified time. Finally, the distribution of all the members of the missile formation could be reached, synthesizing the distribution of all missile members and using the majority of the principle of voting.

4.4.2   Mathematic Models for Task Planning

Task planning mathematical modeling often used methods are parametric and nonparametric method. The parametric method needs to know the characteristic distribution of the target, and Bayes and other methods are used to evaluate the missile's superiority [83]. The nonparametric method mainly constructs the integrated advantage function according to the tactical geometric relation between the missile and the target, including the angle, distance and velocity between the missile and the target. Because the nonparametric method is easy to calculate in real time, it is widely used in practical engineering [84]. Here we use the nonparametric method to construct the synthetically advantage function, according to the angle, distance, velocity between the missile and the target, and missile capture probability to the target.

1. Task load optimization configuration

The optimal configuration of the task load is to improve the integrated combat effectiveness of the MAF aiming at the specific information such as combat task category, tactical program, target information (type, characteristics, quantity, location, combat damage state, etc.), threat information (type, characteristics, quantity, location, treat level, etc.), flight environment and so on. The optimal match of the task load is carried out before the launch, so as to create favorable conditions for the online task planning and the dynamic assignment in the duty shift region in the course of the task.

2. Capture probability

MAF coordinated attack is dependent on the formation support network. Although each missile member has a limited range of detection, it can only detect part of the airspace or track part of the targets, but through the support network can achieve information exchange and sharing, so that each missile member in the formation of a wider range of battlefield situation. They equal to a dynamically distributed seeker composed of a function of the expansion of the "complex eye seeker", through which it can detect a wider range. The missile formation support network inevitably has the problem of delay and packet loss, which leads to the need to obtain target information through the support network when the missile members themselves cannot detect the target. In this condition, the probability of capturing the target is lower than the probability that the sensor itself can capture the target case. The above situation is appropriately simplified, and the construction probability model is constructed as follows:

$$S_c = \begin{cases} p_{sc} & \text{search by seeker} \\ p_{nc} & \text{search by support network} \end{cases} \tag{4.20}$$

p_{nc} is a random variable relying on the support network conditions, $0 \le p_{nc} \le p_{sc} \le 1$.

3. **Angle Advantage**

Assume that the relative motion relationship between the missile and the target is shown in Fig. 4.3. M—stands for the missile. V—is the missile velocity. T—stands for the target. V_T—is the target velocity. LOS—is the line of sight between the missile and the target. η—is the angle between V and LOS, η_T is the angle between V_T and LOS.

When the velocity of the missile closer to the LOS missile is easier to attack the target. In other words, when η equals to zero, the missile has the biggest angle advantages on the target:

$$S_\varphi = e^{-\left(\frac{\eta}{a\pi}\right)^2} \tag{4.21}$$

a is a variable parameter varies with the distance between the missile and the target. Usually a has the proportional relationship with r because the bigger r the smaller the impact on S_φ.

4. **Distance Advantage**

Assume that seeking range of the missile seeker is $(R_{min} \sim R_{max})$, R_{min} and R_{max} are respectively the seeker near and far boundary. Distance advantage is very small when $r \gg R_{max}$ or $r \ll R_{min}$. While the distance advantage is greatest when $r = (R_{min} + R_{max})/2$. Therefore, the distance advantage function can also be constructed as follows:

$$S_r = e^{-\left(\frac{r-R_0}{\sigma_R}\right)^2} \tag{4.22}$$

In the function, $R_0 = (R_{min} + R_{max})/2$; σ_R is a parameter relative to the seeking range of the seeker. For passive seeker, the distance between missile and the target cannot be known by detecting. In this condition, the estimated value of the distance can be obtained by the self-seeker filtering algorithm and the information obtained through the support network.

Fig. 4.3 Relationship of missile and target motion

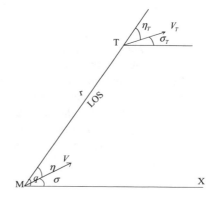

5. Velocity Advantage

In most cases, only when the velocity of the missile is bigger than the target velocity, the missile can attack to the target. Therefore, the following equation could be used to construct the velocity function between the missile and the target:

$$S_v = 1.0 - V_T/V \qquad (4.23)$$

6. The Matching Advantage between Seeker and Target Radiation Source

In the configuration optimizing of the MAF seeker, not only to consider the missile's target damage capability, target type, arrival time and combat capability, but also to consider the multi-seeker collaborative combination of configuration is needed to better play the missile formation seeker of the cooperation detection capabilities and anti-interference capability.

The face of the target to exert a variety of soft and hard defense measures, the detection capacity to the target and anti-interference effectiveness of the missile is related to the seeker guidance system and the target countermeasure type. At the same time the missile formation improve the overall combat effectiveness through the support network for effective information cooperation.

(1) Seeker Optimization Configuration Modeling

Missile set $\{M_n[N_S][N_T]\}$: Integer encoding is used, the length of coding sequence S equals to the size of the missile formation n, M_i is the elements of the set, $1 \leq i \leq n$, each element M_i has two attribute values which respectively represents the type of seeker and the target number to attack. Total amount of seekers' type is N_s.

$M_i[j][k]$: shows the type of the i-th missile seeker is j, this missile will attack the k-th target. $1 \leq i \leq n$, $1 \leq j \leq N_s$, $1 \leq k \leq N_T$, thus, n missiles attack N_T targets (missiles attack the same target make up a small formation). For example, $n = 10$, $N_s = 3$, $N_T = 5$, j has the following values:

$$j = \begin{cases} 1 & \text{Active Radar seeker} \\ 2 & \text{Passive Radar seeker} \\ 3 & \text{Photoelectric seeker} \end{cases}$$

The encoding of the seekers optimized configuration is shown below (Fig. 4.4).

Target defense confrontation set $\{D[N_D]\}$: list the target's a variety of soft and hard defense measures according to the order of time, $1 \leq l \leq N_D$. Common and typical defensive measures the target could take are shown in Table 4.1.

Efficiency matrix E': the element $E'_{k,l}$ is the basic effectiveness for the k-th kind of seeker to attack the l-th kind of defense measures. For example, if the effectiveness for active radar seeker to attack the infrared/smoke interference equals to 1. Thus, $E'_{1,7} = E'_{1,8} = 1$.

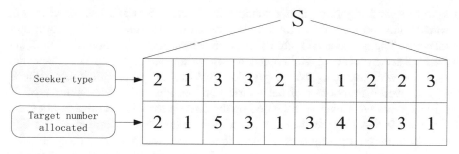

Fig. 4.4 Seeker optimization configuration coding method

Table 4.1 Target defense methods

Number	Target against method	Against object
1	Active bait	Passive radar seeker
2	Remote air defense missile	Missile body
3	Chaff/Angle backflush interference	Active radar seeker
4	Shipborne active jammers	Active radar seeker
5	Mid-range air defense missile	Missile body
6	Outboard active interference	Passive radar seeker
7	Infrared bait play	Infrared imaging seeker
8	Smoke bombs	Visible light seeker
9	Dense array near defense system	Missile body

$$E'[N_T][N_D] = \begin{bmatrix} E'_{1,1} & E'_{1,2} & \cdots & E'_{1,N_D} \\ E'_{2,1} & \ddots & \cdots & \cdots \\ \vdots & \vdots & \ddots & \cdots \\ E'_{N_T,1} & \cdots & \cdots & E'_{N_T,N_D} \end{bmatrix}$$

Missile-target radiation characteristic matching advantage matrix of missile formation S_s:

$$S_s = \sum_{k=1}^{N_T} \omega_i \left(\prod_{l=1}^{N_D} E'_{k,l} \right) \tag{4.24}$$

Weighting factor $0 \leq \omega_i \leq 1$, and $\sum \omega_k = 1$.

(2) Optimization Algorithm

Missile formation seeker optimization configuration is essentially a multi-constrained, strongly coupled complex multi-objective integer optimization problem, whose obvious feature is that the value of the decision variable is discrete.

Particle swarm algorithm is not suitable for discrete field and has the premature shortcoming, so that taboo search is introduced into discrete particle swarm optimization. The global search ability of particle swarm optimization algorithm, and the local search ability of taboo search algorithm are used to improve the calculation precision while reducing the computation time to meet the real-time requirement.

The overall framework of the hybrid optimization assignment strategy of the missile formation seeker is still based on the discrete particle swarm algorithm. The difference is that after the discrete particle swarm algorithm performing an iteration, the algorithm not directly start the next iterative process, but to start a local taboo search in the certain neighborhood of the $P_g(t)$ vector. If a better global optimal position is found within a certain neighborhood of the $P_g(t)$ vector, the vector is replaced with the position vector; If no better global optimal position is found within a certain neighborhood of the vector, the vector is not updated. That is, after each iteration of the particle swarm algorithm, a taboo search for the global optimal vector $P_g(t)$ is added to increase the local search capability of the algorithm. Figure 4.5 shows the flow of the missile formation seeker hybrid optimization algorithm.

(3) Distributed Detection of Missile Formation Seekers

Distributed detection is an important part of cooperation detection of missile formation seekers. It is necessary to establish a mathematical model of distributed detection data fusion.

The use of multiple seekers in different spatial locations to detect and track targets in the same area constitutes a seeker distributed network system. In engineering applications, with the use of distributed signal detection model, the seeker can independently detect and track the target. And the effect on the seeker network is less when any seeker fails.

The distributed signal detection characteristics mainly depend on the detection performance of each seeker and the working mode of the detection center. The working mode of the testing center is its judgment criterion. Usually, the rank K fusion criteria is used.

Rank K fusion criteria: Assuming that there are N seekers, in which there are at least K ($1 \leq K \leq N$) seekers to judge the target exists. The testing center will judge that the target exists. It is called the "OR" principle when $K = 1$. And it is called the "AND" principle when $K = N$.

In the judgement vector $D = (d_1, d_2, \ldots, d_N)$, $d_i = 0$ expresses that the target does not exist according to the i-th seeker's judgement. $di = 1$ expresses that the target exists according to the i-th seeker's judgement. The judgement rule of the rank K fusion criteria is as follow:

$$R(D) = \begin{cases} 1 & \text{(target exists)}, & if \sum_{i=1}^{N} d_i \geq K \\ 0 & \text{(target is not exist)}, & if \sum_{i=1}^{N} d_i < K \end{cases}$$

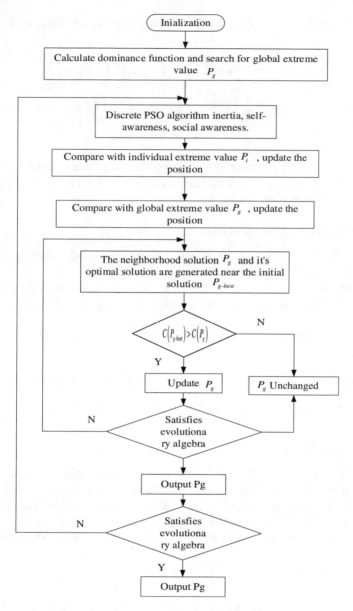

Fig. 4.5 Flow chart of hybrid optimal allocation algorithm for missile formation seeker

The detection probability and false alarm probability of distributed detection center are:

$$P_D = \sum_D R(D) \cdot \prod_{S_0} \left(1 - P_{d,i}\right) \cdot \prod_{S_1} P_{d,i}$$

$$P_{fa} = \sum_D R(D) \cdot \prod_{S_0} \left(1 - P_{fa,i}\right) \cdot \prod_{S_1} P_{fa,i}$$

Σ is all the possible combinations of D. $R(D)$ is the judgement rule of the testing center. S_0 is the seeker group that think the target does not exist. S_1 is the seeker group that think the target exists.

7. Synthetical advantage function

According to the advantages functions above, synthetical advantage function between the missile and the target could be constructed. In the case of a greater distance advantage or velocity advantage, if the angle deviation is very large (the angle advantage is very small), the synthetical advantage is not great.

$$S = S_c \times (C_1 \times S_\varphi \times S_R + C_2 \times S_\varphi \times S_v + C_3 \times S_s) \qquad (4.25)$$

C_1, C_2, C_3 are weight coefficients, whose value are decided by the size of the effect of each advantage to the synthetical advantage. $0 \le C_i \le 1$, and $\sum_{i=1}^{3} C_i = 1$.

8. Advantage matrix

Assume there are n missiles and N_T targets. These two values random dynamic change based on the battlefield situation and network connectivity. According to the mathematical model established in the previous section, the dynamic advantage function value S_{ij} of our i-th missile to the enemy's j-th target could be got by Eq. (4.25). Therefore, the missile-target advantage matrix could be established.

$$S = \begin{bmatrix} S_{11} & \cdots & S_{1N_T} \\ \vdots & \vdots & \vdots \\ & S_{ij} & \\ \vdots & \vdots & \vdots \\ S_{n1} & \cdots & S_{nN_T} \end{bmatrix} \qquad (4.26)$$

The target assignment problem can be described as follows:

$$J_{max} = \sum_{i=1}^{n} \sum_{j=1}^{N_T} S_{ij} X_{ij} \tag{4.27}$$

$$s.t \begin{cases} \sum_{j=1}^{N_T} X_{ij} = 1, & i = 1, 2 \ldots n \\ \sum_{i=1}^{n} X_{ij} \leq T_j, & j = 1, 2 \ldots N_T \\ X_{ij} = \{0, 1\} \end{cases} \tag{4.28}$$

In the equation, the first constraint indicates that a missile can only attack a target. The second constraint indicates that the target j can only be attacked by T_j missiles. Thus, target j needs T_j missiles, and T_j can be obtained from the relevant characteristics of the target. X_{ij} is decision variable and the range is $\{0, 1\}$. X_{ij} shows whether the i-th missile is assigned to the j-th target.

Take the case of a missile formation with a size of $n = 5$ to attack 3 targets. In this case, the time constraint is defined as follow: the maximum number of iterations of the algorithm is 50. Assume that each member of the missile formation is flying at a height of 2000 m. Each member's velocity is 200 m/s. Each member' flight path angle is 0. Each target is flying at the height of 2000 m and the velocity of 20 m/s. The view field parameters of the seeker in the simulation are shown in Table 4.2.

η_u and η_l are the upper and lower bounds of the seeker. q_l and q_r are the azimuth angle of the seeker. R_{max} and R_{min} are the far and near boundaries of the sight distance. The definition of its coordinate system and direction is given in the Ref. [85]. Assume that $p_{nc} = 0.7$, $p_{sc} = 0.95$, $a = 0.003 \cdot r$, $\sigma_R = (R_{min} + R_{max})/25$, $n_c = 1.0$, $n_d = 0.5$, $C_1 = C_2 = 0.35$, $C_3 = 0.3$, $w = 0.3$. The missile-target advantage matrix could be got according to the above mathematical model.

Table 4.2 The field parameters of the seeker in simulation

Number	$\eta_u(°)$	$\eta_l(°)$	$q_l(°)$	$q_r(°)$	R_{max} (km)	R_{min} (km)
1	5	−25	−30	30	40	0.5
2	10	−30	−45	45	50	0.5
3	10	−20	−25	25	35	0.2
4	5	−20	−40	40	60	0.2
5	5	−20	−20	20	60	0.2

$$S = \begin{bmatrix} 0.5538 & 0.9205 & 0.7692 \\ 0.9001 & 0.8410 & 0.8601 \\ 0.5250 & 0.7803 & 0.6801 \\ 0.9503 & 0.5799 & 0.7597 \\ 0.9361 & 0.5301 & 0.7203 \end{bmatrix} \qquad (4.29)$$

Figure 4.6 shows the cooperation assignment result using one round of auction improvement algorithm combined with the majority principles on a computer with a clock speed of 3.6 GHz. In the figure, □ is used to denote the infrared seeker and the infrared heat radiation source. ○ is used to denote the radar seeker and the electromagnetic radiation source. Figure 4.7 shows the search process for the assignment algorithm. It could be seen from the simulation results, the distribution results is reasonable. Missile members ε_1 and ε_2 have the complementary type of seekers, which is conducive to improving the subsequent cooperation combat accuracy. And they are also very suitable to attack target 2 because of their geometric relationship. Likewise, the geometrical relationship between other missile members and the assigned target, its seekers' collocation and radiation source matching relationship both validate the rationality of the results obtained.

In the simulation experiment, the single-missile assignment algorithm takes 0.011 s, and the final assignment time with the synthetical network information is 0.182 s. Considering the update of the advantage matrix elements in each iteration, the objective function of the optimal solution is 3.4996, which is the same as the

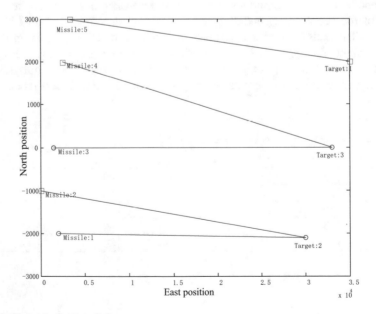

Fig. 4.6 Five missiles attack there targets

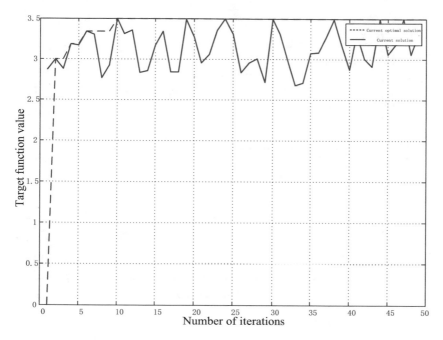

Fig. 4.7 Search progress

ideal objective function obtained by using the exhaustive method. Figure 4.7 shows that 10th iteration is needed to get the optimal solution. The results of multiple simulations show that the average iteration number of the optimal solution need seven times. The ratio of the objective function value of the optimal solution to the ideal objective function value (i.e. the optimal ratio) is between 98.58 and 95%. Without using the initial solution generated by one round of auctions, but to use the stochastic initial solution, the average iteration number of the optimal solution is 68 which will exceed the maximum number of iterations specified by the time constraint. Thus, the use of one round of auction to improve the assignment speed of the algorithm to meet the real-time requirements of the battlefield and has a good optimization performance.

When the missile member ε_5 equips the radar seeker instead, the distribution results shown in Fig. 4.8. As it is shown the figure, since the seeker type of the missile member ε_5 no longer matches the target 1's radiation source and the missile member ε_4's seeker type matches the target 1 radiation source, the assignment scheme is changed. The missile member ε_4 is to attack the target 1, and the missile member ε_3, ε_5 are to attack the target 1.

Figure 4.9 gives the condition of ten missiles attacking eight targets, the time constraint of algorithm is the maximum number of iterations 35 times. Target 3 and target 7 each needs two missiles and other targets each needs 1 missiles. It can be seen that whether from the geometric relationship or from the sensor with the matching relationship, the assignment results are reasonable. The results of multiple

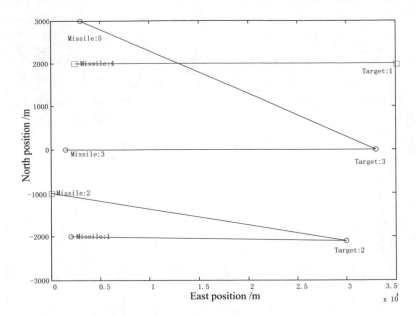

Fig. 4.8 ε_5 changes with radar seeker

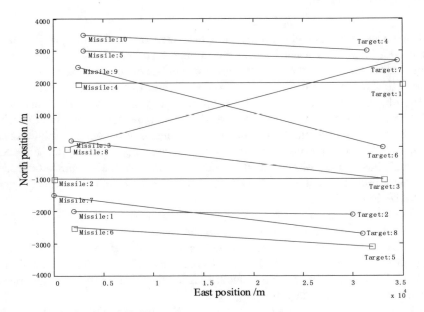

Fig. 4.9 Ten missiles attack eight targets

simulations show that the average iteration number of the optimal solution need 28 times and the probability that there is an optimization ratio more than 95% is 92.25%. It is visible that the algorithm is equally effective for larger scale assignments.

Using support network to provide shared information, taking into account the missile performance, target characteristics and the seeker-target radiation source matching and the relationship between the situations, the missile-target matrix model was established. Using an effective one round of auctions algorithm and an improved algorithm based on taboo search, a single-missile sub-optimal target dynamic assignment scheme for a specified period of time could be got. And then through the information sharing of all single-missile assignment program to obey the majority of the principle of voting, the results of the entire missile formation assignment was got. The simulation results show that the method is reasonable, practical and satisfies the real-time requirement.

4.4.3　Task Static Planning Method

After years of development, the three main solutions for missile formation task static planning are as follows:

(1) Assume that the solution of the discrete case is finite, and find all feasible solutions and then contrast. This method is called exhaustive or enumeration, generally only for small-scale problems.
(2) Firstly ignore the integer requirements, solute according to the continuous situation, and then take an integer processing for the solution. This method includes branch-and-bound method and cutting plane method. The computational complexity of the branch-and-bound method and the cutting plane method is less than that of the enumeration method, but the problem should be linear. The computational complexity increases exponentially with the size of the problem. Therefore these methods is rarely used for large-scale problems.
(3) The biomedical intelligence calculation is to abstract the natural biological community characteristics or phenomena into mathematical model, simulate the biological system of nature, rely entirely on the instincts of the organism, and optimize the survival state of the individual through the unconscious optimization behavior to meet the needs of the environment. Bionic intelligent calculation method has the following characteristics: probabilistic global optimization, not relying on optimization problems strictly mathematical properties, nature of parallelism, self-organization and evolution, and etc.

This section uses the third kind of biomedical intelligence solution. Firstly, the continuous algorithm is discretized, and the velocity and position updating formula of the discrete-time particle swarm algorithm are given. The algorithm is simple and easy to operate under the premise of ensuring its optimal performance. Compared with other group algorithms, the optimal solution of particle swarm optimization is unidirectional, and the algorithm is easy to get into the local optimal. In order to overcome this shortcoming, the taboo search algorithm is used as the supplementary strategy to strengthen the local optimization ability of the algorithm, so that the two algorithms can realize the complementary advantages to improve the search performance.

1. Encoding and Advantage Function of Task Planning Algorithm

The algorithm of the task planning algorithm is encoded by integer (as shown in Fig. 4.10). The length of the coding sequence S is equal to the size of the missile formation n. The elements of it are denoted as s_i, $1 \leq i \leq n$. The positive random number in each element represents the number of the task executed by the missile member. Assume that the amount of the task is N_T, $1 \leq s_i \leq N_T$. For example that the missile formation size is $n = 20$, and there are $N_T = 10$ tasks, there are 20 elements in the figure. Thus the coding sequence of Fig. 4.10 indicates that the missile member ε_1 executes task 2, the missile member ε_2 executes task 1, the missile member ε_3 executes task 10, the missile member ε_4 executes task 3.

The advantage function for the missile formation to execute task is determined by the task advantage table and the coding sequence. Assume the i-th element of the coding sequence is j, thus $s_i = j$. Assume the target sequence exists T. The number of T is N_T. The elements are denoted as $t_i (i = 1, 2 \ldots 10)$. Then the advantage function C is calculated as follows:

$$C = \sum_{i=1}^{N_T} t_i \tag{4.30}$$

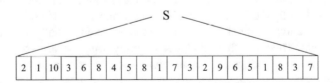

Fig. 4.10 Coding method of task planning algorithm

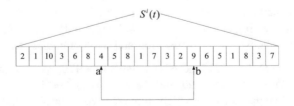

Fig. 4.11 Inertia part of discrete particle swarm algorithm inertia part

where

$$t_i = \begin{cases} \sum_{j=1}^{n} C_{j,i} & if\,(s_j = i, \sum_{j=1}^{n} C_{j,i} < 1) \\ 1 & if\,(s_j = i, \sum_{j=1}^{n} C_{j,i} \geq 1) \end{cases}$$

In this equation, $C_{i,j}$ denotes how well the missile member ε_i executes its duty j. The synthetical advantage function is one of its important variables. $t_i = 1$ denotes that the missile formation has executed the task t_i completely. So the value of advantage function is taken as one and no longer increases.

2. Velocity and location update of discrete particle swarm algorithm

The standard particle swarm algorithm is described below: Located in a M dimensional target search space, there are m particles composed of a group, where the position of the particle i is expressed as $S^i = (s_1^i, s_2^i, \ldots, s_M^i)$, $i = 1, 2, \ldots, m$. Thus, the position of the i-th particle in the M dimensional space is S^i. The position of each particle S^i is a potential feasible solution, which can be substituted into the advantage formula function to obtain its advantage function value, which is used to judge the extent of the advantages and disadvantages of particles X_i. The velocity of the particle i response is expressed as $V^i = (v_1^i, v_2^i, \ldots, v_M^i), i = 1, 2, \ldots, m$. The best position experienced by the particles i is recorded as the individual extreme of the particle i, which is expressed as $P^i = (p_1^i, p_2^i, \ldots, p_M^i), i = 1, 2, \ldots, m$. The best position that all particles of the whole group have experienced are recorded as global extremes, which is expressed as $P_g = (p_{g1}, p_{g2}, \ldots, p_{gM})$. Each particle updates its speed and position based on individual extremes and global extremes, as shown in the following equation:

$$\begin{cases} v_j^i(t+1) = \omega \cdot v_j^i(t) + c_1 \cdot rand1 \cdot \left(p_j^i(t) - s_j^i(t)\right) + c_2 \cdot rand2 \cdot \left(p_{gj}(t) - s_j^i(t)\right) \\ s_j^i(t+1) = s_j^i(t) + v_j^i(t+1), \quad j = 1, 2, \ldots, M \end{cases}$$

$$(4.31)$$

In this equation, t is the evolution generation number, ω is the Inertia weight, which determines the degree of success of the particle's current velocity, the right choice of which allows the particles to have a balanced exploration and development capability. c_1 and c_2 are positive constants, called the learning factor, usually to be random numbers evenly distributed between the [0,1] and could have the ability to be close to its own historical advantages and the group historical advantages.

It can be seen that the standard particle swarm algorithm is suitable for continuous problem, and it is not suitable for solving the problem of integer optimization. Therefore, a discrete particle swarm optimization algorithm based on cross is given.

The essence of particle swarm algorithm is the particles constantly adjust the location and velocity according to their own and their companions' flight experience, so as to the optimal location of flight. The new position of the particle is the result of the interaction between the particle velocity, the individual extremum and the global extremum. Therefore, the position and velocity update formula of the particle swarm algorithm is redefined:

$$S^i(t+1) = f_3\{c_2, P_g(t), f_2[c_1, P^i(t), f_1(\omega, S^i(t))]\} \quad (4.32)$$

Define $S^{i'}$ and $S^{i'''}$ as the intermediate variables, then:

$$S^{i'}(t) = f_1(\omega, S^i(t)) = \begin{cases} g_1(S^i(t)), \ rand() \leq \omega \\ S^i(t), \ rand() > \omega \end{cases} \quad (4.33)$$

Equation (4.33) is the inertial part of the discrete particle swarm algorithm. An evenly distributed random number between [0, 1] was generates by the $rand$ () function. If the random number is greater than ω, the particle $S^{i'}(t) = S^i(t)$ does not change. If the random number is smaller than ω, $S^{i'}(t) = g_1(S^i(t)).g_1(S^i(t))$ represents the difference two random numbers a and b between [1, n]. Then swap $s^i_a(t)$ and $s^i_b(t)$, elements of the vector $s^i(t)$, as Fig. 4.11:

$$S^{i'''}(t) = f_2(c_1, p^i(t), S^{i'}('t)) = \begin{cases} g_2(P^i(t), S^{i'}('t)), \ rand() \leq c_1 \\ S^{i'}(t), \ rand() > c_1 \end{cases} \quad (4.34)$$

Equation (4.34) is the self-cognition part of the discrete particle swarm algorithm. An evenly distributed random number between [0, 1] was generates by the $rand()$ function. If the random number is greater than c_1, the particle $S^{i'''}(t) = S^{i'}(t)$ does not change. If the random number is smaller than c_1, $S^{i'''}(t) = g_2(P^i(t), S^{i'}('t))$. $g_2(P^i(t), S^{i'}('t))$ represents the difference two random numbers a and b between

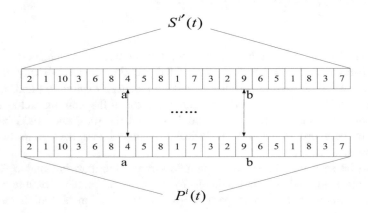

Fig. 4.12 Self-cognition of discrete particle swarm optimization

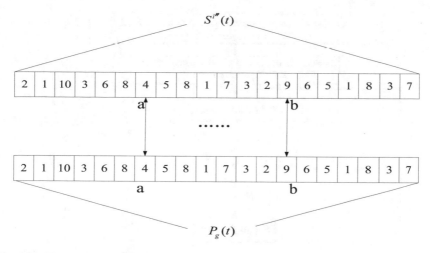

Fig. 4.13 The social cognitive part of discrete particle swarm optimization

$[1, n]$. Then swap the elements between $[a, b]$ in the vector $S^i(t)$ and the corresponding elements in $P^i(t)$, as Fig. 4.12:

$$S^i(t+1) = f_3\left(c_2, P_g(t), S^{i'''}(t)\right) = \begin{cases} g_3\left(P_g(t), S^{i'''}(t)\right), & rand() \leq c_2 \\ S^{i'''}(t), & rand() > c_2 \end{cases} \quad (4.35)$$

Equation (4.35) is the social cognition part of the discrete particle swarm algorithm. An evenly distributed random number between $[0, 1]$ was generates by the $rand()$ function. If the random number is greater than, the particle $S^{i'''}(t+1) = S^{i'''}(t)$ does not change. If the random number is smaller than c_2, $S^i(t+1) = g_3\left(P_g(t), S^{i'''}(t)\right)$. $g_3\left(P_g(t), S^{i'''}(t)\right)$ represents the difference two random numbers a and b between $[1, n]$. Then swap the elements between $[a, b]$ in the vector $S^{i'''}(t)$ and the corresponding elements in $P_g(t)$, as Fig. 4.13. Flowchart of discrete particle swarm algorithm is shown in the Fig. 4.14.

3. **Taboo search algorithm**

Taboo search algorithm has the advantages of fast convergence, strong local search ability and so on. But the convergence result of the algorithm is greatly influenced by the initial solution, and the initial solution is easy to make the algorithm fall into the local optimal [86, 87].

The taboo search procedure used here is as follows:

First set the algorithm parameters. The taboo table is a first-in first-out queue, defined as L_{tabu}, has a length of 10. If the time of the coding sequence staying in the taboo table $n_{tabu} > 5$, the coding sequence is deleted from the taboo table, indicating that the coding sequence is forgotten. Then the initial $S^i(t)$, $t = 0$ is randomly generated, and the current solution $S^i_{cur}(0) = S^i(0)$.

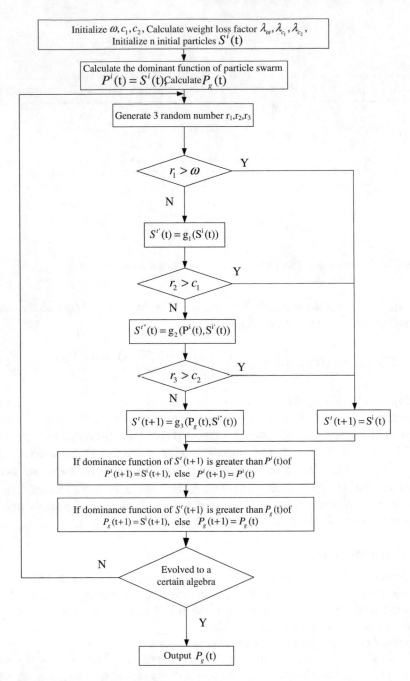

Fig. 4.14 Flowchart of discrete particle swarm optimization

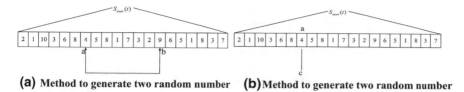

(a) Method to generate two random number **(b)** Method to generate two random number

Fig. 4.15 Taboo search constructs neighborhood

In the vicinity of the initial solution, there are a number of neighbor solutions in the taboo table, and the number of neighbor solution is $n = 10$. Construct the neighbor solution in two ways (as Fig. 4.15): The first is to produce two different random numbers a and b between $[1, n]$, and then swap the element $S_a^i(t)$ and $S_b^i(t)$ of the vector $S^i(t)$. The second is to produce a random number c between $[1, N_T]$ and a random number a between $[1, n]$. Then take the place of element a with the element c and get the maximum solution $S_{best}^i(t)$ of the advantage function in the set of neighbor solutions.

If the advantage function $C(S_{best}^i(t))$ of $S_{best}^i(t)$ is greater than the advantage function $C(S^i(t))$ of $S^i(t)$, then let $S_{cur}^i(t) = S_{best}^i(t)$. And add $S_{cur}^i(t)$ to the taboo table, update the duration of the taboo table, and remove the elements with a duration $n_{tabu} > 5$. If the advantage function $C(S_{best}^i(t))$ of $S_{best}^i(t)$ is smaller than the advantage function $C(S^i(t))$ of $S^i(t)$, then $S_{cur}^i(t)$ does not change. And put $S_{best}^i(t)$ into the taboo table. Therefore, update the duration of the taboo table, and delete the elements whose duration time $n_{tabu} > 5$.

Then, judge whether the algorithm iterates a certain algebra, and if so, end the algorithm and output optimization result; otherwise the algorithm turns to the start step (as Fig. 4.16).

It can be seen from the above process, neighbor function, taboo table and duration time is the key to taboo search, in which the neighbor function follows the idea of local search. The taboo table reflects that the algorithm could avoid the shortcoming of roundabout search; and the duration time is a loosing of the taboo strategy.

The taboo search algorithm has the following shortcomings: (1) it has a strong dependence on the initial solution. A good initial solution allows the taboo search to search for a good solution in the solution space while the bad initial solution will reduce the taboo search convergence rate. (2) The iterative search is serial, only a single state of movement, rather than a parallel search. Aiming at overcoming these shortcomings, the discrete particle swarm optimization algorithm and the taboo search algorithm are combined to form a hybrid integer optimization strategy.

4. Combined Optimization Strategy

The hybrid optimization strategy flow is shown in the Fig. 4.17.

Discrete particle swarm optimization algorithm has the advantages of simple, easy to implement and having strong global search ability. However, the algorithm

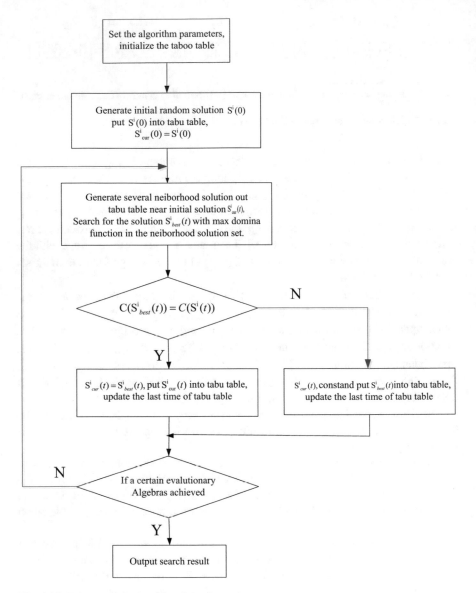

Fig. 4.16 Taboo search algorithm flow chart

has the advantages of low post convergence speed and worse solving precision. The discrete particle swarm algorithm is guided relying on the cooperation and the competitive among the initial random population. Once a particle is found in a current optimal position, the position is passed to the other particles in the social cognitive part of the discrete particle swarm algorithm. The other particles are quickly approaching the optimal position. If the location is locally optimal, the

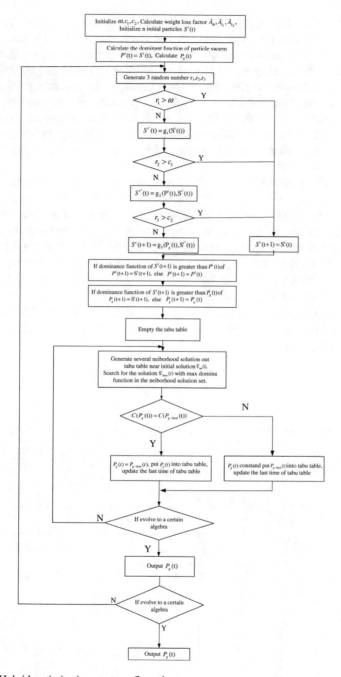

Fig. 4.17 Hybrid optimization strategy flow chart

other particles in the population cannot search for other regions, causing the algorithm to fall into the local optimal. Secondly, due to the existence of inertia weight, so that the particles in the group would fly with a certain inertial, it is possible the particles may miss the optimal solution in the flight process.

In order to overcome the above shortcomings, the discrete particle swarm optimization algorithm and the taboo search algorithm are combined to make full use of the global optimization ability of the discrete particle swarm algorithm and the local search ability of the taboo search algorithm so that the two algorithms can complement each other. This is a current solution that is often used for large-scale optimization problems.

The overall framework of the hybrid optimization strategy of missile formation task is still based on discrete particle swarm optimization. The difference is that the discrete particle swarm algorithm iterations after one time, not directly into the next iterative process, but to take a taboo search in a certain neighborhood of the vector $P_g(t)$. If a better global optimal position is found within a certain neighborhood of the vector $P_g(t)$, the vector $P_g(t)$ is replaced with the position vector, and if no better global optimal position is found within a certain neighborhood of the vector $P_g(t)$, the vector is not updated. That is, after each iteration of the particle swarm algorithm, a taboo search operation for the global optimal vector $P_g(t)$ is added to increase the local search capability of the algorithm. The flow chart of the algorithm is shown in Fig. 4.18.

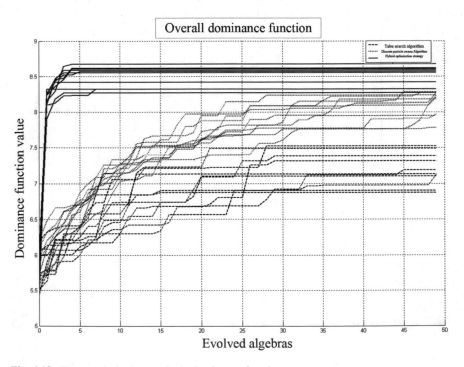

Fig. 4.18 Three optimization methods dominance function curve

5. Simulation

In the experiment consisting of 10 missiles and 10 targets, initial value of ω is 0.9 which eventually decrease to 0.4 as the evolutionary algebra increases. The initial value of c_1, c_2 is 0.6, which eventually decrease to 0.2 as the evolutionary algebra increases. Taboo table length is 10 and taboo time is 5. The computer used has the processor of AMD Athlon64 3800 +, memory 4 GB, the compiler environment for VisualStudio6.0. Figure 4.18 shows the change curve of the overall advantage function of the formation based on three missile formation task planning methods: the discrete particle swarm optimization algorithm, the taboo search algorithm and the combine of the discrete particle swarm algorithm and the taboo search algorithm. Monte Carlo simulations with 50 times of simulations and 50 generations of evolutionary algebra were performed on the three strategies. It can be seen from the Fig. 4.19 that the planning algorithm using the mixed strategy can converge to the optimal solution after 10 generations evolution. And the optimal solution are better than the first two methods in both the convergence speed and the convergence quality.

To sum up, following conclusions could be got:

(1) The discrete particle swarm optimization method presented in this section is to discretize successive algorithms and is based on the velocity and position update formula of the discrete particle swarm algorithm, which makes the

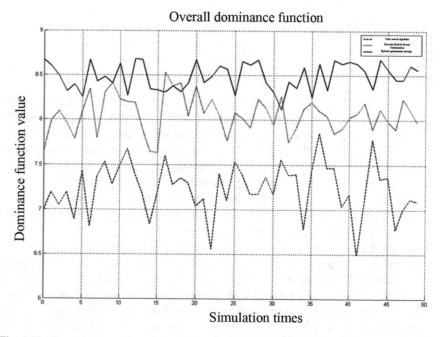

Fig. 4.19 The maximum value of the dominance function of the three optimization methods (20 missiles, 10 targets)

algorithm simple and easy to operate under the premise of guaranteeing its optimizing performance. Simulation results verify the effectiveness of the search method.

(2) The results show that the information sharing mechanism of particle swarm algorithm is different from other algorithms. In the case of genetic algorithms, the information of the chromosomes is shared by the roulette method, that is, the velocity of the whole population moving to the optimal area is more uniform. While the optimal particles of the particle swarm algorithm affect all the particles in the population during each evolutionary process, so the algorithm is easy to get into the local optimum. In view of this shortcoming, a supplementary strategy of taboo search algorithm is adopted. When the inertia weight of the particle swarm algorithm is heavy, the local optimization ability of the algorithm is strengthened so that the original two algorithms could achieve the advantages of complementary. Simulation experiments also validate this feature.

4.4.4 Task Dynamic Planning Method

Missile task dynamic planning method refers to assigning a group or an orderly task (target or spatial location) for each missile member in the process of mission formation, based on certain environmental knowledge and task requirements, according to the number of missiles or load. The purpose is to make the overall efficiency of the missile formation to be optimal at the same time as the maximum possible number of tasks is completed. Dynamic task planning has developed the following two major solutions after nearly a decade of development.

(1) The task planning method based on modern optimization algorithm is to use the method of group rolling optimization to search the best scheme of missile formation cooperative dynamic task planning whose calculation time is long. This method does not apply to real-time strong or uncertain environment conditions. And for this method, most of the problems involved in are single task planning problems. These problems do not reflect the movement of multiple constraints, strong coupling and time-related features. At present, most of these algorithms are used in the static task planning process where the missile information and task information are known.

(2) The multi-agent system method abstracts the task planning problem into the following questions: In multi-agent systems (MAS), there are M different agents to perform N different tasks, how to rationally plan different agents, so as to improve the overall efficiency of the system. Agent usually has two types: one is cooperation, one is self-interest. The auction method based on the market mechanism is the general method to solve the dynamic task planning in the multi-agent system method. The shortcoming is that, in order to achieve optimality, multiple rounds of frequent communication between the agents are

required, which increases the probability of being found for missile formation with missions in the enemy area. To this end, this section will introduce a cooperation mechanism in which missile members could enter the respective mode through only three interactions, to avoid the frequent communication between the missile members.

1. Data Structures Maintained by Missile Members

To complete the cooperation dynamic task planning, each missile member's computer in the formation needs to store the following data:

(1) Missile members own serial number UID, used to confirm the identity of the missile members in the formation. When a mission is broadcasted by a broadcast member of the broadcast mode, the missile member in the response mode confirms the source of the broadcast information by the UID number.

(2) TID number: the number of the task each mission member performs to identify the mode in which the missile member is currently in and the task to be performed.

(3) The task load carried by the missile member itself: $\overline{R}^u_{UID} = [r^u_{UID1}, r^u_{UID2}, \ldots, r^u_{UIDk}, \ldots, r^x_{UIDx}]\overline{R}^u_{UID}$ is expressed in vector form: is the load type. Therefore, $\overline{R}^u_{UID}$ is a $1 \times x$ vector. As the missile members keep on executing the task, the load is constantly consumed until the vector all elements are zeros.

At this time the missile member cannot execute any target task and can only enter the search mode.

(4) The load required by the missile member to perform the TID task is $\overline{R}^t_{TID} = [r^t_{TID1}, r^t_{TID2}, \ldots, r^t_{TIDk}, \ldots, r^t_{TIDx}]$. $\overline{R}^t_{TID}$ is expressed in a $1 \times x$ vector form. The different elements of the vector represent the different kinds of loads required for the task. As the missile members continue to carry out the task, when the vector all the elements are all 0, that the task has been completed. The vector $\overline{R}^t_{TID}$ is empty in search mode.

A background diagram of the dynamic planning of missile formation tasks is given in Fig. 4.20. The pentagon in the figure is a missile; The small box represents the ground task point, which can be a rescue site or a threat to be destroyed. Peripheral solid line indicates the range of detection of each missile member, and the missile member can only find the task point within the detection range; As the missile formation flying into the search area, the number of task points and the task of the required load for the missile formation are unknown, so the task planning is actually determined and distinguished by the location of the task point. For convenience of description, use the task point TID number to represent the location of the task point. Assume that both the task load required by the missile member and the task load required for the target point are three; The number in parentheses in the lower right corner of the box represents the task number TID; The 1×3 vector $\overline{R}^t_{TID} = [r^t_{TID1} \quad r^t_{TID2} \quad r^t_{TID3}]$ in brackets is the load required to complete the task; The

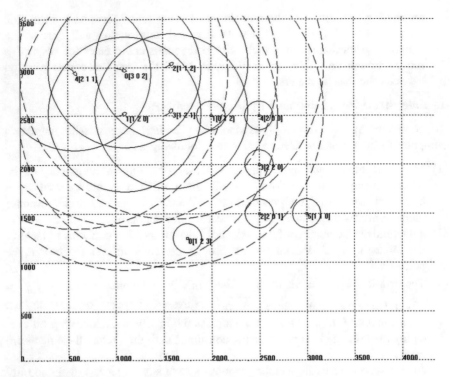

Fig. 4.20 Missile formation task dynamic planning background

different elements of the vector represent different kinds of loads; Numbers indicate the amount of various types of loads. For example, $\overline{R}_1^t = \begin{bmatrix} 0 & 1 & 2 \end{bmatrix}$ indicates that to complete the task in point 1 requires the first load of 0, the second load of 1, and the third load of 2. The amount of load required for the mission point continues to drop as the missile members continue to execute the task until all the elements in the vector equals to zeros. When all elements in $\overline{R}_{TID}^t = \begin{bmatrix} r_{TID1}^t & r_{TID2}^t & r_{TID3}^t \end{bmatrix}$ equal to zeros, the task is completed. The small circle outside of the small box represents the effective radius of the task point, and the missile member needs to detect whether it has entered the effective radius of the task point when executing the task. If a member has entered the effective radius it will put the corresponding task load, then the missile member's task has been completed. The dotted circle indicates the communication radius between the missile members, and the missile members can only communicate with other missile members in the communication radius.

This section uses the circular route as a missile to execute the task route for the following reasons: First, circular route is not easy to cause the missile overload exceeds the limit when the task switches during executing, the actuator is not easy to saturation, so to ensure that the missile flying into the circle with a radius of the effective radius of the task precisely. Second, the circle search makes it easy to expand the search to help find ground task points.

2. The Behavior Mode of Missile Members

The behavior modes of the missile members are divided into search mode and execution mode. The search mode is divided into broadcast mode and response mode. Missile members must be in one of these three modes at any time (Fig. 4.21).

(1) Searching mode ($TID = 1$): missile members do not find executable tasks in the detection area, and no other missile members provide executable tasks within the communication radius. In this condition, the this missile member is in search mode. In this mode, the missile members can execute the regional search according to the predetermined search route, enter the broadcast mode after discovering the task in the search process, share information with other missile members within the communication radius, and complete the task together. In the search mode, if a missile member encounter broadcasts of other missile members within the communication radius, it will enter the response mode and work with the broadcast missile members to complete the target task.

(2) Executing mode: the mode that a missile member finds tasks that can be executed independently within their own detection range and the mode that a missile member receives the broadcast from other missile members and responses.

3. Advantage Function for Missile Members to Execute the Task

Define the UID for the missile member to carry the load as $\overline{R}_i^u = \left[r_{i1}^u, r_{i2}^u, \ldots, r_{ik}^u, \ldots, r_{ix}^u\right]$, The task with the TID of requires a load of $\overline{R}_j^u = \left[r_{j1}^u, r_{j2}^u, \ldots, r_{jk}^u, \ldots, r_{jx}^u\right]$. The advantage vector $A_{ij} = \left[a_{ij1}, a_{ij2}, \ldots\ldots, a_{ijk}, \ldots\ldots a_{ijx}\right]$ calculation formula for the missile with the UID number i to the task with the TID j numberis as follow:

$$a_{ijk} = \begin{cases} 2 & r_{ik}^u \geq r_{jk}^t \ and \ r_{jk}^t \neq 0 \\ 1 & r_{jk}^t > r_{ik}^u \ and \ r_{ik}^u \neq 0 \quad k = 1, 2, 3\ldots x \\ 0 & otherwise \end{cases} \tag{4.36}$$

The equation of advantage function for the missile with the UID number i to execute the task with the TID j is as follow:

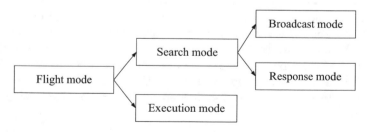

Fig. 4.21 The behavior pattern of the missile member

$$a_{ij} = \sum_{k=1}^{x} a_{ijk} \tag{4.37}$$

In the Eq. (4.36), the reason for the advantage value is as follow: When $a_{ijk} = 2$, the missile member with the UID number i is carrying a type k load, meeting the load needs of the task with the TID number j. In this condition, the missile members do not need other missile members to complete the task, and there is a maximum advantage value of 2. When $a_{ijk} = 1$, the load k of the missile member i is not enough to complete the task point requirements, but still have a certain contribution to the needs of the task point. The member i need to communicate within the radius of the other members by the radio broadcast, to complete the task together. When $a_{ijk} = 0$, the type k load on the missile with UID number i does not contribute to the task demand. It may be that the missile member i does not have a type k load or a task with a TID of j does not require a type k load. In this condition, advantage value of the missile number with the UID number i should not be greater than the advantage value of the missile members called when they find the task themselves. Otherwise it will bring frequent communication between the missile members, increase the probability of ground threat capture.

Combine the dynamic task planning flow chart shown in Fig. 4.22 and take the example to do a step further description for the calculation method of advantage function of missile members in carrying out their tasks: There are three types of load, and the missile member with UID 1 detect the TID 1 and 2 tasks in the detection range. $\bar{R}_1^u = \begin{bmatrix} r_{11}^u & r_{12}^u & r_{13}^u \end{bmatrix} = \begin{bmatrix} 1 & 2 & 0 \end{bmatrix}$, $\bar{R}_1^t = \begin{bmatrix} r_{11}^t & r_{12}^t & r_{13}^t \end{bmatrix} = \begin{bmatrix} 0 & 1 & 2 \end{bmatrix}$, $\bar{R}_2^t = \begin{bmatrix} r_{21}^t & r_{22}^t & r_{23}^t \end{bmatrix} = \begin{bmatrix} 2 & 0 & 1 \end{bmatrix}$.

According to the Eq. (4.36), we can get the advantage value vector for the UID 1 missile to TID 1 task as: $A_{11} = \begin{bmatrix} 0 & 2 & 0 \end{bmatrix}$. The advantage value vector for the UID 1 missile to UID 2 task is written as: $A_{12} = \begin{bmatrix} 1 & 0 & 0 \end{bmatrix}$. The advantage value function for the UID 1 missile to TID 1 task can be gotten using the Eq. (4.37): $a_{11} = 2$. The advantage value function for the UID 1 missile to TID 2 task can be gotten: $a_{12} = 1$.

4. The Dynamic Task Assignment Process

As shown in Fig. 4.22, the onboard detection equipment of the missile member of UID i searches the task point information at a certain period, if some task points were detected, each advantage function for the UID i missile member to execute the tasks will be calculated and the missile member will choose the task (TID is j) which let the value of advantage function be the largest. If the largest value for the UID i UAV to execute the TID j task is given as $a_{ij} = 0$, it means that the load carried by the UAV does not do any contribution to the task and at this time the UAV will keep the search mode, namely the TID is -1. In the other case, namely $a_{ij} \neq 0$, the UAV can do some contribution to the task, and then the UAV can execute this task.

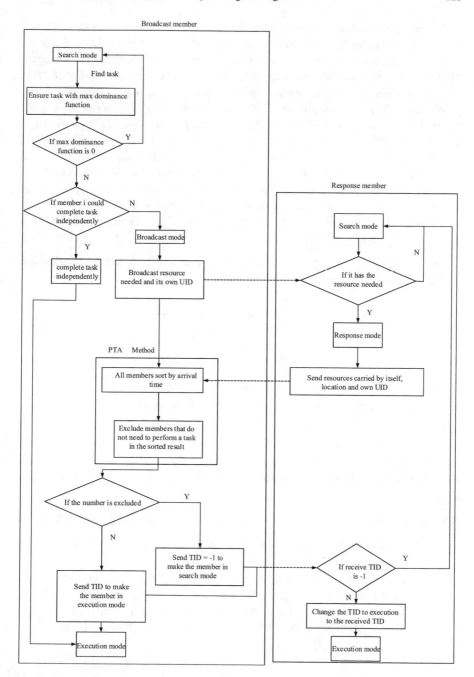

Fig. 4.22 Dynamic task planning flow chart

When the TID j task is determined to be executed, it needs to determine whether the missile member has the ability to independently complete the task. If the advantage function $a_{ij} \geq 2x$, the load carried by the missile member can complete the task alone without broadcasting, and else if $a_{ij} < 2x$, thus the missile member cannot complete independently and need other missile members to execute the TID j task together, so the missile member will get into broadcast mode. Taking the three kinds of load types as an example to further explain: When $a_{ij} = 6$, $a_{ijk} = 2$, $\forall \in N(1, 3)$, $N(1, 3)$ is the set of integers of 1–3. Then the members of the missile can execute the task using its own load. Then this missile enter the execution mode and $TID = j$. When $a_{ij} < 6$, indicating that the missile members do not have the ability to complete the TID j task independently. It need to broadcast calling other missile members as a supplement to complete the TID j task.

In the broadcast mode, the load carried by the missile member with UID i is $\overline{R}_i^u = \left[r_{i1}^u, r_{i2}^u, \ldots, r_{ik}^u, \ldots, r_{ix}^u \right]$. The load needed for TID j task is $\overline{R}_j^u = \left[r_{j1}^u, r_{j2}^u, \ldots, r_{jk}^u, \ldots, r_{jx}^u \right]$. The required load vector to be broadcast is calculated by Eq. 4.38:

$$\overline{S}_{ij}^u = \left[s_{ij1}^u, s_{ij2}^u, \ldots, s_{ijk}^u, \ldots, s_{ijx}^u \right]$$
$$s_{ijk}^u = \begin{cases} r_{jk}^t - r_{ik}^u, & r_{jk}^t > r_{ik}^u \\ 0, & otherwise \end{cases} \tag{4.38}$$

After the UID i member broadcasts its UID and the load vector $\overline{S}_{ij}^u$ that needed, all the missile members in the communication radius, in the search mode and has the elements in the broadcast vector $\overline{S}_{ij}^u$ will response. Other missile members keep in the search mode. The missiles enter the response mode will send its own UID, position and the information of loads carried $\overline{R}_{n_1}^u, \overline{R}_{n_2}^u, \ldots, \overline{R}_{nw}^u, \ldots, \overline{R}_{np}^u$ to the UID i missile member. The UID i missile member will deal with the information collected as follows:

(1) Set $C_{ij} = \{\varnothing\}$, $\overline{R}_j^{C_i} = \{\varnothing\}$ as the initial set. C_{ij} denotes to the set of UIDs of missile members eventually respond. $\overline{R}_j^{C_i}$ denotes to the set of task loads of the missile members that respond of C_{ij}.

(2) Calculate the estimated time of arrival (ETA) for the responding missile members to arrive the task j. In the case of small changes in missile velocity, the ETA for the missile members to fly into the task point can be easily calculated. Assume that the set of estimated time of arrival is $ETAs = \left\{ t_{n_1}, t_{n_2}, \ldots, t_{n_w}, \ldots, t_{n_p} \right\}$. Thus:

$$t_{n_w} \approx \frac{\pi R - r}{V_u} \tag{4.39}$$

where r demotes to the effective radius of the task point, V_u is the velocity of the missile, controlled in 50 m/s by the engine throttle as an example. R is the radius of the route at which the missile member performs the task.

(3) Sort the ETA of the responding missile members: $[N, T] = sort(ETAs)$, where $T = \left\{ t_{n_1^*}, t_{n_2^*}, \ldots, t_{n_w^*}, \ldots, t_{n_p^*} \right\}$ is the result of the ETAs sorted by small to large. $N = \left\{ n_1^*, n_2^*, \ldots, n_w^*, \ldots, n_p^* \right\}$ is the corresponding missile members' UID number.

(4) the set C_{ij} is supplemented in the order of N and $\overline{R}_j^{C_i}$ is supplied with $\overline{R}_{n^*}^u$ in the order of N, until $\overline{R}_j^{C_i} \geq \overline{S}_{ij}^u$ or All elements of N are supplied to C_{ij}. Then the supplementary set of responding missile members is $C_{ij} = \left\{ \hat{n}_1^*, \hat{n}_2^*, \ldots, \hat{n}_w^*, \ldots, \hat{n}_p^* \right\}$, $q \leq p$.

(5) Traverse each element $\hat{n}^*$ in C_{ij} and test whether $\overline{R}_j^{C_i} - \overline{R}_{\hat{n}^*}^u \geq \overline{S}_{ij}^u$ is true. If it is true, remove $\hat{n}^*$ from $\hat{n}^*$ and remove $\overline{R}_{\hat{n}^*}^u$ from $\overline{R}_j^{C_i}$. Finally the set C_{ij} and $\overline{R}_j^{C_i}$ without redundant elements are obtained. The TID of the responding missile member that is not supplied to C_{ij} and removed is set to -1. They are asked to search for other tasks in search mode. At last the TID of the missile members in C_{ij} are set to the task j execution mode.

Through the above method, once one missile members found a task to execute, the formation can determine the missile group to cooperate to execute the task and the missile members to continue to search after three communication interactions.

The reason for implementing the above (3), (4), (5) is to avoid using the Linear Programming method to obtain the C_{ij}. The purpose to express (3), (4), (5) in integer programming idea is shown as following equation:

$$
\begin{cases}
Objective\ Function: & \min_{\hat{T} \subseteq T} \max_{t_{n^*} \in \hat{T}} t_{n^*} \\
subject\ to: & \sum_{n^* \in C_{ij}, C_{ij} \subseteq N} \overline{R}_{n^*}^u \geq \overline{S}_{ij}^u
\end{cases}
\tag{4.40}
$$

where $\hat{T}$ is the subset of T corresponding to C_{ij}. Integer planning belongs to the NP-hard problem. As the number of missile members and the number of missions increases, the calculated amount of broadcast model missile members will produce a combined explosion. So the ETAs Sorting in (3) and elimination of redundant elements in the (4) are used to avoid the combined explosion problem in the integer programming. In this method, the calculation time of the missile members in broadcast mode will be reduced to polynomial time of the sort and remove operation, suitable for occasion with real-time requirements and application. The disadvantage is that the results are often sub-optimal. For the convenience of description, the steps (1) to (5) are referred to as PTA (Polynomial Time Algorithm) method.

5. Guidance law for missile members to execute task

Figure 4.23 shows the guidance of the mission member's mission. Missile members are in a fixed height during the entire search process. So the guidance law of the missile members to the search and execute task is mainly designed to set the track deviation command to ensure that each missile member in the formation can accurately enter the task arc. Assume that the position of a missile member at a certain moment is (x_u, y_u). This missile member determines to execute the task at point O when it is at point A, and then enters into the execution mode. The coordinate of A is (x_A, y_A), and the coordinate of O is (x_O, y_O). The point B is the middle point of A and O, its coordinate is (x_B, y_B). The missile member is hoped to fly into the task point O along the arc with the radius of BA. At this point the missile   member's   route   radius   to   perform   the   task   is $R = \frac{1}{2}\sqrt{(x_A - x_B)^2 + (y_A - y_B)^2}$.

The expected yaw angle instruction of the missile member at (x_u, y_u).

$$\varphi^* = \frac{\pi}{2} - \alpha \qquad (4.41)$$

where, α is the angle between the line of missile member to the task point and the ground coordinate system x axis. $\alpha = a\tan\left(\dfrac{y_u - y_B}{x_u - x_B}\right)$;

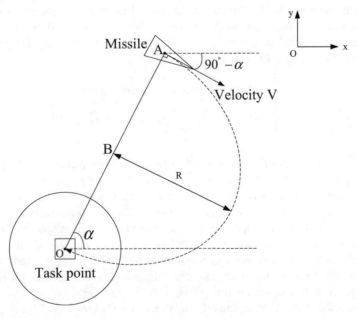

Fig. 4.23 The guidance of the missile member's mission

The feedback of the radius limit is added to the command of the flight yaw angle:

$$\varphi^* = \frac{\pi}{2} - a\tan\left(\frac{y_u - y_B}{x_u - x_B}\right) - z\left(\sqrt{(x_u - x_B)^2 + (y_u - y_B)^2} - R\right) \qquad (4.42)$$

where, Z is the feedback coefficient to be adjusted, here set to 0.002. When the missile member is in the search mode, this guidance law can also be used to make a circular motion along a pre-estimated search center. The above-mentioned guidance law ensures that the missile enters the task point along the arc-shaped trajectory.

6. Simulation experiment

Figure 4.24 is the simulation interface for the formation of dynamic task planning. In the simulation experiment, a five missiles formation is to perform six tasks, the initial state shown in Fig. 4.25. Each missile member has a detection radius of 800 m and a communication radius of 1600 m. When the missile member is in the search mode, the guidance law makes it a circular motion with the center of (2100, 2025), and a radius of 800 m, which is a typical case that the number of mission missiles is less than the number of missions.

When the missile members only choose their own range of the main function of the maximum range of tasks, there is no information between the missile members

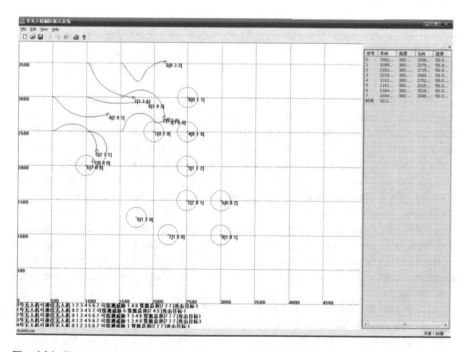

Fig. 4.24 Simulation interface of formation dynamic task planning

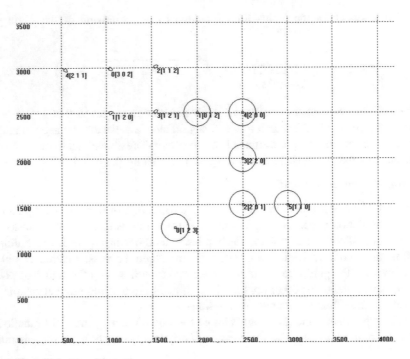

Fig. 4.25   Initial state of formation

of the interactive case, the five missiles to perform six missions of the missile members is shown in Fig. 4.26. Missile initial information and task point initial information are as Fig. 4.25, five missiles to execute all the tasks require time of 200.8 s.

When the missile formation uses the methods given in this section to implement cooperative search and dynamic task planning, the missile members used three communication interaction strategy. The flight trajectories of the missile members are shown in Fig. 4.27. The time required for the missile formation to execute all tasks is 109.2 s. It is seen that the efficiency of the task has been greatly improved.

The above-mentioned task planning method based on modern optimization concept takes into account the optimism and real-time requirements of missile formation, which is better suited to the conditions of environmental uncertainty. From the discovery of the target to enter the implementation mode, the missile members need only three communications, reducing the number of communications to the MAS auction method. The simulation results verify the effectiveness of the proposed method for solving dynamic programming problems in uncertain environments.

In this section, the static task planning method and the dynamic task planning method for missile formation are designed for task planning problems faced by missile formation before and during the task. In the static task planning method, through the improvement of traditional particle swarm algorithm and taboo search

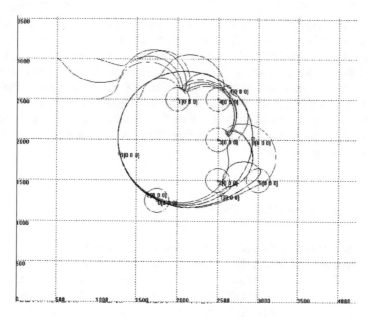

Fig. 4.26 No information interaction between the missile members

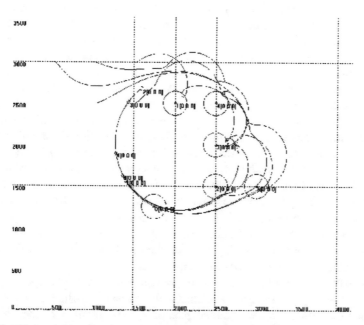

Fig. 4.27 Existing information interaction between the missile members

algorithm, and add taboo search into the discrete particle swarm optimization algorithm, to enhance the local optimization ability of discrete particle swarm algorithm. This method minimizes the calculation time in ensuring the quality of task planning. In the dynamic task planning method, the proposed suboptimal PTA method of polynomial time reduces the number of communication between the missile members and the calculated load of the missile members, which ensures the real-time performance for executing task of the broadcasting missile members and improves the quality for the missile formation to complete the task and the adaptability for application.

4.5   Cooperative Route Planning

Route planning refers to the process of finding a movement trajectory that satisfies a certain performance index from the initial point to the target point under certain constraints. The main purpose of the missile route planning is to find defense penetration routes that can be realized by Terrain Following, Terrain Avoidance and Threat Avoidance, which are optimized for comprehensive combat effectiveness.

The purpose of cooperative planning task of the missile formation is to ensure that the formation is able to reach the specified mission point sequence and target area on time in accordance with the requirements of the task planning to maximize the overall operational effectiveness of the entire missile formation. The task planning of the missile formation is to ensure that the formation is able to reach the specified task point sequence and target area on time in accordance with the requirements to maximize the overall combat effectiveness of the entire missile formation. For example, in order to enable the missile formation to achieve a saturating attack, it is necessary to rationally plan the route of each missile member in the formation to meet the saturation of the target-centered spatial distribution and its arrival time consistency. In the process of missile formation flight, due to factors such as speed and distance constraints, missiles may need to fly grouped or independent, so only through the cooperative route planning to ensure that all the missile members of formation in accordance with the task requirements to reach the target area, to achieve the purpose that entire missile formation has the maximizing overall combat effectiveness. This section will focus on the TF / TA^2 requirements of the missile formation, introducing the combination of global off-line planning and local online real-time planning to solve the problem of contradiction between route planning accuracy and real-time requirement.

Cooperative route planning can be decomposed into two parts: Firstly, single-missile TF / TA^2 integrated optimal route planning, including offline global route planning, online route dynamic planning (local route real-time planning). Secondly, the missile formation cooperative route planning.

4.5.1 Route Planning Methods

There are many route planning methods for single missiles. The route planning technology applied in practical engineering is also more mature. In this book, we introduce only a typical algorithm based on modern optimization principle. Other typical route planning methods can be found in the Refs. [1], [75].

1. Index of route planning

Task survival rate of tactical missiles:

$$P_s = \prod_{i=1}^{N} \left(\overline{P}_{Di} + P_{Di}\overline{P}_{ki}\right)\overline{P}_{Ci} \tag{4.43}$$

where, P_S is task survival probability, $\overline{P}_{Di}$ the probability of not being detected by the enemy (defense penetration probability), $\overline{P}_{ki}$ is the probability of being detected by the enemy but not shot down, $\overline{P}_{ki}$ is the probability of not hit the ground. To facilitate engineering applications, select the following simplified cost targets:

$$J = \int_0^{t_f} \left(\omega_1 C_t^2 + \omega_2 h^2 + \omega_3 f_{TA}\right)dt \tag{4.44}$$

where the first item is the penalty to the too large distance to the ground route to ensure that the missile will not deviate from the specific way point too far, reduce fuel consumption and flight time. The second item makes the flying height of the missile extremely small, to ensure the minimum altitude of the penetration route. The third is used to punish the route too close to the ground threat. Ratio $\frac{\omega_2}{\omega_1}$ and $\frac{\omega_3}{\omega_1}$ is used to coordinate the missile selection over the topographic obstacles and threat points, or choose to bypass them.

When given a flight area, it is usually necessary to first determine a basic reference route (In the offline planning before the launch, the reference route can be selected as the starting point and the target point of the connection; When online real-time planning, the reference route is the route before the launch of the route). Therefore, terrain data and known threat information are used, by making the performance index optimal, the three-dimensional ideal route to meet the requirements could be calculated. In the offline planning, the optimal route is in the large terrain that contains the starting point and the target point. In the real-time planning, the optimal route is located in front of the missile on a terrain. The length of the terrain should be appropriate, generally about 30 s flying missiles, as shown in Fig. 4.28.

When the missile is flying along the previous preferred route, the missile-borne computer will calculate the optimal route on the terrain that the next missile will fly. Therefore, as long as the calculation time of the optimal route is less than or equal

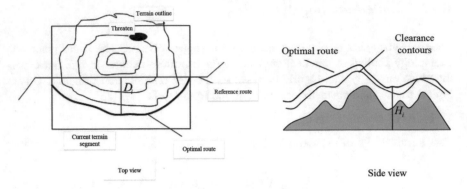

Fig. 4.28 Optimal route based on integrated TF/TA^2

to the optimal route adjustment interval, it can ensure that when the missile enters the next terrain, the optimal route on the terrain has been calculated to ensure that the missile could fly along the optimal route.

Since route planning is based on the digital map after the threat, terrain, digital integration, that threat has been dealt with similar terrain obstacles like the model, so the performance indicators do not have to consider the threat of this factor. So the comprehensive TF/TA^2 performance indicators can be written as:

$$J = \sum_i \left(\omega_1 d_i^2 + \omega_2 h_i^2 \right) \tag{4.45}$$

where, d_i is the deviation of the missile from the reference route; h_i the altitude of the missile. i is the sampling points on the best route, The sampling interval is desirable for the missile to travel at a certain speed within a certain time interval; ω_1, ω_2 is the TF/TA^2 ratio parameter; The scope of the value of the missile will directly affect the choice of flying obstacles to the strategy is to take leap maneuvers over obstacles, or select the lateral maneuver around the obstacles. The value should be based on the actual terrain data characteristics and task requirements to determine. As the performance of the missile deviation from the reference route from the horizontal distance and altitude are to be punished, the best route will be obtained by reference to the low-lying areas around the route.

2. Route Planning Method Based on Modern Optimization Method

In this section, an improved route planning algorithm for single missile is given based on Particle Swarm Optimization (PSO). Because its iterative formula is oriented to continuous space, it is more suitable to solve the non-grid topology of the route planning problem. The standard particle swarm optimization algorithm is prone to precocious phenomenon in the process of optimization. To overcome this flaw, an improved particle swarm optimization algorithm is presented here. The principle is to first select the elite particles and the poor particles according to the corresponding cost function, and adopt the speed update strategy with the kinetic energy compensation for the poor particles, thus avoiding the early maturation phenomenon in the optimization process; In addition, the experience of the worst particles in the movement of individual particles is introduced so that the particles in the population effectively avoid the worst case solution. A large number of simulation experiments show that this improved algorithm in the route planning application has a stronger search ability. The cost of the route planning obtained under the same evolutionary algebra is smaller [4].

(1) Standard PSO algorithm

When using the standard PSO to solve the optimization problem, the solution of the problem corresponds to a 'particle' in the search space, each particle has its own position and velocity, and a cost function determined by the optimized function. Each particle is remembers and follows the current optimal particle. In each iteration, the particle is updated by following two 'extremes' to update its position and velocity: One is the optimal solution $p_{best\,id}^k$ obtained by the particle itself; And the other is the optimal solution $g_{best\,id}^k$ obtained by searching all the particles in the whole particle group. In the standard PSO algorithm, the equation for speed update and position update is shown in Eq. (4.46):

$$\begin{cases} V_{id}^{k+1} = \omega V_i^k + c_1 r_1 \left(p_{best\,id}^k - x_{id}^k \right) + c_2 r_2 \left(g_{best\,id}^k - x_{id}^k \right) \\ x_{id}^{k+1} = x_{id}^k + V_{id}^{k+1} \end{cases} \quad (4.46)$$

where, $i \in N(1, m)$, m the number of particles in the particle group; $d \in N(1, n)$, n is the number of the dimensions of the solution; k is number of iterations; c_1 and c_2 are learning factors; ω is inertia weight; r_1 and r_2 are random number between [0, 1] [88−90].

(2) Improved PSO algorithm

The standard PSO algorithm is prone to precocious phenomena in the process of optimization. The reason is that in the standard PSO algorithm, each particle to obtain the shared information only the global optimal particle information $g_{best\,id}^k$. In the genetic algorithm, chromosomes share all the information through crossover and mutation; The pheromone distribution in the ant colony algorithm is also shared for all ants. Only share information $g_{best\,id}^k$ makes the standard PSO algorithm to accelerate the speed of convergence, and simple operation. But the cost is only the global optimal particle experience to spread to the group, which is a one-way information flow, prone to precocious. In order to overcome this shortcoming, the improved PSO algorithm improves the optimization effect by strengthening the information exchange between the particles within the group and within the particles.

(1) Dynamic velocity update strategy based on kinetic energy

Each particle of the standard PSO algorithm maintains two vectors, velocity vectors V_i, and position vectors x_i during evolution. The velocity vector of the particle determines the rate and direction of the motion. The position vector reflects the position of the solution represented by the particle in the solution space and is the measure of the solution cost (J). In the process of particle swarming, most of the particles will have their own history and the best position in the history of the best movement of the trend, it is easier to fall into the local and global optimal. At this time most of the kinetic energy of particles will be basically lost, from the group point of view entropy in the reduction. The idea of kinetic energy based velocity update strategy is to compensate the kinetic energy of the poor particle swarm with the kinetic energy loss of the elite particle swarm, so that it has a greater speed in the evolutionary process and increase the probability of finding more optimal solutions.

First select m_1 elite particle with smaller J in the m particles of the population. The kinetic energy is:

$$E_k^{m_1} = \sum_{j=1}^{m_1} V_j^T V_j \tag{4.47}$$

where k is the evolutionary algebra of PSO algorithm. The kinetic energy loss of an algebra of elite particle group is:

$$\Delta E_k^{m_1} = E_k^{m_1} - E_{k-1}^{m_1} \tag{4.48}$$

In this group, we select m_1 particles with larger J, and compensate the kinetic energy of the group to obtain the velocity and position of the poor population:

$$\begin{cases} V_{id}^{k+1} = \omega V_i^k + c_1 r_1 \left(p_{bestid}^k - x_{id}^k \right) + c_2 r_2 \left(g_{bestid}^k - x_{id}^k \right) + \Delta_{id} \\ x_{id}^{k+1} = x_{id}^k + V_{id}^{k+1} \\ \Delta_{id} = \Delta E_k^{m_1} / (m_1 \cdot n) \end{cases} \tag{4.49}$$

Equation (4.49) shows that Δ_{id} is essentially the penalty coefficient of the poor particle group induced by the elite particle swarm. In the evolutionary process, due to the poor particles get Δ_{id} compensation, so that the group will always have a part of the poor group in the evolution of algebra under certain conditions in an active state. They are in the search space to increase the scope of movement, is conducive to exploring the unknown optimal solution.

(2) Velocity update strategy based on worst particle

It can be seen from the Eq. (4.49) that for the standard PSO algorithm, a single particle in the process of movement, not only learns from their own individual behavior learning experience, but from the overall behavior of the group. This learning process is based on the optimal or suboptimal particles in the population as a reference. However, the worst case failure experience is also worthy of reference for the group. Based on the experience of the worst particle failure, this part introduces the velocity update strategy based on the worst particle and obtains the improved updating equation as follows:

$$\begin{cases} V_{id}^{k+1} = \omega V_i^k + c_1 r_1 \left(p_{bestid}^k - x_{id}^k \right) + c_2 r_2 \left(g_{bestid}^k - x_{id}^k \right) \\ \quad - \omega_1 r_1 \left(p_{worst\,id}^k - x_{id}^k \right) - \omega_2 r_2 \left(g_{worstid}^k - x_{id}^k \right) \\ x_{id}^{k+1} = x_{id}^k + V_{id}^{k+1} \end{cases} \quad (4.50)$$

where p_{worst} denotes to the position of the worst solution that the particle itself searches for; g_{worst} denotes to the position of the worst solution for the group search. ω_1 and ω_2 are new learning factors used to describe the role of the worst case solution and the global worst solution in the process of particle velocity updating [91].

The velocity and position of the improved particle swarm optimization algorithm are summarized by Eqs. (4.49) and (4.50) are as follows:

$$\text{Poor particles} \begin{cases} V_{id}^{k+1} = \omega V_i^k + c_1 r_1 \left(p_{best\,id}^k - x_{id}^k \right) + c_2 r_2 \left(g_{best\,id}^k - x_{id}^k \right) \\ \quad - \omega_1 r_1 \left(p_{worst\,id}^k - x_{id}^k \right) - \omega_2 r_2 \left(g_{worst\,id}^k - x_{id}^k \right) + \Delta_{id} \\ x_{id}^{k+1} = x_{id}^k + V_{id}^{k+1} \\ \Delta_{id} = \Delta E_k^{m_1} / (m_1 \cdot n) \end{cases} \quad (4.51)$$

$$\text{Other particles} \begin{cases} V_{id}^{k+1} = \omega V_i^k + c_1 r_1 \left(p_{best\,id}^k - x_{id}^k \right) + c_2 r_2 \left(g_{best\,id}^k - x_{id}^k \right) \\ \quad - \omega_1 r_1 \left(p_{worst\,id}^k - x_{id}^k \right) - \omega_2 r_2 \left(g_{worst\,id}^k - x_{id}^k \right) \\ x_{id}^{k+1} = x_{id}^k + V_{id}^{k+1} \end{cases}$$

(3) Indications determination of route planning

The traditional route planning algorithm must first describe the way of the missile in the form of a discrete network map. Each point on the network is associated with cost information (height information, offset information with reference routes, no-fly zone and threat information, etc.). The problem of finding the lowest-cost route problem for the lowest-cost route of the given starting point and the end-point network map. In order to solve the real-time problem of route planning algorithm in practical application engineering, when the influence of weather and other factors is not taken into account, usually make the following assumptions: The height of the missile during the flight from the ground is the height of the terrain and the height of the flight safety and is not introduced as a variable in the optimization process. This can simplify the three-dimensional route planning problem into two-dimensional programming problem, save the corresponding calculation time.

In summary, the traditional route planning algorithm in the form of discrete network map description of the waypoint information is not complete. Because in the topology of this discrete network map process will lose the route between the grid information, and the real optimal solution may be contained between the grid. The size of the grid and the calculation of the algorithm are contradictory. When the area of the route planning is the same, the larger the grid, the smaller the computational complexity of the algorithm, but the greater the probability of the optimal solution between the loss grids. The smaller the grid, the smaller the probability of loss of the optimal solution, the calculation of the algorithm will surge. Particle swarm algorithm has the inherent advantage in dealing with such problems. From the Eq. (4.46) we can see that the iterative equation and the iterative process of the particle swarm algorithm are continuous, but it shows the immaturity in dealing with the discrete problem, so choose the way point as shown in Fig. 4.29.

In Fig. 4.29, the connection between the starting point and the end point is the reference route, and the horizontal deviation a is given, which is convenient for planning. b denotes to the waypoint departs from the reference route, which is a continuous value, can be set to a random number within a range $[-range, +range]$. The corresponding optimization indicators are as follows:

$$\min(J) = \int_0^L (k_1 h + k_2 l + k_3 r) \tag{4.52}$$

where k_1, k_2 and k_3 denote respectively to, the missile's flight height, deviation from the reference route distance and no-fly zone factors. L is the route length; h, l and r denote respectively to the cost of the flight altitude of the waypoint, the distance from the reference route and the no-fly zone.

(4) Planning process

Step 1: Initialize and set the learning factor c_1, c_2, ω_1, ω_2, and inertia weight ω. Generated a particle composition of the initial m particle group in the solution space randomly. The position of the particle x represents the distance at which the

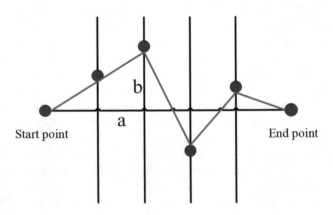

Fig. 4.29 Waypoint topology diagram

waypoint deviates from the reference route. The initial velocity V of each particle is randomly generated. The maximum value of $|V|$ is 15% of the maximum distance from the reference route.

Step 2: Calculate the value *pBest* and *gBest* of the initial group according to the Eq. (4.53).

Step 3: Update the velocity and position of the particles in the group according to the Eq. (4.52). To prevent the particle's position from crossing the set range $[-range, +range]$, the saturation function is used to limit the motion range of the particle, the equation is as follow:

$$x = \begin{cases} range & x > range \\ x & |x| < range \\ -range & x < -range \end{cases} \qquad (4.53)$$

Step 4: Evaluate each particle according to Eq. (4.53).

Step 5: Update the *pBest* according to the particle value.

Step 6: Update the group *gBest* according to each particle *pBest*.

Step 7: Check the end condition, if satisfied, end the optimization. Otherwise, go to step 3, and $k = k + 1$. The end condition is the optimal evolutionary algebra achieved by the optimization, or the difference between the two generations of the evaluation values is less than the given error e.

Some of the results of the route planning simulation experiment are given in Fig. 4.30. The simulation of the computer processor used for the AMD Athlon64 3800 + processor, memory 4 GB, the compiler environment for VisualStudio6.0. Figure 4.31 is the average cost curve of the standard PSO algorithm and the improved PSO algorithm over 30 contrast simulations. The average run time is 90.82 and 120.63 s respectively. Compared with the standard PSO algorithm, the

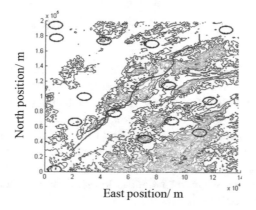

Fig. 4.30 Route plan result

Fig. 4.31 Average cost curve

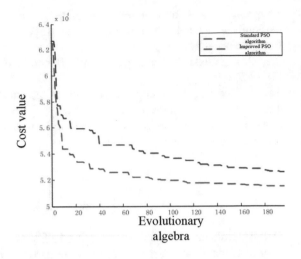

Fig. 4.32 Height and terrain

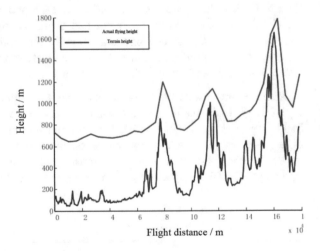

evolution of the PSO algorithm is accelerated on the basis of the experience of the worst particle failure. When evolved to about 20 generations, its generation value has dropped to 5.36×10^4. In the later stages of evolution, due to the kinetic energy compensation for the poor particle swarm, the velocity of the poor particles is increased and the solution to speed up to find of the solution more in line with the requirements. When evolved to about 200 generations, the value of its generation down to 5.15×10^4. As can be seen from Fig. 4.32, using the improved PSO algorithm to select the waypoint height is lower, more conducive to missile low altitude defense penetration flight.

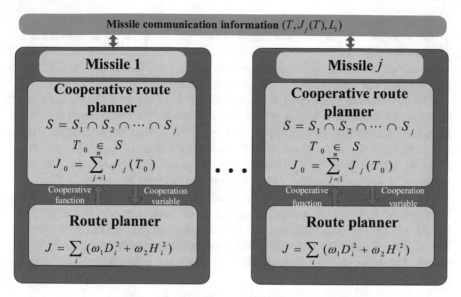

Fig. 4.33 Schematic diagram of cooperative route planning

4.5.2 Cooperative Route Planning Methods

Different from the single missile route planning problem, In addition to taking into account the requirements of each missile member's route planning, it is more important that the three basic principles of the formation should be followed to ensure that the overall operational effectiveness of the formation is maximized. In the cooperative route planning, there will be more optimization constraints and complexity and greater computational complexity. For example, when the missile formation is performing a saturating attack, according to the battlefield situation, flight environment information and missile formation condition, for each member of the missile in a timely manner to meet the requirements of the route, it is need to ensure that the formation of the maximum probability of survival at the same time to reach the target area to complete the saturation attack task. Missile formation in the course of the flight may fail to form formation, or damage so that the missile members can only separate fly due to network failure, electronic confrontation and other reasons. In some other conditions, missile formation need to be divided into a number of small teams to carry out combat task due to some missile members of the range, speed and other factors or the specific requirements of combat task. These circumstances will undoubtedly have a direct impact on the process and results of cooperative route planning, which will complicate the associated algorithm for cooperative route planning.

Typical cooperative route planning methods typically take both centralized and decentralized structures. The centralized planning is to solve the cooperative route planning by a missile member, and then the planned waypoint sent to other missile members. This method requires a lot of information to be passed, and the robustness of the algorithm is poor. While the decentralized planning is to overcome the shortcomings of each of the missile members have their own route planner, independently calculate their own optimal cooperative route. This section will focus on the decentralized cooperative route planning approach.

The main task of cooperative route planning is to ensure that the missile groups launched from different launch platforms are attacking the same target area, each missile in the missile cluster can plan an optimal or suboptimal route according to the given requirements and the missile formation in a coordinated manner at the expected arrival time T_f at the same time to reach the target area. Structural frameworks for decentralized cooperative route planning is shown in Fig. 4.33, each missile member has its own cooperative route planner and route planner. Route planer in accordance with the algorithm of the integrated optimal TF / TA^2 route to plan the optimal route and sub-optimal routes of multiple alternative routes. Through the cooperative route planner, the optimal expected arrival time T_0 for the missile formation, and the corresponding planning route for each missile member could be obtained.

The cost function of choosing cooperative route planning is:

$$J = \sum_{j=1}^{n} \sum_{i=1}^{m_j} J_j(X_i, V_i) \tag{4.54}$$

where n is number of Missile members planned for cooperative route planning (formation size), m_j is the number of routes planned by missile member ε_j, X_i is the number i route planned by missile member ε_j, V_i is the velocity of the missile member ε_j to fly along the number i route X_i, J_j is the cost of the route planned by missile member ε_j. These could be obtained by Eq. (4.45).

Each missile member can obtain multiple routes X_i to the target area by the TF/TA^2 optimal route planning method, Including the optimal route and sub-optimal route. Route X_i has a length of L_i. Find the estimated time range $s_i \in \left(\frac{L_i}{V_{jmax}}, \frac{L_i}{V_{jmin}} \right)$ for the missile members to fly to the target area along the planned route X_i, depending on the range of the missile ε_j's own velocity $V_j \in \left(V_{jmin}, V_{jmax} \right)$. The set S_j of expected times for the flight of the missile member ε_j to the target area is:

$$S_j \in \left(s_1 \quad s_2 \quad \ldots s_i \quad \ldots s_{m_j} \right) \tag{4.55}$$

Whereby the intersection of the expected time sets of all the missile members reaching the target area can be obtained:

$$S = S_1 \cap S_2 \cap \cdots \cap S_j$$

Expected arrival time of missile formation $T \in S$.

If the intersection S is not an empty set, take the time T_0 in the intersection makes the cooperative route planning cost function J minimum as the formation coordination time. The minimum cooperative route planning cost is $J_0 = \sum_{j=1}^{n} J_j(T_0)$. The velocity of each missile member corresponding to the coordination time T_0 is the flight velocity command of the missile member. Missile members fly in accordance with the corresponding planning route to the target area, to achieve the purpose of collaborative operations. If the intersection is empty, the second and third routes of suboptimal are selected by the TF/TA^2 optimal route planning method until the intersection of time sets is not empty.

In the process of cooperative route planning, the cost function of the route planning is the cooperative function, and the expected arrival time T of the missile members is the cooperative variable. In simple terms, the cooperative route planning task is to choose the appropriate T to minimize the sum of the synergistic functions of the missile members, but also to ensure that the selected T satisfies the performance constraints of the missile members.

1. Local route real-time planning

The local route real-time planning of missile members is mainly refers to: the missile members in the flight process, for some reason, cannot be planned in accordance with the best route flight off-line, according to the current situation need to re-plan a new optimal route to the target.

(1) In the formation duty shift area (mid-guidance to final guidance shift stage), the formation will be split into a number of small-scale formation or a single missile member need to plan the optimal route through the local route real-time planning algorithm to reach the target according to the selection result or the maximum spread of the target group.
(2) In the formation flight area (mid-guidance stage), when the use of lead-follow form, if the leader missile has failed, the new leader need to implement local route real-time planning.
(3) In the formation flight area (mid-guidance stage), the target location from the target indicating system or the relay information system changes greatly, the entire formation need to carry out local route real-time planning.
(4) In the course of the formation flight, due to some failures (such as network failure, etc.) or other reasons leading to the departure of the missile members, they need to implement real-time planning of local routes in order to get the most suitable route, as far as possible to continue to carry out combat task

As the local route real-time planning requires a certain amount of time $T(T \leq T_l,$ Tl is the maximum time required for offline planning), in time T the missile members need to fly along the transition route. A transitional route with a length of $L(L = T_l \cdot V)$ is set as a transition route, as shown in Fig. 4.34a in order to facilitate the engineering implementation and reduce the calculation time. When the missile member travels along the route section AC, the optimal route from the waypoint C to the target area B is planned. When a member of the missile moves along the

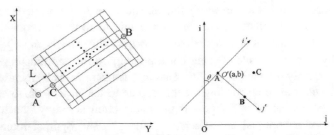

(a) **Real-time planning of local route** **(b)** **Map coordinate transformation diagram**

Fig. 4.34 Local route real-time planning and map coordinates conversion diagram

transition route AC, if the obstacle is detected, the missile member obstructs the obstacle avoidance method. The method of obstruction control for autonomous formation will be described in detail in Sect. 5.3.4.

Local route real-time planning requires: topological rectangular block size, shape can be arbitrarily selected. And it is need to establish a new coordinate system, through the coordinate transformation algorithm, to plan route between any two points in real-time.

The coordinate conversion method is shown in Fig. 4.34b. Planning the route between any two point A, B, it is necessary to take AB connection as a reference route and A point as the origin to establish a new coordinate system $i'\,O'\,j'$. Coordinates of θ and O' could be calculated by the coordinate of A and B. Thus, the new coordinate of any point C in the coordinate system $i'\,O'\,j'$ is (x,y). And the coordinate (x',y') of C in the inertial coordinate system iOj is calculated by Eq. (4.56):

$$\begin{cases} x' = (x - a)\cos\theta + (y - b)\sin\theta \\ y' = -(x - a)\sin\theta + (y - b)\cos\theta \end{cases} \tag{4.56}$$

2. Cooperative route planning technology of MAF

Task Planning/Target Dynamic Assignment is the prerequisite and foundation for cooperative route planning of MAF. The result of task planning/target dynamic assignment is the main constraint and input of cooperation route planning. In the process of optimization of integrated operational effectiveness, the find of the TF/TA^2 route of ensuring that the formation can reach the designated task point sequence and the target area of the route, could achieve the purpose of the largest integrated combat effectiveness of the entire missile formation.

In the following, a practical application example is introduced to apply the above-mentioned task planning/target dynamic assignment method and cooperative route planning method. In this example, solution for the cooperation route planning problem of MAF from the gather area to the duty shift area such as: the modeling of the planning problem, the reasonable extraction of the target assignment cost function, the optimal matching of the emission point-assembly area, the planning of the cooperative route, the simulation verification experiment, etc. The application of cooperative route planning is developed and the effectiveness of the above method is verified by practical application.

(1) Problem description

Take a certain region as a research area, the map format is USGS DEM, as shown in Fig. 4.35. Planning area for 200 km × 140 km, a total of 24 launch points, the specific requirements of the current battlefield situation given randomly. Each firing point can launch a missile, a total of 24 missiles can be launched. Given seven gather areas, each for the three-dimensional area, divided into several layers, the distance between the layers of at least 100 meters to ensure that the actual ballistic trajectory in the planning route within the error cone. There are limits on the number of missile accesses in the gather area, as shown in Table 4.3. There are a number of no-fly zones in the above-mentioned areas. The no-fly zone has a certain radius of protection, set at 5 km, representing the core areas of important cities on the map, etc. The tasks of the initial task planning and route planning are arbitrarily designated at the launch point, each launch point requires the launching of one missile and to meet the number of missiles entering the gather area and the nearest distance between routes. In the course of the missile flight no-fly zone cannot be crossed, can only be bypassed. In order to ensure flight safety, in the course of flight only when the flight distance is greater than a certain distance the next maneuver could be conducted and cannot conduct continuous maneuver. Route planning time should meet the tactical requirements within a limited time.

The figure shows the terrain of the planning area, the locations of the launch points, gather area locations and no-fly zone locations.

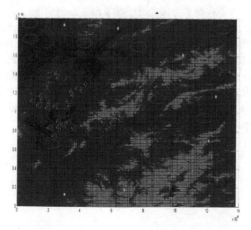

Fig. 4.35 Cooperative route planning background map of MAF

Table 4.3 From the launch point to the assembly area of the task requirements

Gather area	1	2	3	4	5	6	7
Missiles number	3	4	3	4	4	3	3

According to the knowledge of permutations and combinations, we know that the number of feasible solutions satisfying Table 4.3 in the solution space of the task planning from the emission point to the assembly area is:

$$n = C_{24}^3 C_{21}^4 C_{17}^3 C_{14}^4 C_{10}^4 C_6^3 C_3^3 = 3.4631 * 10^{16} \tag{4.57}$$

In such a large solution space to search for the optimal solution, using of traditional heuristic algorithm, the time required is unacceptable. We should use the intelligent calculation method to get the matching result between the launching point and the gather area to meet the requirement of the index, and then plan the specific route for each group matching result in the specified planning time. This requires the route planning algorithm to have short calculating time, easy to engineering practice. Since the dynamic programming approach has the advantage in dealing with small-scale route optimization, the method will be used here.

(2) Coordinate planning method of emission point and gather area

The problem of coordination between the launch point and the gather area is solved by genetic algorithm. The weighting of two cost functions is used as a performance index to describe the assignment of targets. Cost 1 is the sum of the distance from all launch points to the gather area; Cost 2 is the cost of describing whether there is a no-fly zone between the launch point and the assembly area and the distance between the no-fly zone to the gather area and the launch point. The smaller the cost 1, the smaller the probability that the projection of a plurality of missile ballistic projections intersect, the more safe of the flight. The smaller the cost 2, the less the no-fly zone on the connection point between the launch point and the gather area. If there is a no-fly zone on the connection line between the launch point and the gather area, the closer the no-fly zone to the midpoint of the connection, the less the cost 2. The reason for setting cost 2 is to reduce the pressure on the second phase of dynamic planning. If the no-fly zone is close to the launch point and the gather area, in the dynamic planning process, no matter how the topology tree is generated, it will intersect the no-fly zone, resulting in the failure of dynamic planning. In summary, the performance index function of the target assignment is as follows:

$$J = \sum_{i=1}^{24} Dist_{i-k} + \sum_{i=1}^{24} Is\, Cross_{i-k} \cdot \Delta_{i-k} \tag{4.58}$$

where

$$Is\,Cross_{i-k} = \begin{cases} 1 & if \quad \text{the } i-k \text{ connection intersects the no-fly zone} \\ 0 & if \quad otherwise \end{cases} \quad (4.59)$$

$$\Delta_{i-k} = \begin{cases} \alpha_2 \cdot Dist(midpoint\_i - k, prohibitedCircleCenter\_k) & IsCross_{i-k} = 1 \\ 0 & IsCross_{i-k} = 0 \end{cases}$$

$$(4.60)$$

where i is the launch point number, k is the gather area number, $Dist_{i-k}$ is the distance from launch point i to gather area k. $Dist(Midpoint\_i - k, Prohibited\,Circle\,Center\_k)$ is the distance between the midpoint of the $i - k$ connection line and the center of the intersection no-fly zone. α_2 is the adjustable parameter, reflecting the importance of different costs, preferably between the value of 1–10.

The fitness F is taken as the reciprocal of the index function:

$$F = \frac{1}{J} \quad (4.61)$$

The algorithm that determines the intersection of line segments and circles can refer to the sweep method in computer graphics.

(3) Dynamic planning method

The multi-stage decision process with n stages has numbers of $K = 0, 1, 2, \ldots, N - 1$. Allowing strategy $P^*_{0,n-1} = (x^*_0, x^*_1, \ldots, x^*_{n-1})$ is a sufficient condition for the optimal strategy. For any $k(0 < k < n - 1)$ and initial state $s_0 \in S_0$:

Where $P_{0,n-1} = (P_{0,k-1}, P_{k,n-1})$, $\tilde{s}_k = T_{k-1}(s_{k-1}, x_{k-1})$, It is the k-stage state determined by the initial state s_0 and sub-strategy $P_{0,k-1}. T_{k-1}(s_{k-1}, x_{k-1})$ is the state transition equation.

Fig. 4.36 Route and node diagram

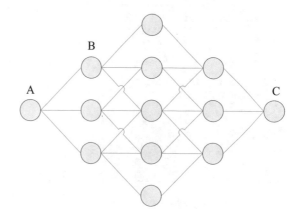

In order to facilitate engineering applications, take the cost function of a route as shown in Eq. (4.44). The tree topology is shown in Fig. 4.36. AC is the connection line between the emission point and the assembly area. The cost of point B is composed of three parts, the height of point B, the distance between point B and AC line, and whether there is no-fly zone f_{TA} on AB connection. Thus, a single route planning problem was changed into a given starting point of the end of the weighted map of the shortest road problem. The optimal route of single missile can be obtained by solving the shortest path of the graph by dynamic programming.

There may be a problem that distance between the routes is less than the set route safety distance, the following gives a way to determine the existence of such a route. The two endpoints of the route L_1 in the space are $A(x_a, y_a, z_a)$, $B(x_b, y_b, z_b)$; The two endpoints of the route L_2 in the space are $C(x_c, y_c, z_c)$, $D(x_d, y_d, z_d)$, The coordinates of the points X_1 on the route L_1 can be expressed as:

$$X_1(x_a + p \cdot (x_b - x_a), y_a + p \cdot (y_b - y_a), z_a + p \cdot (z_b - z_a)) \tag{4.63}$$

The coordinates of the points X_2 on the route L_2 can be expressed as:

$$X_2(x_c + q \cdot (x_d - x_c), y_c + q \cdot (y_d - y_c), z_c + q \cdot (z_d - z_c)) \tag{4.64}$$

Thus the distance $f(p, q)$ between X_1 and X_2 could be expressed as:

$$f(p,q) = Ap^2 + Bq^2 + Cpq + Dp + Eq + F \tag{4.65}$$

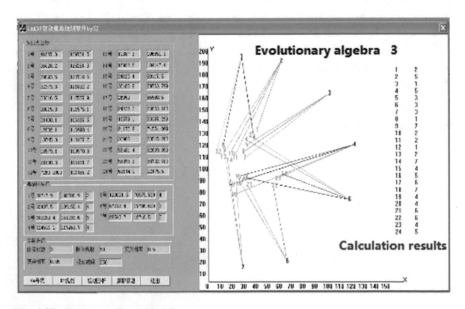

Fig. 4.37 Planning software interface

where A, B, C, D, E, F are respectively the function of (x_a, y_a, z_a), (x_b, y_b, z_b), (x_c, y_c, z_c), and (x_d, y_d, z_d). Calculate the $f(p, q)$ partial derivative of p and q. Let it to be 0, corresponding value of p and q and $f(p, q)_{min}$ could be obtained. The minimum distance of the two routes is as follows:

$$d_{min} = min\{f(0, 1), \quad f(1, 0), \quad f(1, 1), \quad f(0, 0), \quad f(p, q)_{min}\} \qquad (4.66)$$

(4) Simulation experiment

In the experiment, the route planning software uses Visual C ++ 6.0 IDE and ISO/ ANSI-98C ++ standard compiler. As shown in Fig. 4.37, the main interface displays the target assignment results of the genetic algorithm. Dynamic programming message response function call MATLAB engine, the use of MATLAB to draw out the USGS DEM format file digital elevation. The function of the planning analysis is to analyze whether the distance between the trajectories meets the requirements for the safety distance between the traps. The final match results and the final planning results are shown in Figs. 4.38 and 4.39, respectively.

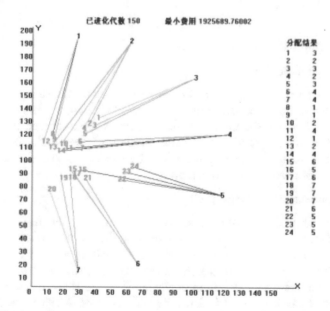

Fig. 4.38 Match result

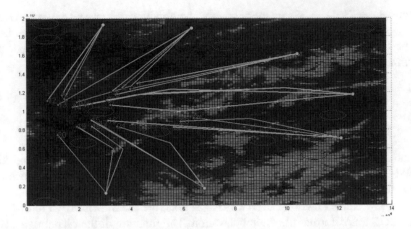

Fig. 4.39 The results of dynamic route planning

4.5.3   *Route Smoothing and Instruction Transformation for TF/TA^2*

1. TF/TA^2 **Optimal Route Smoothing**

(1) Optimal route fitting

The horizontal and vertical optimal route obtained by the above algorithm is two polylines. Due to the maneuverability of the missile members, the resulting way-point command causes the missile member to have supersaturated flight control problems. Therefore, it is necessary to fit the optimal route point by fitting the algorithm, that is, to fit the planned optimal route into a curve. For example, polynomial fitting can be used to achieve.

(2) Simulation experiment example

The given area size is 100 km × 140 km and the digital map resolution is 200 m × 200 m. Assuming that three missiles are fired at the same time from the launch area, the initial positions of the missiles are (100 m, 2500 m), (100 m, 100 m), (100 m, −2500 m), the target area is (100 m, 1000 m), through the cooperative route planning algorithm to obtain the co-variables (that is expected to reach the target area time) for 544.2 s, missile flight speed is 193.5 m/s, 202.5 m/s, and 195.2 m/s. Simulation based on VC ++ environment, using a randomly generated digital map, in Fig. 4.40 gives some of the results of the

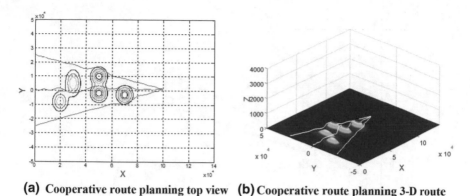

(a) Cooperative route planning top view   **(b)** Cooperative route planning 3-D route

Fig. 4.40 Two-dimensional and three-dimensional maps of three missile cooperation

simulation experiment. The following conclusions can be drawn from the simulation results:

(1) In the cooperative route planning algorithm, in order to enable the missile formation to reach the target area at the same time, the missile members near the target area will use the maneuvering strategy of the obstacle avoidance to increase the flight range. On the other hand, missile members who are far from the target area, in order to reduce the time to reach the target area, make the intersection of time with other missile members reach the target area is not empty, and the planned route is with less maneuvering.

(2) The route proposed by the cooperative route planning algorithm not only ensures that missiles fired from different launch zones can reach the target area simultaneously to achieve saturation attacks, but also have a TF/TA^2 function for each missile member

(3) The vertical route given by the cooperative route planning algorithm not only has better terrain tracking function, but also satisfies the maneuvering constraints of each missile member

2. **Route guidance technology in mid-guidance stage**

(1) angle command method

In the guidance section of the route guidance using the angle command method [92, 93], the principle shown in Fig. 4.41.

In the figure, a is the height of the ground set by the missile. At the point A, the terrain following radar can measure the inclination and the slope R of the distance C from the front of the mountain. If the current inclination of the missile is used, thus:

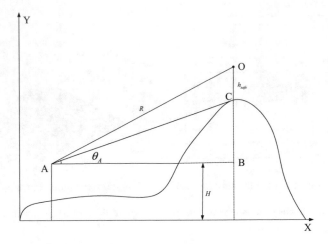

Fig. 4.41 Angle command method

$$\theta^* = \theta_A + \frac{h_{safe}}{R} \tag{4.67}$$

In the case of a vertical plane, assuming that the coordinates of the current point of the missile are (x_M, y_M) and the selected target point coordinates are (x_T, y_T), the track inclination command is:

$$\theta^* = \arctan\left(\frac{z_T - z_M}{x_T - x_M}\right) * 57.3(^\circ) \tag{4.68}$$

The concept of the reaction distance is introduced. The reaction distance is: after the missile detects obstacles, the minimum distance to response to this obstacle and ensure that the collision does not occur, as L_0. In practical applications, the missile's reaction distance is usually less than the farthest detection distance L_R of the seeker, and the greater the $L_R - L_0$, the more difficult the missile to implement terrain avoidance. In the progress the missile flies over the obstacles, the greater L_0 the missile pull up sooner, and the higher the gap between the way point, the worse the hidden. The smaller the L_0, the smaller the gap with the last waypoint, the better the concealment. After crossing the obstacle, the larger the L_0, the smaller the track overshoot, the better the missile concealment. The smaller the L_0, the larger the track overshoot, the worse the concealment. It can be seen that the requirements L_0 before and after crossing the obstacle are just the opposite, so the adaptive angle method with the suppression function is given to solve this contradiction.

(2) Adaptive angle method with suppression function

Although the command method allows the missile to successfully cross the obstacle, but the route cannot be very close to the terrain contour, will reduce the probability of penetration. In this part, the suppression function F_s and the scale factor K are introduced into the angle instruction method to form the adaptive angle algorithm. The function of the suppression function is that the missile does not immediately make a climb action when the missile encounters an obstacle (the distance from the obstacle is L_R), but continues to approach the obstacle. When the distance is less than a certain value (that is, the distance from the obstacle L_0), with the control command greater than θ^* to take the maximum normal overload climbing, the equation is as follows:

$$\theta^* = K(\theta - F_s) \qquad (4.69)$$

The traditional suppression function has been positive during the rise of the missile, although it can delay the pull-up time of the missile, but it may also produce a large overtop time (Overtop time is defined as: in the process over the obstacles, time during which the missile height is greater than the $h + h_{safe}$, h_{safe} is the missile height from the ground safety height, the general to take 50 m for plain, 100–150 m for mountain), shown as Fig. 4.41.

The form of the suppression function used in curve 1 in Fig. 4.42 is:

$$F_s = 57.3 \arctan\left(\frac{z_T - z_M}{x_T - x_M}\right) - 57.3 \arctan\left(\frac{z_T - z_M}{x_T - x_M}\right) \sin\left(\frac{2\pi}{4L_R}(x_M - x_T)\right)$$
$$(4.70)$$

where L_R is the optical seeker detection distance of 5000 m, curve 2 suppression function is zero. It can be seen from the figure that the traditional suppression

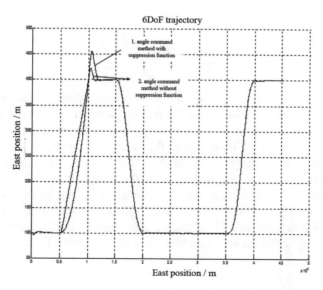

Fig. 4.42 Comparison of the effect of suppressing function and sine suppression function

function can delay the pull-up time of the missile, but it will be extended the overtop time, easily captured by the enemy radar after the top, and the fixed height flight over the top will have a greater overload. This is due to the rise rate when the top is too large.

(3) Vertical navigation strategy

The feedback of the height and path angle plays a decisive role in the process of climbing and descending of the missile, but in the process of climbing and descending, the height feedback should be turned off, otherwise the high degree of feedback will affect the role of the track angle feedback, this is not conducive to play the effect of adaptation angle method.

Based on the angle command method, the suppression function is optimized and improved. That is, the positive suppression function is adopted in the initial stage of the missile climb, the negative suppression function is used in the middle section, and the zero suppression function is used in the final section. This suppression function is essentially a decision simulating man to manipulate an airplane when flying over the top. When the distance from the obstacle is very far, the appropriate delay in used to pull up the aircraft (this time is equivalent to the suppression function is positive). When the distance from the obstacle reaches the reaction distance of the aircraft, in order to have a small overtop time, raise the aircraft ahead of time (equivalent to the suppression function is negative). The purpose is to make the aircraft at a certain distance from the obstacle, the flight height has reached a predetermined height, so that there is a small flight path angle need when overtop, not only reduces the track overshoot, but also prepares for further operations after the pilot crosses the obstacle. At this point because the track angle has been close to zero, even the subduction after crossing the obstacle will reduce the corresponding overload. Suppose that the form of the suppression function is:

$$F_s = 57.3 \arctan\left(\frac{z_T - z_M}{x_T - x_M}\right) \cos\left(\frac{2\pi}{4L_R \cdot \frac{4}{3}}(x_M - x_T)\right) \qquad (4.71)$$

The simulation results are shown in Fig. 4.43. The vertical suppression function can also take the form of a piecewise function, as shown in Fig. 4.44. The area of its triangle ABC represents the effect of the angle command on the delayed pull, and the area of the triangular CDE represents the advance pull-up effect set to reduce the overtop time. The advantage of this method of setting the suppression function is that the C and D points are adjustable, and if the point C is shifted to the right, the effect of the delay pull is enhanced, and a better terrain tracking effect is obtained in areas where the terrain is not very intense. C point to the left or D point down in advance to increase the role of enhanced, which in the relatively complex terrain area can get better results, but the additional cost is the need for missile overload.

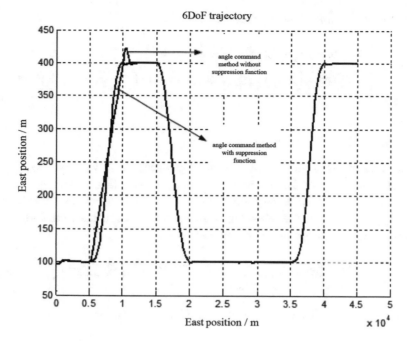

Fig. 4.43 Comparison of the effect of suppressing function and cosine suppression function

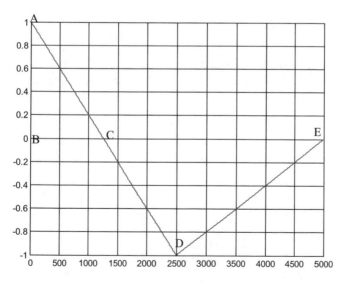

Fig. 4.44 Line segment suppression function diagram

The specific form of the suppression function is:

$$
F_s = \begin{cases} 57.3 \ \arctan\left(\frac{z_T - z_M}{x_T - x_M}\right)\left(\frac{y_D - y_A}{x_D}(x_T - x_M) + 1\right) & , \quad x_T - x_M \in [0, x_D] \\ 57.3 \ \arctan\left(\frac{z_T - z_M}{x_T - x_M}\right)\left(\frac{y_D}{x_D - L_R}(x_T - x_M - L_R)\right) & , \quad x_T - x_M \in [x_D, L_R] \end{cases}
$$

$$(4.72)$$

(4) Horizontal and lateral guidance strategy

Figure 4.45 shows the horizontal and lateral guidance. (x_{pre}, y_{pre}) and (x_T, y_T) are two waypoints to pass. The purpose of the lateral navigation strategy is to design the command of the trajectory skew and roll angle to guide the missile from (x_{pre}, y_{pre}) to (x_T, y_T). The sinusoidal trigonometric function is used to fit the curve to obtain the form of the sinusoidal curve:

$$
y = \frac{y_T - y_{pre}}{2} \sin\left(\frac{\pi}{x_T - x_{pre}}\left(x - \frac{x_T + x_{pre}}{2}\right)\right) + \frac{y_T + y_{pre}}{2} \tag{4.73}
$$

Both sides of the equation seek differential at the same time:

$$
y' = \frac{y_T - y_{pre}}{2} \cdot \frac{\pi}{x_T - x_{pre}} \cos\left(\frac{\pi}{x_T - x_{pre}}\left(x - \frac{x_T + x_{pre}}{2}\right)\right) x' \tag{4.74}
$$

$$
\varphi^* = \arctan\left[\frac{\pi}{2} \cdot \frac{y_T - y_{pre}}{x_T - x_{pre}} \sin\left(\frac{\pi}{x_T - x_{pre}}(x - x_{pre})\right)\right] \tag{4.75}
$$

The horizontal and lateral direction of the missile is controlled by the track angle, and the feedback amount is the track inclination angle, the roll angular

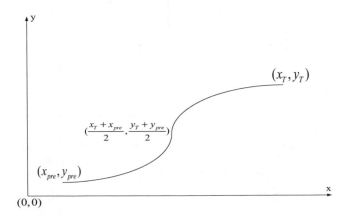

Fig. 4.45 Horizontal and lateral guide

velocity, the yaw rate, the yaw angle and the roll angle, respectively. The coordination direction is used as follows:

$$\begin{cases} \phi^* = 8.0(\varphi^* - \varphi) \\ \varphi^* = \arctan\left[\frac{\pi}{2} \cdot \frac{z_T - z_M}{x_T - x_M} \sin\left(\frac{\pi}{x_T - x_{pre}}(x_M - x_{pre})\right)\right] \end{cases} \quad (4.76)$$

where φ^* is the flight path angle command, ϕ^* is the roll angle command, ϕ is the roll angle. Desirable flight path angle command could also use the following form:

$$\begin{cases} \phi^* = 8.0(\varphi^* - \varphi) \\ \varphi^* = K \cdot \arctan\left(\frac{z_T - z_M}{x_T - x_M}\right)\left[1 - \cos\left(\left(\frac{2\pi}{x_T - x_{pre}}\right)(x_M - x_{pre})\right)\right] \end{cases} \quad (4.77)$$

The simulation results are shown in Fig. 4.46.

When the missile in the vertical side of the simultaneous maneuver, with the above-mentioned guidance law will produce larger error of the missile from the navigation point. The reason is: when the missile takes longitudinal and lateral maneuver at the same time, as the missile roll angle in this process is not zero, the elevator will be negative deflection to increase the angle of attack, to make up for the loss of lift, and then make some impact to the horizontal and lateral of the missile.

In order to solve this problem, the use of control instructions: in the vertical, lateral simultaneous maneuver, the rudder and aileron flight deflection angle command is zero, relying on rolling turn for lateral maneuver, longitudinal lift loss in the case of less lateral maneuverability could be made up by the robustness of the flight control system itself. Due to low-altitude missile penetration cannot be a large roll of maneuver, it is need to prevent the too large height loss that lead to increase the probability of collision with ground. The command form is:

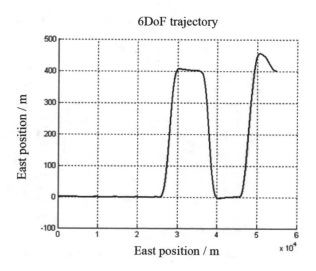

Fig. 4.46 Horizontal and lateral guidance law simulation

$$\begin{cases} \theta_{hope} = \arctan\left(\frac{y_T - y_M}{x_T - x_M}\right) \\ \theta_{sup} = \arctan\left(\frac{y_T - y_M}{x_T - x_M}\right) \cdot \cos\left[\frac{2\pi}{L_R \frac{4}{3}} \cdot \left(x_m - x_{pre}\right)\right] \\ \theta^* = \theta_{hope} - \theta_{sup} \\ \varphi^* = 0 \\ \varphi_1 = -0.8 \arctan\left(\frac{z_T - z_m}{x_T - x_m}\right)\left[1 - \cos\left(\frac{2\pi}{L_R}\left(x_m - x_{pre}\right)\right)\right] \\ \phi^* = 8.0(\varphi_1 - \varphi) \end{cases} \tag{4.78}$$

The corresponding simulation results are shown in Fig. 4.47.

This section studies the single-missile route planning method, cooperative route planning method and middle guidance low altitude penetration route guidance method for missile formation, targeted to the contradiction between the requirement of high penetration probability and the low probability of collision in the low altitude penetration of missile formation. For the single-missile route planning, the velocity update strategy based on kinetic energy and worst particle is given, and the route planning index and process based on continuous particle swarm optimization are introduced. In the cooperative route planning of MAF, the genetic algorithm and the dynamic programming method are combined to solve the multi-launch point, multi-gather area and multi-constrained route planning problem. The adaptive angle method with the suppression function is applied to the low-altitude penetration guidance, which can better coordinate the requirement of the probability of penetration and the probability of collision to ground.

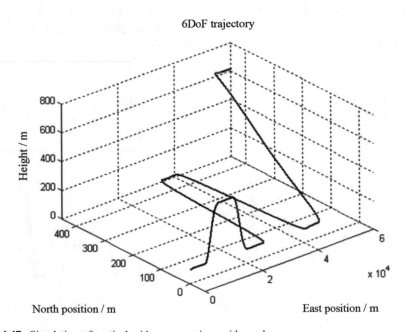

Fig. 4.47 Simulation of vertical with maneuvering guidance law

4.6   Handover Guidance and Cooperative Guidance

The cooperative electronic confrontation, cooperative attack and saturation attack tactics are important means for missiles formation to realize the system confrontation and improve the comprehensive combat effectiveness. At the same time, these need high level of coherence guidance time consistency.

When the missiles formation engaged in collaborative combat environment, the topology of the sensor and the support network is also dynamically changed due to the movement of the missile and the failure of the communication channel, and the calculation is highly distributed without a fixed and defined centralized controller. When the missile formation is in a high dynamic uncertainty or strong antagonistic environment, centralized control structures have great limitations, and there are two aspects of reasons: firstly, this structure requires a lot of information flow only from the perspective of bandwidth, so it is no doubt very slow and expensive; secondly, in a strong antagonistic environment, a large number of data transmission depends on the channel is unreliable and dynamic changes, but also easy to be intercepted and detected. It can be seen that the centralized control structure is very fragile and lacks the necessary robustness and concealment. Therefore, it is necessary to study the distributed information processing and its guidance technology with minimum information flow requirements for missiles autonomous formation.

4.6.1   Design and Analysis of Guidance Handover Boundary

Guidance handover is to solve the problem of effective switching between the different guidance laws in the midcourse and terminal guidance, and to ensure that the seeker is able to capture the target in the final stage of the midcourse guidance to smoothly transfer to the terminal guidance stage. In the Refs. [94, 95], the main sources of error in the guidance handover are analyzed and the method for instruction handover between different guidance laws is given, but the random perturbation problem is not taken into account.

Due to the existence of stochastic factors, the missiles formation will be transferred to the terminal guidance from midcourse guidance at the handover boundary at random time. Therefore, the random disturbance factors will be considered in this section, and the design method of the optimal handover interface will be given combining with the modern optimization algorithm, in order to meet requirements of the minimum miss distance and variance.

1. System Model

In order to adjust the position and attitude of the missile in the final stage of the midcourse guidance, and ensure the missile can successfully capture the target, the flight control system will use attitude control in the midcourse guidance, and the guidance law will use attitude tracking, while using the procedural instructions or virtual target guidance for the missile to reach the boundary. In the terminal guidance mode, in order to improve the maneuverability and precision of the missile, the flight control system adopts the overload control method [1]. The guidance law adopts the proportional guidance law. The midcourse guidance mode is the first structure, and the terminal guidance mode is the second structure. When the missile flight state reaches the handover boundary, the system will transform from the first structure to the second structure at a certain probability. According to the definition in Ref. [77], this is a two-structure system with a centralized transition of a sequential transformation structure.

Without loss of generality, considering the random disturbance factor, the system state equation is expressed as:

$$\dot{X} = F(X,t) + BU(s,t) + \xi(t) \tag{4.79}$$

where the $F(X, t)$ is the vector function of the corresponding dimension, and B is a known constant matrix, and the control quantity $U(s,t) = \delta_{zc}$ is the control rudder deflection instruction, $s = \overline{1,2}$ is the state label of the system structure. $\xi(t)$ is the Gaussian white noise vector of the corresponding dimension whose intensity matrix is $Q(t)$, the correlation function matrix is $Q(t)\delta(t - t')$, and $\delta(t - t')$ is the Dirac function, and other related variables are defined in Ref. [77].

Owing to different guidance laws used in the midcourse guidance and the terminal guidance, control instructions are described below in different modes:

$$\begin{cases} U(1,t) = k_\theta(\theta^* - \theta) + k_\vartheta(\vartheta^* - \vartheta) + k_\varepsilon\varepsilon + k_{\omega_z}\omega_z + k_{\delta_z}\delta_z \\ U(2,t) = k_{n_y}\left(n_y^* - n_y\right) + k_\eta\eta + k_{\omega_z}\omega_z + k_{\delta_z} \end{cases} \tag{4.80}$$

where all the k are the state feedback coefficients, and ϑ^*, θ^* and n_y^* are the pitch angle command, the trajectory inclination command and the overload command respectively, and $\theta^* = \vartheta^* - \alpha_0$. α_0 is the equilibrium angle of attack for the flat flight state. ϑ^* and n_y^* are determined by the guidance law adopted by the midcourse and terminal guidance mode, respectively. $\varepsilon = \int_0^t [\vartheta^*(\tau) - \vartheta]dt$, $\eta = g \int_0^t \left[n_y^*(\tau) - n_y(\tau)\right]dt$ are the integral compensation introduced in order to eliminate the steady-state error of the command response, where g is the acceleration of gravity, and other related variables are defined in Refs. [1, 75].

2. System Probability Analysis

According to the definition in Ref. [77], it can be seen that in the process of centralized transition, the transition of the structure occurs only when the state reaches the boundary described by the deterministic function $X^{(rs)}(t)$. The probability analysis of the handing over midcourse to terminal guidance is to solve the conditional probability density function $f(X, t|s, t)$ and the state probability $p(s,t)$ of the state vector, see the Ref. [77] for the correlation variable definition and the system probability analysis method.

3. Probability Moments and Index Functions of Miss Distance

After obtaining the conditional probability density function $f(X, t|s, t)$ and the state probability $p(s, t)$ of the state vector, we can calculate the probability moment of the current target mirror error (miss distance) $\Delta_d(X, t)$.

Assume that x_t and h_t represent the coordinates of the target centroid in the ground coordinate system respectively, and let $x_r = x_t - x$, $y_r = h_t - h$. The expressions for $\Delta_d(X, t)$ is as follows [77]:

$$\Delta_d(X, t) = -\frac{R(t)^2}{\dot{R}(t)}\dot{\varphi}(t) \tag{4.81}$$

In the formula, $R(t) = \sqrt{x_r^2(t) + y_r^2(t)}$ is the straight line distance between the missile and the target, $\varphi(t) = \arctan[y_r(t)/x_r(t)]$ indicating the altitude angle. The mathematical expectations and variance of $\Delta_d(t)$ are as follows:

$$m_{\Delta_d}(t) = \sum_{s=1}^{2} p(s, t) \int_{-\infty}^{+\infty} \Delta_d(X, t)f(X, t|s, t)dX \tag{4.82}$$

$$D_{\Delta_d}(t) = \sum_{s=1}^{2} p(s, t) \int_{-\infty}^{+\infty} [\Delta_d(X, t) - m_{\Delta_d}(t)]^2 f(X, t|s, t)dX \tag{4.83}$$

In order to assess the accuracy of the missile hit target, let t in the above formula be $t = t_a = t_k - t_1$, where $t_k = \frac{R_0}{V_m - V_t}$, $t_1 = \frac{R_1}{V_m - V_t}$, and R_0 is the straight line distance between the missile and the target, R_1 is the minimum distance flying without control depend on the control system and the target type, V_m and V_t are the average values of the missiles and targets speed. The value of t_k and t_1 can also be estimated by Monte Carlo.

In order to improve the accuracy of the missile hitting the target, we can use the modern optimization algorithm to optimize the following objective function to obtain the optimal handover switching surface and minimize the index function as follows:

$$J(\Omega) = k_m m_{\Delta_d}^2(t_a) + k_D D_{\Delta_d}(t_a) \tag{4.84}$$

where Ω is the handover surface, and k_m and k_D are weight coefficients.

Many of the computational processes of the previous variables involve the infinite integral problem of function for the independent variables among $(-\infty, \infty)$, which cannot be given analytically. Therefore, it needs to be solved by means of Gauss-Hermite integral algorithm.

4. Simulation

Assuming that the mathematical expectation initial value of the component in the missile state vector X is $m_x(0|s,0) = 170,000$ m, $m_h(0|s,0) = 3000$ m, $m_V(0|s,0) = 200$ m/s, the available longitudinal load for the missile is 5 g, and $g = 9.81$ m/s^2. The initial value of the variance matrix is $\Theta(0|,s,0) = \mathbf{0}$, $s = \overline{1,2}$. Noise intensity matrix is

$$Q(t) = diag[(1.0\,\text{m/s})^2, (0.01\,\text{rad})^2, (0.01\,\text{rad/s})^2, (0.01\,\text{rad})^2, (10.0\,\text{m})^2, (10.0\,\text{m})^2, (0.2°)^2],$$

In which the data corresponds to the noise intensity of each variable in the state vector $X = [V, \theta, \omega_z, \vartheta, x, h, \delta_z]^T$ in turn. The initial position of target is $x_t = 182,000$ m, $h_t = 0$ m, $V_t = 20$ m/s. The maximum detection distance of the seeker $R_{max} \leq 6$ km, altitude angle of detection is $\eta \in [-35°, 10°]$.

The virtual guidance law is adopted in the midcourse guidance, and the virtual target is set according to the handover switching surface. The proportional guidance law is adopted in the terminal guidance. First, we use the *Monte-Carlo* method to get the approximate constraint boundary value of each variable in Ω, and then use the taboo search algorithm to optimize the objective function (4.84) within this constraint boundary value range.

Let $k_m = 0.65, k_D = 0.35$, and then the optimal handover shifting surface is gotten as $\Omega_h = (-1146°, 177965.23\,\text{m}, 987.45\,\text{m})$, and simultaneously we have $m_{\Delta_d}(t_a) = 2.156\,\text{m}$, $D_{\Delta_d}(t_a) = 1.623\,\text{m}^2$, where $t_a = 68s$. Figures 4.48, 4.49 and 4.50 show some of the results obtained using the simulation of switching surface Ω_h.

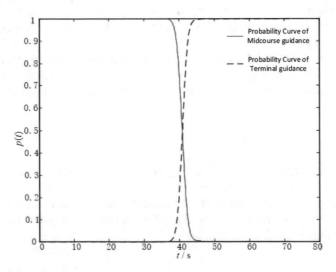

Fig. 4.48 Probability curve of two modes

Fig. 4.49 Mathematical expectations of longitudinal trajectories

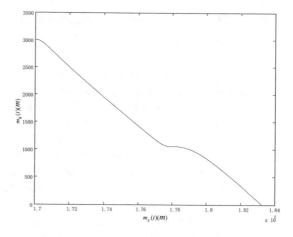

Fig. 4.50 Mathematical expectations of pitching angles

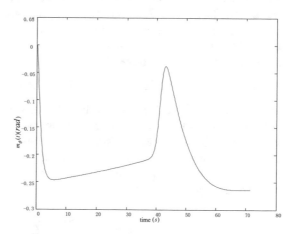

Figure 4.48 shows the probability curve of the system structure during the handover process. It can be seen from the figure that in the initial stage, the system is in the midcourse guidance structure with the probability of almost one, The absorption intensity that the system is converted to the terminal guidance structure from the midcourse guidance structure is significantly increased, namely the system transforms from the first structure to the second structure, the probability that the system is in the terminal guidance structure of is gradually increased, which finally is almost one. Figure 4.49 shows the unconditional mathematical expectations for longitudinal trajectories, and Fig. 4.50 gives unconditional mathematical expectations for pitch angles. From the two figures, it is can be seen that in the process that the missile converts from the midcourse guidance to the terminal guidance process, flight trajectory and attitude angle has been adjusted and is smooth. It can be seen that the resulting handover shifting surface Ω_h can make the final miss distance minimum, and the trajectories link up smoothly, so the simulation results show that the method is effective.

4.6.2 Time Consistence of Distributed Cooperative Guidance

In the strong antagonistic environment, the formation support networks will lead to state changes of its topology and network connectivity due to a variety of uncertain factors and these changes for the missiles formation executing cooperative combat task is not negligible.

Refs. [96−98] discussed the distributed cooperative guidance time consistent problem for missiles formation when the network topology is under a strong coherence/balance or a weakly connected/unbalanced conditions. In Ref. [96], a method of analyzing the cooperative guidance problem under the condition of random topology of the network topology and communication noise is generated, based on the structural random transition system theory, Here, we will focus on the cooperative guidance time consistency problem in the presence of communication noise and delay, and the occurrence of random transition/unbalanced network topology, and give the corresponding network adjustment and control strategy to achieve the optimal balance design between the cooperative guidance time consistency and penetration probability.

1. Guidance Time Adjustment and Coordination Consistency Issues

In order to achieve saturation attack tactics that the missiles hit the target at the same time, the guidance law is required to have the ability to adjust the time to guide. In Refs. [99, 100], a guidance law with a guidance time limit (ITCG) is discussed, which would allow a missile to hit a target at a predetermined time. It is based on the pure proportional guidance law (PNG), by adding a correction to achieve the adjustment of the guidance time, in the form of the following:

$$a = a_p - \frac{60V^5}{a_p R_{go}^3}(\bar{T}_{go} - \hat{T}_{go}) \tag{4.85}$$

where the first term $a_p = NV\dot{\lambda}$ is the amount of control given by the PNG, N is the navigation ratio, λ is the angle of sight. The second term is a correction term for adjusting the guidance time, where $\bar{T}_{go}$ is the specified guidance time, $\hat{T}_{go} = \left(1 + \frac{(\theta-\lambda)^2}{10}\right) R_{go}/V$ is the remaining time estimation of PNG guidance. However in the literature, in order to make multiple missiles hit the target at the same time, it is needed to specify the guidance time in advance, rather than through the dynamic information interaction between the missiles, obviously this is not a real sense of cooperation, and cannot meet the ever-changing requirements of the battlefield environment [101, 102].

2. Topology Transition Problem of Formation Support Network

The use of directed or undirected graphs to establish the information interaction model of the formation is a very effective way. In this way, the transition of the formation support network structure can be described using the example diagram shown in Fig. 4.51. The figure shows three different network topology structures $\{G_a, G_b, G_c\}$ between the three missiles. Figure 4.51d shows the process of transition with the states $\{G_a, G_b, G_c\}$. In the example, the network structure is in a discrete state G_a at the beginning and converts to the next state in the order shown in Fig. 4.51d. Similarly, the multi-topology formation support network transition in other cases, can be described using a similar map.

3. Time Consistency of Distributed Cooperative Terminal Guidance

(1) Double-layer cooperative guidance structure

In order to solve the problem of distributed cooperative guidance time consistency with ITCG in the case of formation support network topology transition, the following double-layer cooperative guidance and control structure is adopted, as shown in Fig. 4.52.

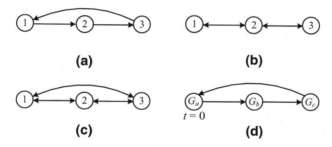

Fig. 4.51 Network topology and transition process

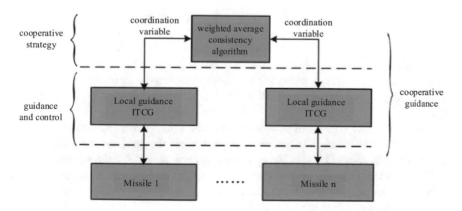

Fig. 4.52 Double-layer cooperative guidance structure

In this structure, the upper layer cooperative strategy adopts the distributed weighted average consistency algorithm, which solves the expected guidance time of the missile group according to the expected guidance time provided by each missile. The underlying ITCG guides the missile members to the target at the same time based on the resulting missiles formation expectations guidance time. The cooperative guidance and control structure is simple and clear, with analytical solution and conducive to engineering.

(2) System modeling

The weighted average consistency algorithm cooperative terminal guidance is presented below, and a hybrid system model is established to describe the dynamic characteristics of the expected guidance time of each missile under the network topology transition. In the Refs. [100, 101], with the centralized cooperative strategy, using the expected guidance time as the coordination variable, and then the sum of the control energy of all the missile members as the centralized coordination function, we will have the suboptimal value of the coordination variables ξ:

$$\xi = \sum_{i=1}^{n} \alpha_i^2 \hat{T}_{goi} \bigg/ \sum_{i=1}^{n} \alpha_i^2 \tag{4.86}$$

Which make the centralized coordination function $J = \sum_{i=1}^{n} u_i^2$ to achieve the minimum, where is n the formation size, $\alpha_i = 60 V_i^5 / (a_{pi} R_{goi}^3)$ and i is the missile member index number, u_i is defined as the ith missile's guidance law in Eq. (4.85). ξ has a very significant physical meaning: the missiles formation expected guidance time value obtained through consultation is the weighted average of the estimated remaining time value of each missile member.

In order to further study the existence of the network topology transition, mixed system model is established as follows with continuous node state and discrete graph G state:

$$\dot{\xi}(t) = -CL(G_k)\xi(t), k = s(t), G_k \in \Gamma_n \tag{4.87}$$

where $\xi = (\xi_1, \xi_2, \cdots, \xi_n)^T$, $C = diag(c_1, c_2, \ldots, c_n)$, $L(G_k)$ denotes the Laplacian matrix of graphs G_k, $\Gamma_n = \{G = (v, \varepsilon, A) : rank(L) = n - 1, 1^T L = 0\}$ represents a finite set of n-rank strong connected equilibrium directed graphs. $s(t) : R_{\geq 0} \to I_{\Gamma_n}$ is a jumping signal, where $I_{\Gamma_n} \subset N$ is the set of indexes associated with the elements in Γ_n.

The distributed consistent algorithm requires infinite time to converge to the decision value in theory. In practical applications, the problem of infinite time convergence is solved by specifying a convergence period for the system.

The following describes the cooperative terminal guidance time consistency problem in the presence of communication noise and delay, and the occurrence of random transition/unbalanced network topology.

(1) System model

Structural random transition system model with communication noise and network topology random transition is as follows:

$$\dot{\xi}(t) = diag[-L(G_k)]\xi(t) + [-L(G_k) - diag(-L(G_k))]\xi(t-\tau) + B(s,t)\varsigma(t), \; k$$
$$= s(t), G_k \in \Gamma_n$$

(4.88)

where $k = s(t) = \overline{1,M}$ the state label of the system structure, M is is the number of deterministic network topology. $B(s, t)$ is the matrix of the corresponding dimension. $\varsigma(t)$ is the corresponding dimension of the Gaussian white noise vector, whose intensity matrix is $Q(t)$, the correlation function matrix is $Q(t)\delta(t-t')$, and $\delta(t-t')$ is the Dirac function, and other related variables are defined in Ref. [77].

(2) System probability analysis

The mathematical expectation $m(t|s,t)$ and covariance matrix $\Theta(t|s,t)$ of the state vector for each structure in the system (4.88) has the following form:

$$\dot{m}(t|s,t) = F(s,t)m(t|s,t) + H(s,t)m(t-\tau|s,t-\tau)$$
$$- \sum_{r=1\neq s}^{M} v^{(sr)}(t)\frac{p(r,t)}{p(s,t)}[m(t|s,t)-, m(t|r,t)]$$

(4.89)

$$\dot{\Theta}(t|s,t) = F(s,t)\Theta(t|s,t) + \Theta(t|s,t)F^T(s,t) + H(s,t)\Theta(t-\tau|s,t-\tau)$$
$$+ \Theta^T(t-\tau|s,t-\tau)H(s,t) + B(s,t)Q(t)B^T(s,t) - \sum_{r=1\neq s}^{M} v^{(sr)}(t)\frac{p(r,t)}{p(s,t)}$$
$$\cdot \{\Theta(t|s,t) - \Theta(t|r,t) + [m(t|s,t)][m(t|r,t) - m(t|s,t)]^T$$

(4.90)

where

$$\left.\begin{array}{c} F(s,t) = diag[-CL(G_k)] \\ H(s,t) = [-CL(G_k) - diag(-CL(G_k))] \\ k = s(t) = \overline{1,M} \end{array}\right\}$$

(4.91)

When the network topology is weakly connected/unbalanced, C is the unit matrix of the corresponding dimension. When the network topology is strong connected/balanced, the matrix C is the same as defined in (4.87), so that the strong continuity/balanced network topology is embodied in the system (4.88) as a special case of weakly connected/unbalanced topologies, and the advantage of using the weighted average consistency algorithm for the network topology with strong connectivity.

(3) Simulation example

In the simulation, not only the support network topology between the missile members is different, but also the communication delay $\tau = 0.5\,\text{s}$ is joined.

Assuming that the support network topology between the missile members is shown in Fig. 4.53, the strong connectivity network topology is also considered as a structural state. We can get the Laplacian matrix of each network topology at this time, we can see that only graph G_a is a strong connectivity balance and the rest is weak connectivity and unbalanced, and there is no network connection in graph G_d:

$$L(G_a) = \begin{bmatrix} 1 & 0 & -1 \\ -1 & 1 & 0 \\ 0 & -1 & 0 \end{bmatrix}, \ L(G_b) = \begin{bmatrix} 1 & -1 & 0 \\ 0 & 0 & 0 \\ 0 & 0 & 0 \end{bmatrix}, \ L(G_c)$$
$$= \begin{bmatrix} 0 & 0 & 0 \\ 0 & 0 & 0 \\ -1 & 0 & 1 \end{bmatrix}, \ L(g_d) = \begin{bmatrix} 0 & 0 & 0 \\ 0 & 0 & 0 \\ 0 & 0 & 0 \end{bmatrix}$$

Design the probability of the system being each structure at the initial moments and the transfer strength matrix between the structures as follows:

$$p(1,0) = 0.5, p(2,0) = 0.1, p(3,0) = 2.0, p(4,0) = 0.2,$$
$$v = \begin{bmatrix} 0.1 & 0.1 & 0.2 & 0.3 \\ 0.1 & 0.1 & 0.1 & 0.2 \\ 0.1 & 0.1 & 0.1 & 0.2 \\ 0.1 & 0.1 & 0.1 & 0.2 \end{bmatrix},$$

Set the convergence cycle $\Delta T = 1$ in the simulation, the following are the simulation results obtained based on the given conditions.

It is can be seen from the ballistic curves under cooperative guidance shown in Fig. 4.54a and the remaining guidance time shown in Fig. 4.54b, the missiles formation does not hit the target at the same time. The missile member ε_3 first hit the target and the missile member ε_1 and ε_2 arrived after 1.3 s and about 0.7 s to hit

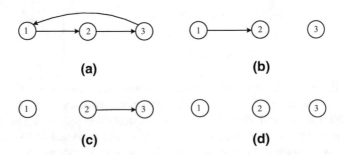

(a)   **(b)**

(c)   **(d)**

Fig. 4.53 The four different network topologies between three missiles

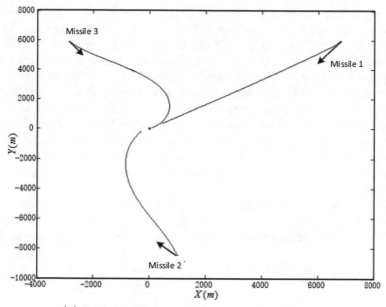

(a) Ballistic trajectory under cooperative guidance

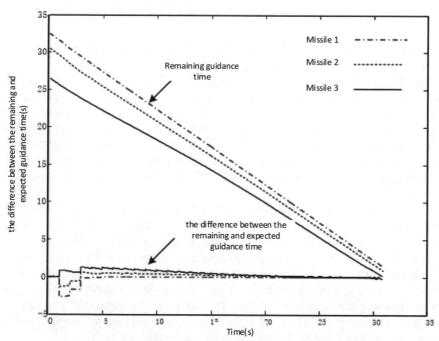

(b) the difference between the remaining and expected guidance time

Fig. 4.54 Cooperative guidance simulation results in poor network state

the target. The reason is that the system is in a weakly connected non-balanced network topology under a great probability. As can be seen from Fig. 4.55a, the steady-state probability of the system in structure 1 is only 0.14, that is, the probability that the system is in a strong-connected balanced network topology is very low and the steady-state probability of the fully-unconnected communication structure 4 up to 0.43. The steady state probabilities of weakly connected unbalanced communication structures 2 and 3 are 0.20 and 0.23, respectively. Figure 4.55b shows the structural state probability and the coordinated variable of each network topology, that is, the transition curve for the unconditional mathematical expectation of the expected guidance time. It can be seen from the figure that in this network connectivity, the expected guidance time of the missiles failed to reach consistency.

The following is the simulation under a better network state to compare with the above example. Design the probability of the system being each structure at the initial moments and the transfer strength matrix between the structures as follows:

$$p(1,0) = 0.1, p(2,0) = 2.0, p(3,0) = 0.3, p(4,0) = 0.4,$$

$$v = \begin{bmatrix} 1.0 & 1.0 & 0.9 & 0.8 \\ 0.5 & 0.6 & 0.4 & 0.5 \\ 0.4 & 0.5 & 0.6 & 0.7 \\ 0.1 & 0.1 & 0.1 & 0.1 \end{bmatrix}$$

It is can be seen from the ballistic curves shown in Fig. 4.56a and the remaining guidance time shown in Fig. 4.56b, three missiles hit the target at the same time. From the specified guidance time shown in Fig. 4.56b, it can be seen that the expected guidance time is quickly agreed, and the normal overload performance is better. The network connectivity is better by analyzing the structure state probability shown in Fig. 4.57a, and then Fig. 4.57b gives the variation curve of the unconditional mathematical expectation of the coordinated variable and the main diagonal elements of the unconditional covariance matrix. It can be seen that the convergence process of the coordinated variables is consistent and the system is stable.

Compared results of simulation under the above two different conditions, it can be seen that the network connectivity condition has a great influence on the consistency of the coordinated terminal guidance time, and the stochastic transition system theory can be used to quantitatively analyze the influence. When the network connectivity is controllable, on the basis of this quantitative analysis, we can develop a control strategy for network connectivity in cooperative terminal guidance, so that the missile can hid itself by controlling the connectivity in order to improve the anti-probability, and at the same time to achieve the purpose of time consistency. This issue is studied and analyzed below.

Fig. 4.55 The structure and coordinated variables simulation results in poor network state   ▶

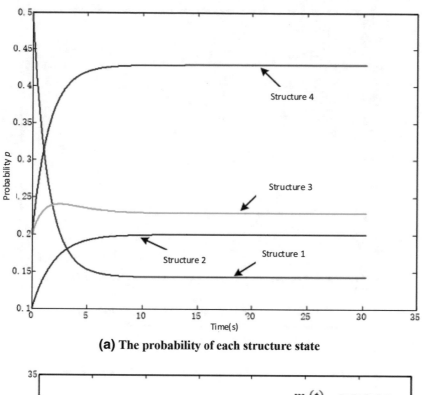

(a) The probability of each structure state

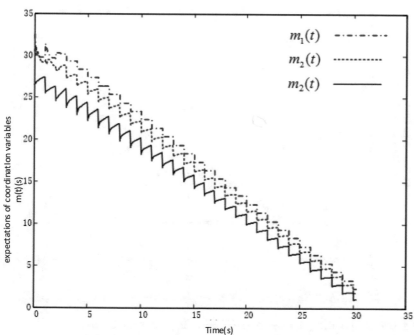

(b) Unconditional mathematical expectations of coordinated variables

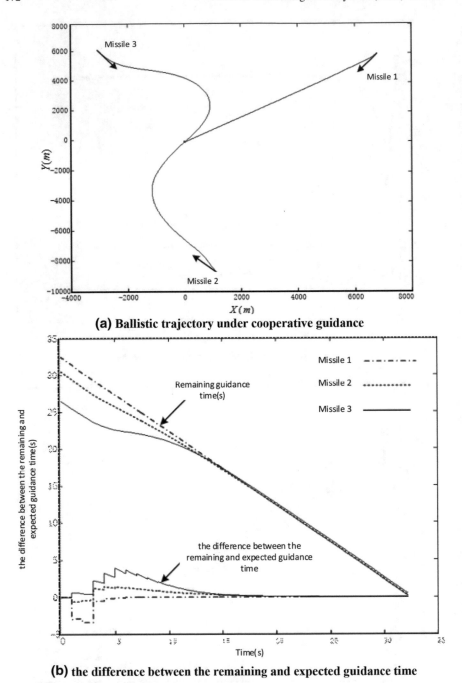

(a) Ballistic trajectory under cooperative guidance

(b) the difference between the remaining and expected guidance time

Fig. 4.56 Cooperative guidance simulation results in good network state

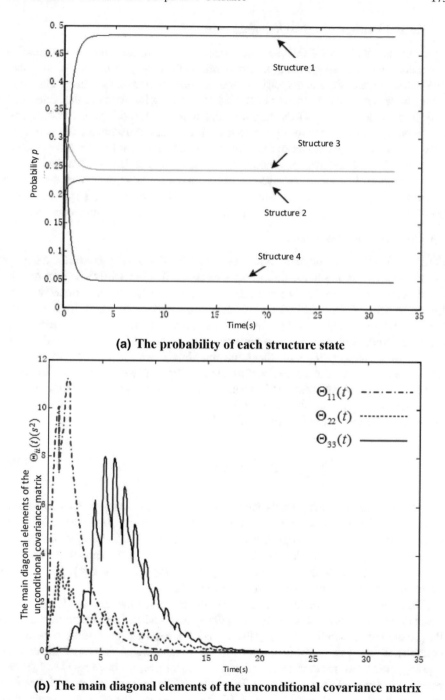

(a) The probability of each structure state

(b) The main diagonal elements of the unconditional covariance matrix

Fig. 4.57 The structure and coordinated variables simulation results in good network state

(3) Control strategy of network connectivity

Through the analysis of the previous sections, we can see that in the missiles formation cooperative combat, the network connectivity situation is the main influencing factor of the cooperative terminal guidance time consistency. The better the network connectivity, the better the effect of hitting the target at the same time, and the missiles cannot hit the target at the same time under poor network connectivity. But in the electronic confrontation combat environment, the better the network connectivity also means the worse the missile's own hidden and the lower the probability of penetration. Then it is necessary to put forward a network connectivity control strategy to address the above contradictions. The systematic quantitative analysis of the system using the stochastic transition system theory provides the basis for the design of the network connectivity control strategy.

(1) Objective function model

Firstly, the objective function model of network connectivity control strategy is established. According to the foregoing, we can see that there are three factors to be considered in the objective function: first, the influence of network connectivity on the probability of the missile being discovered, followed by the influence of network connectivity on the cooperative terminal guidance time consistency result, and the third is effect of network connectivity on miss distance. As mentioned earlier, the network connectivity will affect the missile's normal overload performance, while the poor normal overload performance will produce a large miss distance.

Without loss of generality, it is assumed that the relationship between network connectivity and the probability of the j-th missile member being discovered can be represented by the following function model:

$$p_{fj}(t) = \frac{1}{n-1} \sum_{s=1}^{M} \sum_{i=1}^{n} a_{ij}(s,t) p(s,t), \quad j = 1, \ldots, n \qquad (4.92)$$

where $p_{fj}(t)$ indicates the probability that the j-th missile member is discovered at time t, s indicates the state label of the system structure, M is the number of deterministic network topologies, and n is the number of missile members participating in cooperative guidance, and $a_{ij}(s,t) \in \{0,1\}$ is the element in the adjacency matrix $A(s,t)$ of the s-th network topology graph $G = (v, \varepsilon, A)$, $p(s,t)$ is the probability of the network structure state [77, 96].

The above equation indicates that when the j-th missile member sends a signal to other missile members, it is likely to be discovered and the more members sent to the greater the probability of being discovered. For example, according to the above equation, if the j-th member of the formation in each of the possible network topology sends information to the other $n-1$ members, we have $p_{fj}(t) = 1$, if any message is not send, then $p_{fj}(t) = 0$. When the adjacency matrix A of each possible network topology is time-invariant, $p_{fj}(t)$ tends to be stabilized as the steady state value while the state probability of each structure tends to steady state. Based on

Eq. (4.92), the probability that the entire missiles formation is discovered can be expressed as follows:

$$p_f(t) = \sum_{j=1}^{n} p_{fj}(t) \tag{4.93}$$

The results of network connectivity and cooperative terminal guidance time coincidence and its relationship with final miss distance can be obtained according to the previous quantitative analysis method.

Assuming Δt_{ij} indicates that the time difference between the i-th missile member hits the target and the moment the j-missile member does, it can be seen from the foregoing that it is a function of the Laplacian matrix $L(G_k)$ and the probability of each state of the structure $p(s,t)$, and can be defined as follows:

$$\Delta t_{ij} = f_t[L(G_k), p(s,t)] \tag{4.94}$$

where $k = s(t) = \overline{1,M}$ the state label of the system structure and M is is the number of deterministic network topology. According to the previous system probability quantitative analysis method based on the structure random jump system theory, the specific value of Δt_{ij} can be obtained. According to Eq. (4.94), the sum of time difference among each missile can be obtained in following form:

$$\Delta t = \sum_{j=1, j \neq 1}^{n} \Delta t_{ij}, i \in \{1, \ldots, n\} \tag{4.95}$$

Similarly, assuming that Δd_i indicates the miss distance of the i-th missile and then it can be obtained similar to Eq. (4.95) as follows:

$$\Delta d_i = f_d[L(G_k), p(s,t)] \tag{4.96}$$

In the same way, the sum of miss distances is given:

$$\Delta d = \sum_{i=1}^{n} \Delta d_i, i \in \{1, \ldots, n\} \tag{4.97}$$

According to Eqs. (4.93), (4.94) and (4.96), the objective functions are as follows:

$$J(p) = k_1 p_f(t) + k_2 \Delta t + k_3 \Delta d \tag{4.98}$$

where, k_1, k_2 and k_3 are the weighting factors. In order to improve the penetration probability of missiles and the performance of time coincidence, the modern optimization algorithm can be used to find the optimal value of Eq. (4.98), in order to obtain the optimal stability of each structural state $p^*(s)$ when the possible network topology is given in advance, so that the value of the objective function is minimized. It is possible to control the network connectivity by determining the mutual transfer strength between the structures based on the initial probabilities of the structural states.

(2) Simulation example

It is to be noted that, when the possible network topologies are given in advance, it is needed to take full account of the physical characteristics limit of the missile, to avoid terminal guidance time inconsistence, which is because of a missile member's speed limit.

The objective function (4.98) is optimized by using the taboo search algorithm in the probability value range [0, 1] of each structural state. Let $k_1 = 0.7, k_2 = 0.15, k_3 = 0.15$, and then we have the optimal steady state values of structural states $p^*(1) = 0.194, p^*(2) = 0.20, p^*(3) = 0.210, p^*(4) = 0.396$, and $p_f = 0.495, \Delta t = 0.527\text{s}, \Delta d = 3.361\,\text{m}$.

Accordingly, when the network state is good, we will have $p_f = 0.959, \Delta t = 0.463\,\text{s}, \Delta d = 3.245\,\text{m}$, so it can be seen that the optimal probability of each structural state can guarantee the cooperative time coincidence and the miss distance requirement, and the probability of being discovered is the lowest, which proves the validity of the optimization result. At this time, according to the initial probability of each structure and the obtained optimal probability steady-state value of each structure state, the transfer intensity matrix between the structures can be obtained as

$$v = \begin{bmatrix} 0.12 & 0.16 & 0.23 & 0.13 \\ 0.21 & 0.28 & 0.17 & 0.12 \\ 0.16 & 0.14 & 0.26 & 0.18 \\ 0.31 & 0.32 & 0.22 & 0.24 \end{bmatrix},$$

which can be used as a network connectivity control strategy.

In this chapter, we establish a hybrid system model with continuous node state and discrete state of topology, and give a distributed cooperative guidance method for missiles formation in random transition and non-equilibrium network topology. Based on the structural random jump system theory, the system model is established and the quantitative analysis of system probability and probability moments are given, in the presence of communication noise and delay, and the occurrence of random transition network topology. In order to ensure that the missiles formation satisfies the requirement of high penetration probability and the minimum terminal time difference and miss distance of the saturation attack, the adjustment and control strategy of the support network is given. The effectiveness of the above-mentioned system model and distributed cooperative guidance method and its network regulation and control strategy are verified by the cooperative guidance simulation experiment of missiles formation in the terminal guidance area.

4.6.3 Simulation of Cooperative Guidance

This section will use the CGCS of IPCLab digital simulation system (described in detail in Chap. 8) based on a type of anti-ship missile model data [1, 103, 104] for the missiles autonomous formation cooperative guidance system simulation and verifying the validity of the proposed target dynamic assignment algorithm and the design of guidance handover boundary and the network connectivity control strategy, and gives the cooperative terminal guidance time consistency analysis of the missiles autonomous formation.

1. **Cooperative terminal guidance simulation system architecture**

Cooperative terminal guidance simulation system is subsystem of the CGCS of IPCLab digital simulation system, and the system structure is shown in Fig. 4.58.

Missile members of the neighbor missiles of the missile ε_i are defined as N_j $(j \in \{1, \ldots n\}, i \neq j)$. As shown in the figure, when the missiles formation executes the dynamic target assignment, the neighbor missile members N_j send the state information and the obtained target information to the missile ε_i through the support network, and then the missile ε_i use the information and the target information obtained itself to participate in the dynamic target assignment. Missile ε_i and its neighbor missiles N_j will do a network vote with the results of their respective target assignment, and at last the missiles formation will get the missile-target matching result.

The missiles formation is transferred from the midcourse guidance mode to the terminal guidance mode according to the missile-target matching relationship and the optimal handover surface. The missile members each lock the target and enter into the cooperative terminal guidance section. The missile ε_i share the expected guidance time information with the neighbor missile members N_i who attack the

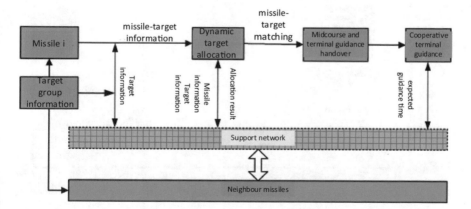

Fig. 4.58 Cooperative terminal guidance simulation system of MAF structure graph

same target, realize the high penetration probability and the minimum terminal time difference and miss distance under the cooperative terminal guidance time consistency and the network connectivity control strategy.

2. Cooperative terminal guidance simulation conditions and settings

Assuming the scale of missiles formation $n = 16$, and the formation executes collaborative attack task on the 8 targets. The initial state of the missiles formation entering the handover area is dynamically generated by the cooperative route planning. Assuming that all missiles' flying speed is $V_m = 230$ m/s, flight height is 2000 m, and the moving height of all targets is 0 m, with a velocity $V_s = 20$ m/s, and the path declination is 0 degrees. Since the target movement speed is much smaller than the missile flight speed, so in the final cooperative attack, we approximate that the target is still. Assuming the initial position of the missiles formation and the target group is shown in Fig. 4.59, and □ denotes the infrared seeker and the infrared heat radiation source, while ○ denotes the radar seeker and the electromagnetic radiation source. The number in parentheses of the target label indicates the amount of missiles needed to destroy the target. The field parameters of each missile's seeker in the simulation are basically the same as those of the five

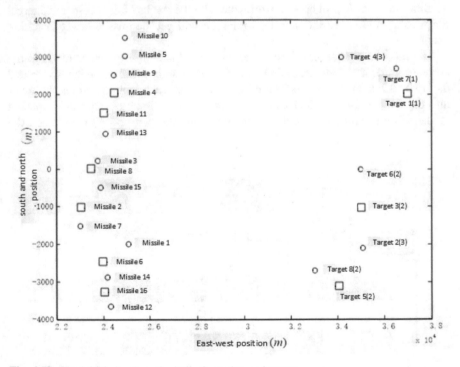

Fig. 4.59 The initial situation of missile formation and targets

missiles listed in Table 4.2. In the dynamic target assignment process, 20 s are needed for assignment once and a total of 5 times are needed to meet the changes in the battlefield situation.

3. **Guidance and control system of the missile member**

A type of anti-ship missile for example uses three-channel robust PID overload flight control system [1, 75, 77, 105, 106]. In the terminal guidance section, to adjust the guidance time, the guidance law (ITCG) with guidance time limit is used:

$$
n_y^* = \begin{cases} 5g, & n_y^* > 5g \\ a_{py} - \frac{60V_m^5}{a_{py}R_{ms}^3}(\bar{T}_{go} - \hat{T}_{goy}) + \cos\theta_m, & -5g \le n_y^* \le 5g \\ -5g & n_y^* < -5g \end{cases} \tag{4.99}
$$

$$
n_z^* = \begin{cases} 5g, & n_z^* > 5g \\ a_{pz} - \frac{60(V_m\cos\theta_m)^5}{a_{pz}R_{ms}^3}(\bar{T}_{go} - \hat{T}_{goz}), & -5g \le n_z^* \le 5g \\ -5g & n_z^* < -5g \end{cases} \tag{4.100}
$$

where n_y^*, n_z^* are the longitudinal and lateral overload instructions, R_{ms} is the distance between the missile and target, θ_m is the trajectory inclination of the missile, and $\bar{T}_{go}$ is the specified guidance time, $\hat{T}_{goy}$ and $\hat{T}_{goz}$ are the longitudinal and lateral guidance remaining time estimation using PNG, the expression as follows:

$$
\hat{T}_{goy} = \left(1 + \frac{(\theta_m - q_\alpha)^2}{10}\right)R_{ms}/V_m \tag{4.101}
$$

$$
\hat{T}_{goz} = \left(1 + \frac{(\varphi_m - q_\beta)^2}{10}\right)R_{ms}\cos q_\alpha/(V_m\cos\theta_m) \tag{4.102}
$$

where q_α and q_β are respectively the angular altitude and horizontal angle of the sight line, and φ_m is the missile ballistic angle. The control quantities a_{py}, a_{pz} are given by the longitudinal and lateral PNG, respectively, are as follows:

$$
a_{py} = k_y|\dot{R}_{ms}|\dot{q}_\alpha \tag{4.103}
$$

$$
a_{pz} = -k_z|\dot{R}_{ms}|\cos\theta_m\dot{q}_\beta \tag{4.104}
$$

where k_y, k_z are the longitudinal and lateral guide proportional coefficient, whose value are taken as 4.0.

4. **Formation support network parameters setting**

The formation support network data link terminal consists of a spread spectrum radio station and an embedded data link protocol. A single physical channel is used to provide half duplex service and each network node can communicate with each other according to the dynamic time division multiple access (MAC), what's more, in the mobile process the nodes can communicate with each other and send data to the terrestrial monitoring system and can receive all uplink data. The node sends its own information in the time slot it occupies, and one slot can only be occupied by one node. The slot structure is shown in Fig. 4.60. With 200 ms as a frame, three frames form a super frame, and each frame is divided into seven time slots, then each time slot is 200 ms/7 ≈ 28.58 ms.

As a result of the use of single-duplex radio, there is a switching delay T_d in the conversion process of transmission and reception mode, and according to the test we get that Td = 15– 20 ms and it is random. When the above slot allocation structure is used, the random delay can be converted to a fixed delay of 28.58 ms. The support network data link system for transmitting high-speed information in high dynamic environments is described in detail in Chap. 7.

5. **Simulation Result**

The target assignment results for the missiles formation are shown in Fig. 4.61. Due to limited space, only a representative target 2 is selected as an example to illustrate the description here. It can be seen from the figure that the missile 2 need 3 missiles and is an electromagnetic radiation source, so the missiles ε_1, ε_2 and ε_7 are allocated to attack it, the missile ε_1 and ε_7 are equipped with active radar seekers, while the missile ε_2 with an infrared seeker, in this way, the three missile seeker

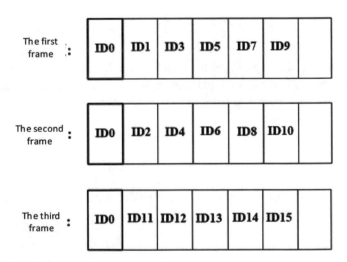

Fig. 4.60 The network slot structure

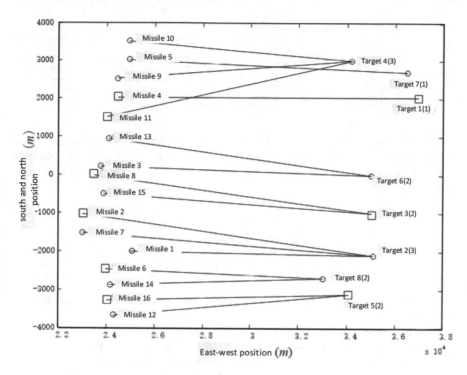

Fig. 4.61 The result of dynamic target assignment

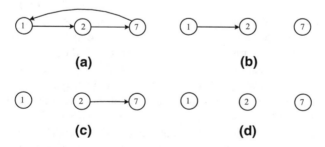

Fig. 4.62 Four different network topologies used between the missiles $\varepsilon_1, \varepsilon_2$ and ε_7

types complement each other, and is conducive to improving the accuracy of combat. The distribution results are reasonable and effective, seen from the missile-target matching results.

According to the calculated optimal handover surface of the missile member $\varepsilon_1, \varepsilon_2$ and ε_7 (shown in Table 4.4), the missile formation entered into the cooperative terminal guidance stage after 10 s of dynamic target assignment. The network topology used between the missile members shown in Fig. 4.62.

Table 4.4 Handover surface of the missiles $\varepsilon_1, \varepsilon_2$ and ε_7

Missile	x_f (m)	h_f (m)	z_f (m)	ϑ_f (°)	ψ_f (°)
1	25026.2	1773.6	−1063.4	−3.4	−5.3
2	24187.8	1986.5	−1493.9	−3.6	−0.6
7	24509.7	1989.4	−1984.7	−2.1	−0.5

Let the probability the system be in each structure at the initial time: $p(1,0) = 0.5, p(2,0) = 0.1, p(3,0) = 2.0, p(4,0) = 0.2$. The optimal transfer intensity matrix between the structures is obtained by using the network connectivity control strategy:

$$
v = \begin{bmatrix}
0.18 & 0.26 & 0.37 & 0.22 \\
0.33 & 0.37 & 0.28 & 0.26 \\
0.24 & 0.21 & 0.35 & 0.27 \\
0.45 & 0.42 & 0.31 & 0.35
\end{bmatrix}.
$$

At this point the probability of the missile formation was discovered is $p_f = 0.537$.

Figure 4.63 shows the midcourse-terminal guidance handover ballistic trajectories and cooperative terminal guidance ballistic trajectories of the missile $\varepsilon_1, \varepsilon_2$ and ε_7, and the missile trajectory and guidance time (not including the first 10 s of the midcourse-terminal guidance handover accommodation time) of each missile member are also given in Fig. 4.63a, using the PNG guidance and cooperative guidance. The difference between the longest and shortest guidance times when using PNG guidance is 6.38 s, and there is a relatively large difference in the guidance time between missile members. In the case of cooperative terminal guidance, three missiles hit the target almost simultaneously, and the time is relatively closer to the maximum guidance time of the PNG guidance (namely the guidance time of the missile member ε_2, 45.21 s). With PNG guidance, the missile ε_2 using longest time have the most straightforward trajectory and the missile ε_1 and ε_7 wait for the missile ε_2 by longitudinal and lateral movements to achieve the goal of simultaneously hitting the target. In Fig. 4.64, the curves of the normal acceleration and the network topologies probability are given, which is the result of the control of the network connectivity. The convergence curve of the coordinated variables (i.e. the expected guidance time) is shown in Fig. 4.65a, and reflects the convergence process of expected guidance time, where the convergence period is 2.5 s. It can be seen from the figure that although the initial differences in the coordination variables are large, when the network connectivity conditions are guaranteed, it can converge quickly, and the conclusion can be drawn from the specified guidance time shown in Fig. 4.65b. Figure 4.65c gives the remaining guidance time estimate and its difference from the specified guidance time. It can be seen that under ITCG, the error of the remaining time and the coordinated variable of each missile eventually converges to zero, the sum of the time difference $\Delta t = 0.67$, while the miss distance $\Delta d = 4.15$ m.

Fig. 4.63 the trajectory curve using cooperative guidance

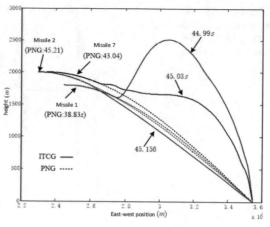

(a) Longitudinal ballistic trajectories

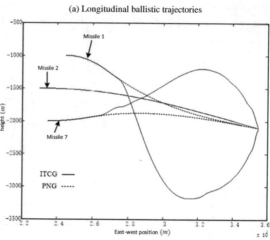

(b) Lateral ballistic trajectories

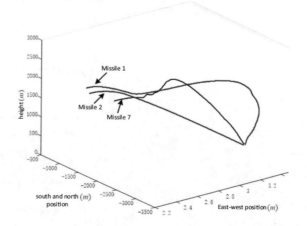

(c) Three-dimensional space ballistic trajectories

Fig. 4.64 Normal
acceleration and structural
state probability of
cooperative guidance

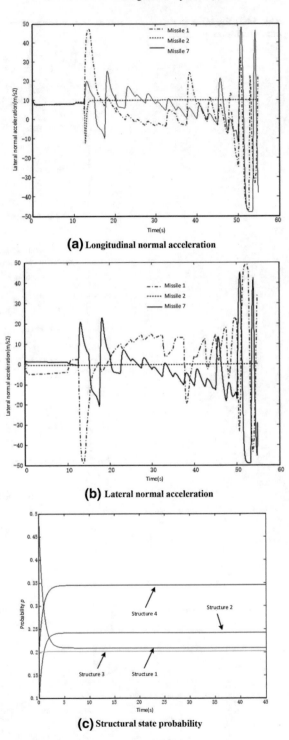

(a) Longitudinal normal acceleration

(b) Lateral normal acceleration

(c) Structural state probability

Fig. 4.65 The curve of
each time value in the
process of cooperative
guidance

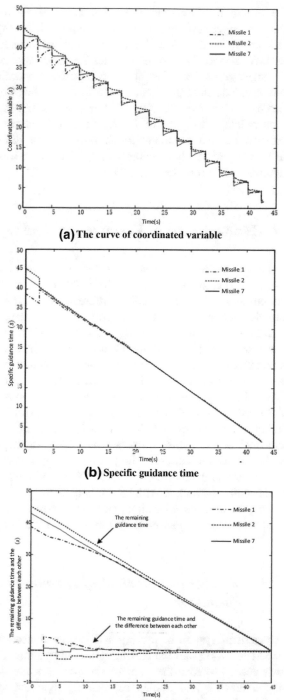

(a) The curve of coordinated variable

(b) Specific guidance time

(c) The remaining guidance time and the difference between each other

The simulation results show that the missiles formation has completed the dynamic target assignment with the support of the support network, and successfully completed the midcourse-terminal guidance handover according to the optimal handover surface Ω, and the missiles formation realized the cooperative attack on the target group by using the time coherence coordination algorithm and ITCG guidance. The mission is completed in a high level by the formation, the missile trajectory of the missiles formation is straight, and the coordinated variables converge quickly; through the implementation of network connectivity control, we achieved the goal that the minimum terminal time difference and miss distance and the lowest probability of being found, which improves the comprehensive combat effectiveness of missiles formation.

4.7   Management of Leaving and Joining Formation

During the mission of the missiles formation, the missile members often leave or join the formation flight due to the malfunction of the missile member or being shorten down, the controllability difference of the missile member's speed and the battlefield situation, thus the formation size will changed accordingly. Therefore, it is necessary to propose a management strategy in a timely manner by management system of leaving and joining formation, in order to adjust the formation principle/performance index, task planning/target assignment, route planning/co-guidance of the missiles formation, and to complete the formation control by the formation flight control system, so that the new formation can maintain stability. When some missile breakdown, formation fault tolerance control is a special case of leaving and joining management. At present, only the fault tolerance control research of unmanned aerial vehicle formation and robot formation can be seen in references. As the missiles autonomous formation mainly uses decentralized control structure and the formation control is not hierarchical, so in this chapter diagram theory based on the adjacent matrix will be applied to the leaving and joining management method, so that the formation can quickly form a new formation to keep stable, when some missile join or leave the missile formation.

4.7.1   Management Method of Leaving and Joining Formation

Management system of leaving and joining formation is based on the formation support network, and the leaving and joining formation detection system continuously detects whether there are missile members have joined or left the formation; if so, the leaving and joining formation judgement system is used to determine whether the missile members can be stable to join the support network or completely leave the support network. Furthermore, it needs to determine the role of missile members who have joined or left the formation by leaving and joining formation member role judgement module.

1. Leaving and joining formation detection: in the process of missile formation, the support network continuously detects whether the size of the network node has changed and which nodes have changed. If there is a change, it means that a missile member will join or leave the missiles formation.
2. Leaving and joining formation judgement: start a timer since the beginning of the leaving and joining formation detection to the missile members leaving or joining the formation, and in the set time T_Δ:

   (1) If the formation size and the composition of the missile members did not change again, it can be determined that the missile members had left or joined, and then execute leaving and joining formation member role judgement and give management strategy.
   (2) If the formation size or missile component composition changes again, then time again.

Here, the set time T_Δ is to avoid to miscarriage of justice and adjust frequently the management strategy of leaving and joining formation due to support network packet loss, delay and system instability and other factors, so as to improve the missiles autonomous formation cooperative and control system.

(3) Leaving and joining formation member role judgement: autonomous formation support network structure is divided into two types, planar structure and hierarchical structure (described in detail in Chap. 7). When there is a difference in the role of the missile members in the formation, it is necessary to determine the role of the missile member who leaves or joins the missile formation. For example, in the "leader-follower" formation, it needs to determine whether the missile is served as a leader, or a follower.
(4) Management strategy of leaving and joining formation: it determines how to adjust the formation after leaving and joining formation detection, leaving and joining formation judgement and leaving and joining formation member role judgement. If the adjustment was decided, then the formation principle/performance indicators, task planning/target assignment, route planning/cooperative guidance of the three major parts should be appropriately adjusted based on specific circumstances. Figure 4.66 shows the structure of management system of leaving and joining formation.

4.7.2 Management Strategy of Leaving and Joining Formation

In the following, we give the relationship between management system of leaving and joining formation and the adjustment of the formation control (see in Fig. 4.67)

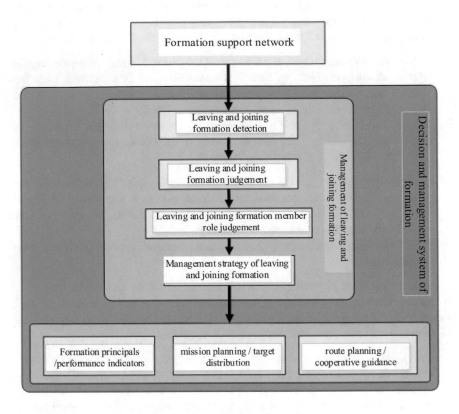

Fig. 4.66 Management system of leaving and joining formation

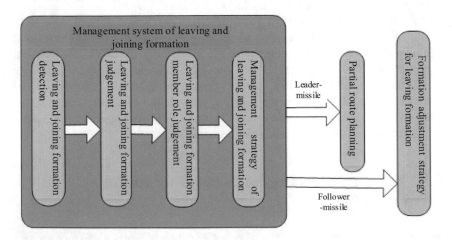

Fig. 4.67 Management system of leaving and joining formation

through the route planning/cooperative guidance, and the relationship between management systems of leaving and joining formation and the other parts can also be so pushed.

As can be seen from Fig. 4.67, after leaving and joining formation detection, leaving and joining formation judgement and leaving and joining formation member role judgement, management system of leaving and joining formation has made a decision whether to adjust or not, and give an adjustment strategy by determine the role of the missile. If it is a follower, then local route planning is needed; the missile members who left the formation independently fly through the local route planning; then, the missiles quickly generate a new formation by formation strategy of leaving and joining formation, and the missiles formation continue fly stably.

1. Formation adjustment strategy for leaving formation

(1) Adjacent matrix of formation

The formation of the formation is a directed graph $G = (\varepsilon, E, D)$, where $\varepsilon = \{ \varepsilon_1, \ldots \varepsilon_n \}$ is the node set of the missile formation; edge set E is the relationship matrix between the nodes in the formation, D is the formation hierarchical structure of the missiles formation, that is, formation depth.

Now, formation composed of six missiles as an example, as shown in Fig. 4.68, the missile ε_1 is the formation's leader missile, and the formation depth is zero; the remaining missile members follow the missile members ε_1 flying, so the formation depth of the missile member ε_2 and ε_3 is one and that of missile member $\varepsilon_4, \varepsilon_5$ and ε_6 is two; with the adjacent matrix of formation, the formation relationship is denoted as follows:

$$E = \begin{bmatrix} 0 & 1 & 1 & 0 & 0 & 0 \\ 0 & 0 & 0 & 1 & 1 & 0 \\ 0 & 0 & 0 & 0 & 1 & 1 \\ 0 & 0 & 0 & 0 & 0 & 0 \\ 0 & 0 & 0 & 0 & 0 & 0 \\ 0 & 0 & 0 & 0 & 0 & 0 \end{bmatrix}$$

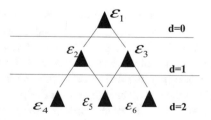

Fig. 4.68 Graph for formation depth of the missiles formation

According to the above formation rule of missiles formation, the adjacent matrix is defined as $\varepsilon_{ij} \in E$, and $\varepsilon_{ij} = 1$ indicates that the missile ε_i is the parent node of the missile ε_j, $n_{ai}(q)$ is the child node set for the node i.

(2) Heuristic search method of formation

No information search method is a blind search method to find a path to a target node, regardless of the width-first algorithm or the depth-first algorithm. For many tasks, it is possible to use task-related information to help reduce the search. This type of information is often referred to as heuristic information, and the search process using heuristic information is called a heuristic search method. It is often possible to specify some of the heuristic information that reduces the workload of the search without sacrificing the guarantee of finding the minimum length path.

The evaluation function in the heuristic search method: heuristic information can be used to arrange the nodes on the OPEN table in the h-th step of the search process so that the search extends along those sections that are considered to be the most promising. Assuming that the evaluation function is marked with $f(n)$, which means the value of the function on the node, and is an estimated dissipated value of the minimum dissipation path from the originating node to the target node.

The selection method of the evaluation function is: $f(n) = g(n) + h(n) + n_{an}$, where $g(n)$ is the path dissipation of the node n, and $h(n)$ is the minimum value of the estimated value for all nodes in the child node set $n_{an}(q)$ of the node n, n_{an} is the number of the elements in $n_{an}(q)$. The path dissipation of each path can be set as one, and the leaf nodes have no path dissipation values, because of that they are target nodes needed to find, so let the estimated value of leaf nodes be zero.

If a missile member leaves the formation for some reason, the formation needs to be readjusted. Using the heuristic search algorithm, search from the leaving missile for members whose positions are needed to adjust, and put these follower missiles' numbers into the directed order list A in turn. The algorithm is based on the principle of small formation disruption and short time of disorganization, to achieve the formation restoration. Using the heuristic evaluation function defined earlier ensures that the search results are the smallest adjustment path for the formation. At the same time, the algorithm ensures that no matter any missiles in the formation leave the formation, other missile members can get the same formation adjustment strategy.

(3) Replacement algorithm for leaving formation

By the above heuristic search algorithm, the numbers of the missile members' need to adjust can be obtained, and then through the replacement algorithm for leaving formation, the missile members needed position adjustment are adjusted with the shortest disorganization time to improve the stability of the missiles formation.

Auxiliary matrix: According to the definition of the previous adjacent matrix, we can get that in the adjacent matrix, the i-th column reflects $n_{ai}(q)$ of any missile member ε_i, the i-th line reflects $n_{ac}(q)$ of any missile member ε_i. The problem of the formation adjustment is attributed to the fact that the rows and columns in the table

A in the adjacent matrix are exchanged in turn, i.e. the result of the heuristic search algorithm is used to adjust the formation. It is necessary to introduce an auxiliary matrix P, which is left or right multiplied by the adjacency matrix to realize the exchange of these specific row and column vectors in the adjacent matrix, thus restoring the formation's formation.

Assuming that the size of the missiles formation is n, then the adjacent matrix E of the formation:

$$E = \begin{bmatrix} e_{11} & \cdots & e_{1i} & \cdots & e_{1j} & \cdots & e_{1k} & \cdots & e_{1n} \\ \vdots & \cdots & \vdots & \cdots & \vdots & \cdots & \vdots & \cdots & \vdots \\ e_{i1} & \cdots & e_{ii} & \cdots & e_{ij} & \cdots & e_{ik} & \cdots & e_{in} \\ \vdots & \cdots & \vdots & \cdots & \vdots & \cdots & \vdots & \cdots & \vdots \\ e_{j1} & \cdots & e_{ji} & \cdots & e_{jj} & \cdots & e_{jk} & \cdots & e_{jn} \\ \vdots & \cdots & \vdots & \cdots & \vdots & \cdots & \vdots & \cdots & \vdots \\ e_{k1} & \cdots & e_{ki} & \cdots & e_{kj} & \cdots & e_{kk} & \cdots & e_{kn} \\ \vdots & \cdots & \vdots & \cdots & \vdots & \cdots & \vdots & \cdots & \vdots \\ e_{n1} & \cdots & e_{ni} & \cdots & e_{nj} & \cdots & e_{nk} & \cdots & e_{nn} \end{bmatrix} \tag{4.105}$$

P is the unit matrix $I_{n \times n}$ If there is a missile fault in the formation, it is necessary to update the P by using the heuristic search algorithm to get the directed order table A. The specific update method is as follows [4, 91]:

Definition 4.1 $\exists A = \begin{bmatrix} n & m & \cdots & s \end{bmatrix}$ and the matrix P is adjusted according to the following rule (namely CP rule):

$$p_{nn} = 0, p_{nm} = 1, p_{mm} = 0, \ldots, p_{ss} = 0, p_{sn} = 1$$

According to the definition 4.9, update the auxiliary matrix P using CP rule, and then observe the adjusted auxiliary matrix $P, \forall Q_{n \times n}$, and it has the following properties:

(1) $\forall \mathbf{P}_i = \begin{bmatrix} 0 \ldots 0 \, p_{ij} \, 0 \ldots 0 \end{bmatrix}, (p_{ij} = 1, i \neq j)$

Right multiplied by Q, namely $\mathbf{Q}' = \mathbf{PQ}$, then the result is $\mathbf{Q}'_i = \mathbf{Q}_j$ (both are row vectors)

(2) $\forall \mathbf{P}_i = \begin{bmatrix} 0 \ldots 0 \, p_{ij} \, 0 \ldots 0 \end{bmatrix}^{\mathrm{T}} (p_{ij} = 1, i \neq j)$

Left multiplied by Q, i.e. $\mathbf{Q}'' = \mathbf{QP}_i$, then the result is (both are row vectors) And then update the adjacent matrix according to Proposition 4.1.

Proposition 4.1 for $\forall E_{n \times n}, \exists \mathbf{P}_{n \times n}$ (the adjusted auxiliary matrix P). Then, after the replacement for leaving formation, the formation of the formation meets the adjacent matrix E'' : $\mathrm{E}'' = \mathbf{PEP}^T$

2. **Formation adjustment strategy for joining formation**

The formation adjustment strategy for joining formation can be mainly divided into the following steps:

(1) Generate new adjacent matrix E with the relative position of the joining missile and the other missiles in the missiles formation;

(2) Using the heuristic search algorithm described above, search from the joining missile for members whose positions are needed to adjust, and put these follower missiles' numbers into the directed order list A in turn.

(3) The new adjacency matrix E'' is generated by the same replacement algorithm as the principle of the replacement algorithm for leaving formation. The missiles continue to fly to the target area according to the new formation.

4.7.3 Simulation for Management of Leaving and Joining Formation

In the above simulation environment, with the same initial conditions, experiments for leaving and joining formation are completed. In the simulation process, a missile in the formation is randomly selected at random time and loses communication links with the formation, and leaves the formation; after a period of time, the missile join the formation after the communication recovery. This experiment is used to detect the formation effectiveness of the management of leaving and joining formation.

Figure 4.69 shows the flight route information recorded by the missile member (follower-missile 1) that has left or joined the formation. It can be seen from the figure, the missile member leaves the formation and fly to the target area independently, according to the local route real-time planning algorithm; when the missile member resumed communication with the formation support network, the missile member joined the formation of three other missiles, and then the formation of the missiles formation reconstructed, the formation changed from wedge to diamond. The simulation results show that the local route real-time planning algorithm and management algorithm for joining formation are reliable and effective.

Figure 4.70 shows the flight path of the missiles formation recorded by missile members who have not left the missiles formation. It can be seen from the figure that when a missile left the formation network, the formation size was reduced from 4 to 3, the missile members reconstructed the formation in time, and the formation changed from diamond to wedge, then the formation continued to fly stably; when the missile member recovered to rejoin the support network, the formation size was

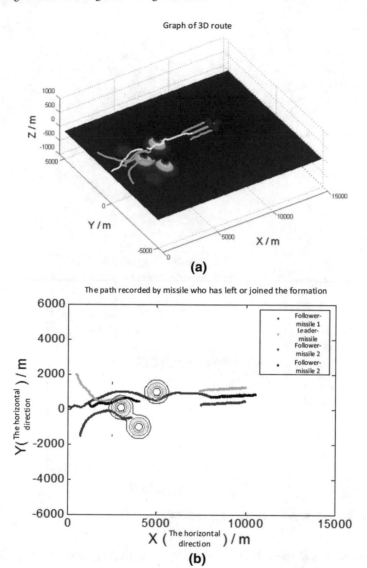

Fig. 4.69 Routes recorded by missile members who has left or joined the formation

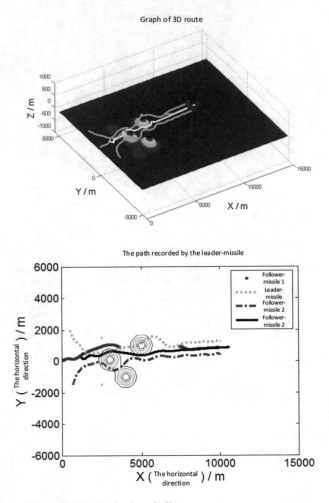

Fig. 4.70 The path recorded by the leader-missile

expanded to 4, and the missiles formation reconstruction carried out through the management algorithm for joining formation. The formation was converted from wedge to diamond and stable to the target area. The simulation results show the effectiveness of the formation fault tolerance control algorithm and the management algorithm of leaving and joining formation.

Chapter 5
Flight Control System of MAF (FCSM)

5.1 Compositions of FCSM

FCSM includes two aspects: one is formation generation and guidance, the other is formation configuration control and holding. Its responsibility is that: according to the formation optimization index and formation requirements generated by the formation decision and management system (DMS), optimize and generate in real-time the instructions of formation guidance, control and holding, and ensure the realization of nodes collision avoidance maneuvers control and high-quality formation configuration through member flight control system (MFCS). The structure of FCSM is as shown in Fig. 5.1.

5.2 Formation Generation and Guidance

In Chap. 2, most commonly used basic formation types used in the missile formation, namely column and horizontal formation, wedge-shaped formation, diamond-shaped formation, shuttle-shaped formation and polygon formation, has been defined and detailed analyzed. No matter how large the formation scale is, it can be made up of these four basic types in principle. It can be seen from the function structure Fig. 1.13 and architecture Fig. 2.19 of MAF's cooperative guidance and control that, the formation decision and management system should apperceive the information obtained and formation information according to the situation, and generate the required target formation q_e composed of basic formations in real-time. The characteristics of target formation q_e are determined by the definitions of α—adjacent formation, σ—adjacent formation and Quasi-σ—adjacent formation, the Definitions 2.12–2.14 give the characteristics of these target formations q_e. The target formation q_e sets the expected safety distance allowance $\Delta\mu_{eij} = E\{\Delta d_{eij}(t)\} = \mu_{eij} - d_{si}$ for each missile member ε_i. How to guide the formation to quickly and accurately form the required target formation q_e is the focus of this section.

© National Defense Industry Press, Beijing and Springer Nature Singapore Pte Ltd. 2019
S. Wu, *Cooperative Guidance & Control of Missiles Autonomous Formation*,
https://doi.org/10.1007/978-981-13-0953-3_5

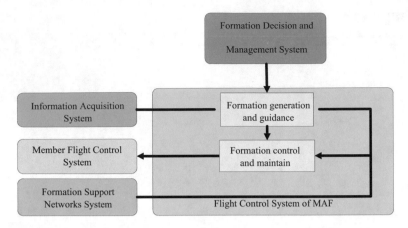

Flight Control System of MAF

Fig. 5.1 The structure of FCSM

5.2.1 Flocking for Multi-agent Dynamic Systems

In this section, we present a set of distributed algorithms for flocking in free-space, or free-flocking [107–109]. We refer to a physical agent with dynamics $\ddot{q}_i = u_i$ as α-agent. Later, we introduce virtual agents called β-agents and γ-agents which model the effect of "obstacles" and "collective objective" of a group, respectively. The primary objective of an α-agent in a flock is to form an α-lattice with its neighboring α-agents. In free-flocking, each α-agent applies a control input that consists of three terms:

$$u_i = f_i^g + f_i^d + f_i^\gamma \tag{5.1}$$

where $f_i^g = -\nabla_{q_i} W(q)$ is a gradient-based term, f_i^d is a velocity consensus term that acts as a damping force, and f_i^γ is a navigational feedback due to a group objective. We propose two distributed algorithms that can be used for creation of flocking motion in $\mathbb{R}^m$ [107–109].

Algorithm 5.1 $u_i = u_i^\alpha$ with

$$u_i^\alpha = \sum_{j \in n_{ic}} \phi_\alpha(\|q_j - q_i\|_\sigma)\boldsymbol{n}_{ij} + \sum_{j \in n_{ci}} a_{ij}(q)(p_j - p_i) \tag{5.2}$$

where $\boldsymbol{n}_{ij} = \sigma_e(q_j - q_i) = (q_j - q_i)/\sqrt{1 + e\|q_j - q_i\|^2}$ is a vector along the line connecting q_i to q_j and $e \in (0, 1)$ is a fixed parameter of the σ-norm. Algorithm 5.1 has no group objective and is known as the (α, α) protocol of flocking, because it states the interaction rule between two α-agents.

Algorithm 5.2 $u_i = u_i^\alpha + u_i^\gamma$ or

$$u_i = \sum_{j \in n_{ci}} \phi_\alpha(\|q_j - q_i\|_o) n_{ij} + \sum_{j \in n_{ci}} a_{ij}(q)(p_j - p_i) + f_i^\gamma(q_i, p_i, q_r, p_r) \qquad (5.3)$$

where u_i^γ is the navigational feedback and is given by

$$u_i^\gamma = f_i^\gamma(q_i, p_i, q_r, p_r) = -c_1(q_i - q_r) - c_2(p_i - p_r), \quad c_1, c_2 > 0 \qquad (5.4)$$

The pair $(q_r, p_r) \in \mathbb{R}^m \times \mathbb{R}^m$ is the state of a γ-agent. A γ-agent is dynamic/static agent that represents a group objective and can be viewed as a moving rendezvous point. Let (q_d, p_d) be a fixed pair of vectors that demote the initial position and velocity of a γ-agent. A dynamic γ-agent has the following model:

$$\begin{aligned} \dot{q}_r &= p_r \\ \dot{p}_r &= f_r(q_r, p_r) \end{aligned} \qquad (5.5)$$

With $(q_r(0), p_r(0)) = (q_d, p_d)$. A static γ-agent has a fixed state that is equal to (q_d, p_d) for all time. The design of $f_r(q_r, p_r)$ is part of tracking control design for a group of agents. Based on the expression of u_i^γ, a secondary objective of an α-agent is to track a γ-agent.

We shall see that despite the similarities between certain terms in these protocols, the collective behavior of a group of agents that use Algorithm 5.1 is drastically different than a group of agents applying Algorithm 5.2. Algorithm 5.1 leads to flocking behavior only for a very restricted set of initial states. For generic set of initial states and large number of agents, Algorithm 5.1 most likely fails to produce flocking behavior and instead leads to regular fragmentation. In contrast, Algorithm 5.2 never leads to fragmentation. The importance of Algorithm 5.1 is due to its fundamental role in forming lattice-shape structures during flocking as a key element of Algorithm 5.2.

The collective dynamics of a group of α-agents applying Algorithm 5.2 is in the form:

$$\begin{cases} \dot{q} = p \\ \dot{p} = -\nabla W(q) - \widehat{L}(q)p + f_\gamma(q, p, q_r, p_r) \end{cases} \qquad (5.6)$$

where $W(q)$ is a smooth collective potential function, and $\widehat{L}(q)$ is the m-dimensional Laplacian of the proximity net $G(q)$ with a state-dependent adjacency matrix $A(q) = [a_{ij}(q)]$.

In Algorithm 5.1, $f_\gamma \equiv 0$, then system (5.6) is a dissipative particle system with Hamiltonian.

$$H(q, p) = W(q) + \sum_{i=1}^{n} \|p_i\|^2 \qquad (5.7)$$

This is due to $\dot{H} = -p^T \hat{L}(q)p \leq 0$ and the fact that the multi-dimensional graph Laplacian $\hat{L}(q)$ is positive semidefinite matrix for all q. The key in stability analysis of collective dynamics is employing a correct relative motion system. We propose the use of a moving frame to analyze the stability of flocking motion.

Consider a moving frame that is centered at q_c. Let $\text{Ave}(z) = (1/n)\sum_{i=1}^{n} z_i$ denote the average of the elements z_i in $z = \text{col}(z_1, \ldots, z_n)$. Let $q_c = \text{Ave}(q)$ and $p_c = \text{Ave}(p)$ denote the position and velocity of the origin of the moving frame. Then $\dot{q}_c(t) = p_c(t)$ and $\dot{p}_c(t) = \text{Ave}(u(t))$. The position and velocity of agent ε_i in the moving frame is given by

$$\begin{cases} x_i = q_i - q_c \\ v_i = p_i - p_c \end{cases} \tag{5.8}$$

The relative positions and velocities remain the same in the moving frame, i.e. $x_j - x_i = q_j - q_i, v_j - v_i = p_j - p_i$, thus $W(q) = W(x), \nabla W(q) = \nabla W(x)$. The (α, α) protocol in the moving frame can be expressed as

$$u_i^{\alpha} = \sum_{j \in n_{ci}} \phi_{\alpha}(\|x_j - x_i\|_{\sigma})\boldsymbol{n}_{ij} + \sum_{j \in n_{ci}} a_{ij}(q)(v_j - v_i) \tag{5.9}$$

with $a_{ij}(x) = \rho_h(\|x_j - x_i\|_{\sigma}/\|d_{imax}\|_{\sigma})$. Our first result is a decomposition lemma [107–109] that is the basis for posing a structural stability problem for the motion of flocks.

Lemma 5.1 *Suppose that the navigational feedback $f_{\gamma}(q, p)$ is linear, i.e. there exists decomposition in the following form:*

$$f_{\gamma}(q, p, q_r, p_r) = g(x, v) + \boldsymbol{1}_n \otimes h(q_c, p_c, q_r, p_r) \tag{5.10}$$

Then, the collective dynamics of a group of agents applying Algorithm 5.2 can be decomposed as n second-order systems in the moving frame

$$\begin{cases} \dot{x} = v \\ \dot{v} = -\nabla W(x) - \hat{L}(x)v + g(x, v) \end{cases} \tag{5.11}$$

and one second-order system in the reference frame

$$\begin{cases} \dot{q}_c = p_c \\ \dot{p}_c = h(q_c, p_c, q_r, p_r) \end{cases} \tag{5.12}$$

where $g(x, v) = -c_1 x - c_2 v$, $h(q_c, p_c, q_r, p_r) = -c_1(q_c - q_r) - c_2(p_c - p_r)$ and (q_r, p_r) is the state of the γ-agent. Equation (5.11) is called structural dynamic, and Eq. (5.12) is called translational dynamic.

According to the decomposition lemma, we are now at the position to define stable flocking motion as the combination of the following forms of stability properties: (1) stability of certain equilibria of the structural dynamics, and

(2) stability of a desired equilibrium of the translational dynamics. The challenge in analysis of flocking behavior is to establish stability property (1).

The significant differences in group behaviors created by Algorithms 5.1 and 5.2 are due to the considerable differences in the structural dynamics induced by the two algorithms. Given Algorithm 5.1, the following structural dynamics:

$$\sum_1 : \begin{cases} \dot{x} = v \\ \dot{v} = -\nabla W(x) - \hat{L}(x)v \end{cases} \tag{5.13}$$

In comparison, the structural dynamics of a group of agents applying Algorithm 5.2 is in the form

$$\sum_2 : \begin{cases} \dot{x} = v \\ \dot{v} = -\nabla U_\lambda(x) - D(x)v \end{cases} \tag{5.14}$$

where $U_\lambda(x)$ is called the aggregate potential function and is defined by

$$U_\lambda(x) = W(x) + \lambda J(x) \tag{5.15}$$

The map $J(x) = \frac{1}{2}\sum_{i=1}^{n} \|x_i\|^2$ is the moment of inertia of all particles and $\lambda = c_1 > 0$ is a parameter of the navigational feedback. Moreover, the damping matrix $D(x) = c_2 I_m + \hat{L}(x)$ is a positive definite matrix. Define the structural Hamiltonian of systems Σ_1, Σ_2 as follows:

$$\begin{aligned} H(x, v) &= W(x) + K(v) \\ H_\lambda(x, v) &= U_\lambda(x) + K(v) \end{aligned} \tag{5.16}$$

where $K(v) = \frac{1}{2}\sum_{i=1}^{n} \|v_i\|^2$ is the velocity mismatch function, or the kinetic energy of the particle system in the moving frame. We also need to define what we mean by "cohesion of a group" and "flocks" [107–109].

Definition 5.1 Let $(q(\cdot), p(\cdot)) : t \mapsto \mathbb{R}^{mn} \times \mathbb{R}^{mn}$ be the state trajectory of a group of dynamic agents over the time interval $[t_0, t_f]$. We say the group is cohesive for all $t \in [t_0, t_f]$ if a ball of radius $R > 0$ centered at $q_c(t) = \text{Ave}(q(t))$ that contains all the agents for all time $\exists R > 0 : \|x(t)\| \le R, \forall t \in [t_0, t_f]$ exists.

Definition 5.2 A group of α-agents is called a flock over the internal $[t_0, t_f]$ if the proximity net $G(q(t))$ is connected over $[t_0, t_f]$.

The group is called a quasi-flock if the largest component of the proximity net is highly populated. The following lemma provides a geometric characterization of the set of local minima of the collective potential and plays a critical role in establishing the spatial-order of self-organizing flocks.

Lemma 5.2 *Every local minima of $W(q)$ is an α-lattice and vice versa.*

Theorem 5.1 [107–109]: *Consider a group of α-agents applying Algorithm 5.1 with structural dynamics Σ_1 [See formula (5.13)]. Let $\Omega_c = \{(x, \mathrm{v}) : H(x, \mathrm{v}) \leq c\}$ be a level-set of the Hamiltonian $H(x, \mathrm{v})$ of Σ_1 such that for any solution starting in Ω_c, the agents form a cohesive flock $\forall t \geq 0$. Then, the following statements hold.*

   (i)   Almost every solution of the structural dynamics converges to an equilibrium point $(x^*, 0)$, with a configuration x^* that is an α-lattice.

   (ii) All agents asymptotically move with the same velocity.

  (iii) Given $c < c^* = \psi_\alpha(0)$, no internal collisions among agents occur for all $\forall t \geq 0$.

Theorem 5.2 [107–109]: *Consider a group of α-agents applying Algorithm 5.2 with $c_1, c_2 > 0$, and structural dynamics Σ_2 [See formula (5.14)]. Assume that the initial velocity mismatch $K(\mathrm{v}(0))$ and inertia $J(x(0))$ are finite. Then the following statements hold.*

   (i)   The group of agents remain cohesive for all $t \geq 0$.

   (ii) Almost every solution of Σ_2 asymptotically converges to an equilibrium point $(x_\lambda^*, 0)$ where x_λ^* is a local minima of $U_\lambda(x)$.

  (iii) All agents asymptotically move with the same velocity.

  (iv) Assume the initial structural energy of the particle system is less than $(k+1)c^*$ with $c^* = \psi_\alpha(0)$ and $k \in \mathbb{Z}_+$. Then, at most k distinct α-agents could possibly collide ($k = 0$ guarantees a collision-free motion).

Theorem 5.2 establishes some critical properties of collective behavior of a group of agents applying Algorithm 5.2 including cohesion, convergence, asymptotic velocity matching, and collision-avoidance. But it is difficult to give a rigorous proof of the geometric characterization of local minima of $U_\lambda(q)$. But it can be inferred that the local minima of $U_\lambda(q)$ should have the following two conjectures.

Conjecture 5.1 *Any local minima q_λ^* of $U_\lambda(q)$ for $\lambda > 0$ induces a connected proximity net, which means that under control law of Algorithm 5.2 a flock of α-agents is asymptotically self-assembled.*

Conjecture 5.2 *When $\lambda > 0$, for any fixed n, d, d_{imax}, satisfying $d_{imax}/d = 1 + \varepsilon(\varepsilon \ll 1)$, there exists a $\lambda^* \ll 1$ so that (1) any local minima q_λ^* of $U_\lambda(q)$ with $\lambda \in (0, \lambda^*)$ is a quasi α-lattice. (2) q_λ^* induces a planar graph $(G(q_\lambda^*), q_\lambda^*)$ in dimensions $m = 1, 2, 3$.*

Theorem 5.3 [107–109]: *Let q be an α-lattice of scale $\mathrm{d} > 0$ and ratio $\kappa > 1$ with n nodes at distinct positions. Then*

   (i)   The proximity structure $(G(q), q)$ is a planar graph in dimensions $m = 2, 3$.

   (ii) The proximity net $G(q)$ has at most $3n - 6$ links in $\mathbb{R}^2$.

  (iii) The proximity net $G(q)$ with $n > m + 1$ nodes cannot be a complete graph in $\mathbb{R}^m (m = 1, 2, 3)$.

According to Theorem 5.3, the planarity of proximity structure in 2-D space is that the total number of interaction terms for maintaining flocking motion is $O(n)$. This is a substantial reduction is computational complexity due to use of a distributed flocking algorithm compared to an $(O(n^2))$ cost of implementation of all-to-all interaction topologies in some existing models of swarms. Moreover, the planarity of graphs imposes a restriction on maximum ratio of the interaction range $\kappa = d_{imax}/d$ to desired distance, i.e. $1 < \kappa < \sqrt{m}$.

The equilibrium state of flocking motion satisfies $p_1 = p_2 =, \ldots, = p_n = p_r$, then all α-agents satisfy the following dynamics:

$$c_1^\alpha \sum_{j \in n_{ci}} \phi_\alpha(\|q_j - q_i\|_\sigma) n_{ij} = c_1^\gamma (q_i - q_r) \tag{5.17}$$

It can be seen that the parameters c_1^α, c_1^γ have a decisive effect on the final equilibrium state. The larger the parameter c_1^α is, the average distance d' between α-agent and its neighbors is closer to d, and α-agent is father away from γ-agent. The larger the parameter c_1^γ is, the average distance d' between α-agent and its neighbors is further less than d, and the α-agent is closer to γ-agent.

5.2.2 Simulation for the Airbreathing Hypersonic Vehicles Formation

A certain type of airbreathing hypersonic vehicle is in cruise flight course with a constant height. Aiming at the problems of formation generation and guidance in the lateral plane, a controller is designed based on the flocking theory of multi-agents. Because of the strict restriction of the angle of attack, the airbreathing hypersonic vehicle is not suitable for high speed maneuvering and is more suitable for cruise flight. Consider n airbreathing hypersonic vehicles in cruise flight course, the dynamic equation is illustrated in Refs. [1, 75].

Let $q_i = (x_i, z_i)$ be the position of vehicle i in the lateral plane, in the equation x represents the forward position of the vehicle, z represents the lateral position of the vehicle, and y represents the height. Let

$$p_i = (V_i \cos \theta_i \cos \varphi_i, -V_i \cos \theta_i \sin \varphi_i)$$

represent the velocity in the lateral plane of vehicle i, where φ_i, θ_i respectively represent trajectory deflection angle and trajectory inclination angle. Formation of vehicles is built up using the Algorithm 5.2, let Ac_i be the acceleration vector instruction through the Algorithm 5.2, then

$$Ac_i = c_1^\alpha \sum_{j \in n_{ci}} \phi_\alpha(\|q_j - q_i\|_\sigma) n_{ij} + c_2^\alpha \sum_{j \in n_{ci}} a_{ij}(q)(p_j - p_i) - c_1^\gamma(q_i - q_r) - c_2^\gamma(p_i - p_r) \tag{5.18}$$

where (q_r, p_r) is the position and velocity of the reference point, in this section it is set to cruise flight uniformly and in a straight line.

However the practical airbreathing hypersonic vehicle adopts velocity control and attitude control, so the acceleration instruction is required to convert to velocity instruction and attitude instruction. The velocity vector instruction can be obtained by integrating the acceleration vector instruction. Let p_i^* be the velocity vector instruction after integration, δt represents the integration step, then $p_i^* = p_i + Ac_i \cdot \delta t$.

We may as well let $p_i^*(1), p_i^2(2)$ respectively represent the instruction of forward and lateral velocity. Then the velocity instruction V_i^* and the trajectory deflection angle instruction φ_i^* are obtained as follows:

$$V_i^* = \sqrt{p_i^*(1)^2 + p_i^*(2)^2}$$
$$\varphi_i^* = -\arcsin(p_i^*(2)/V_i^*)$$

(5.19)

The formation control is realized through the control of velocity and trajectory deflection angle.

A flight controller structure [1, 75] is given as follows. The vertical cruise control adopts the coordinated control method of throttle-elevator, and the controller structure is $[\delta_{tc}\ \delta_{zc}]^T = K[V^* - V\ \theta^* - \theta - \omega_z\ \vartheta^* - \vartheta\ h^* - h - I_h - \delta_t - \delta_z]^T$, and the control law is $K = \begin{bmatrix} 0.03 & -49.4 & -4.33 & -9.7 & -0.01 & -0.003 & 0.7 & 0.03 \\ -0.88 & -173.4 & -16.9 & -36.2 & -0.04 & -0.01 & 2.7 & 0.1 \end{bmatrix}$.

The lateral control adopts the method of roll-turning, and the controller structure is:

$$[\delta_{yc}\ \delta_{xc}]^T = K[\varphi^* - \varphi\ -\omega_x - \omega_y\ \psi^* - \psi\ \phi^* - \phi\ -\delta_y - \delta_x]^T$$

The control law is

$$K = \begin{bmatrix} 5.7523 & -3.3397 & 1.8355 & -3.1745 & -3.112 & 0.0411 & 0.0 \\ -2.3235 & -1.5806 & -8.2392 & -0.2675 & -1.294 & 0.0137 & 0.0 \end{bmatrix}$$

Because that the vehicle adopts roll-turning, the rotation angle instruction can be obtained by empirical equation:

$$\phi^* = \frac{-V(\varphi^* - \varphi)}{g \cdot dt}$$

(5.20)

where dt is the simulation step size and the range of rotation angle instruction is $-60° \sim 60°$.

A simulation example is given for the formation of 5 aircrafts [107]. The initial position of five aircrafts are randomly distributed, the lateral plane positions are respectively $(-4000\ m,\ -4000\ m)$, $(0,\ -1500\ m)$, $(-2000\ m,\ 0)$, $(-3000\ m,\ 2500\ m)$, $(-1000\ m,\ 3500\ m)$, the initial heights are all 24 km, the initial Mach numbers are all 6 Ma, the initial trajectory deflection angles and trajectory inclination angles are all 0. Assume that the communication delay of the formation support network is τ, so the Algorithm 5.2 is rewrited as

$$u_i(t) = \sum_{j \in n_{ci}} \phi_\alpha(\|q_j(t-\tau) - q_i(t)\|_\sigma) n_{ij}(\tau)$$
$$+ \sum_{j \in n_{ci}} a_{ij}(q(\tau))(p_j(t-\tau) - p_i(t)) + f_i^\gamma(q_i(t), p_i(t), q_r, p_r) \tag{5.21}$$

$$n_{ij}(\tau) = \left((q_j(t-\tau) - q_i(t))\right) \bigg/ \sqrt{1 + e\|q_j(t-\tau) - q_i(t)\|^2} \tag{5.22}$$

$$a_{ij}(q(\tau)) = \rho_h(\|q_j(t-\tau) - q_i(t)\|_\sigma / \|d_{imax}\|_\sigma) \in [0, 1], \ j \neq i \tag{5.23}$$

That is at time t, every vehicle received the information from its adjacent vehicles is information at time $t - \tau$. Assume the delay is $\tau = 0.1$ s, select parameters $c_1^\gamma = 0.05, c_1^\alpha = 7$. The other parameters are respectively restrained distance (formation spacing) $d = 1000$ m, $d_{imax} = 1.4d(\kappa = 1.4)$, in the σ—norm $e = 0.1$, in $\phi(z) a = b = 5$ and in $\phi_\alpha(z) h = 0.2$.

It can be seen from the simulation results of Fig. 5.2 and 5.3, in the presence of communication delay, vehicle formation still can form a stable flocking formation. It should be noted that, why the formation is able to tolerate the delay is related to the given restrained distance $d = 1000$ m and the Mach number. In the process of formation flocking, the error of the position information of the adjacent vehicles due to communication delay is related to the Mach number. The larger the Mach number is, the larger the error. When the Mach number is 6 Ma, the error is about 180 m which is acceptable for the formation with 1000 m formation spacing. But when the distance is reduced to 200 m, the error will have a great impact on the formation. Therefore, the selection of restrained distance (formation spacing) d is particularly important for the formation generation and guidance, and the effect of position error due to the communication delay under a hypersonic flight condition should be considered. Because if the formation is a subsonic vehicle formation, the position error due to the communication delay of 0.1 s is at most about 30 m which is acceptable for the subsonic vehicle formation with a distance of 200 m. So the hypersonic vehicle formation is much larger than that of the subsonic or supersonic vehicle formation.

The simulation results show that: although the vehicles can flock in a stable formation, the subsequent cooperative terminal guidance is affected because of the influence of different initial positions. Because the reduction of the horizontal distance between vehicles will make the terminal lateral maneuvering increase, make the front and rear distance between vehicles increase, and make the terminal time of hitting the target increase. These all will have a certain impact on the comprehensive combat-effectiveness. The simulation experiment results show that: the formation guidance strategy based on the flocking theory in the different support network states can realize the formation generation and guidance of missiles

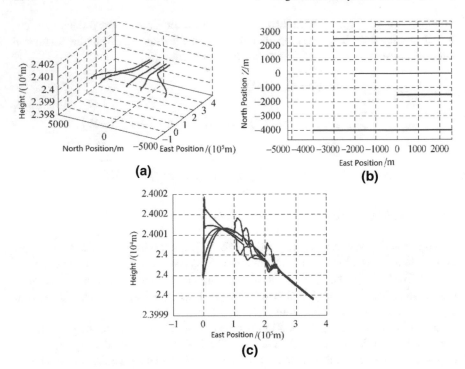

Fig. 5.2 Hypersonic vehicle formation trajectory, **a** 6DOF trajectory, **b** lateral horizontal trajectory, **c** longitudinal vertical trajectory

autonomous formation in accordance with the principles of cooperativity. For the airbreathing hypersonic vehicle whose Mach number and flight attitude are controllable, the flocking and stabilization of the formation have a certain degree of flexibility and robustness.

5.3   Loose Formation Control

This section will focus on the flight control problems of missiles in loose formation, that is, the design of formation spacing controller of the formation. It is the research basis of dense formation control problem, and the dense formation control problem will be researched in the next section.

According to the Definition 2.15, the loose formation means that the mathematical expectation of safe distance allowance $\Delta\mu_{ij} > \sigma_i$ for all missile members ε_i, the chance of collisions between missile members in the formation flight is little, the formation tightness and denseness are at a low state, and its potential value $E(q)$ is large. In this case, the performance requirement of cooperative guidance controller for missile formation is not high. Therefore, the method of route planning or scheduling planning (common in the air traffic control and ground robot formation)

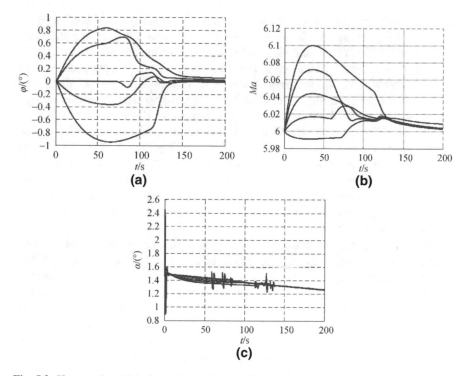

Fig. 5.3 Hypersonic vehicle formation flight state diagram, **a** trajectory deflection angle, **b** mach number, **c** angle of attack

is generally adopted to solve the collision avoidance problem of loose formation. The method can be summarized as follows: through the route planning, the nodes are exclusive in the spatial and time domains. That is, to ensure that the nodes in the same time domain do not appear in the same spatial domain and the nodes in the same spatial domain do not appear in the same time domain. Usually the route planning is offline and centralized, and often does not need the formation to implement and hold the formation configuration. Although the holding of formation configuration is required, because the distance between nodes is larger than the maximum safe distance, so factors such as information uncertainty, complex maneuvering and non-expected maneuvering of the nodes will be filtered out from the formation motion system with a large scale relative distance. Therefore, the collision avoidance problem between nodes in loose formation situation is not prominent and can be avoided by proper offline planning and scheduling, the online real-time collision avoidance control is less stressful. From the point of view, to a certain extent, the loose formation can be considered to realize collision avoidance through setting the node spacing which is far greater than the safe distance.

Considering combat task requirements and engineering implementation of the formation, the three-dimensional missiles autonomous formation configuration can be decomposed into two two-dimensional formation configurations that are horizontal and vertical formation configuration. The basic types of horizontal formation configuration can be divided into: column and horizontal formation, wedge-shaped formation, and diamond-shaped formation, as shown in Fig. 5.4. According to the requirements of battlefield situation, tactical application, obstacle and collision avoidance maneuvering and other factors, based on the determined horizontal formation configuration, the basic types of vertical formation configuration can be divided into: low altitude penetration for the whole formation, one missile as a high altitude navigator in the formation, and multi missiles as high altitude navigators in the formation, as shown in Fig. 5.5 [3, 74]. Because the vertical formation configuration only aims at the control of flight altitude based on the determined horizontal formation configuration, and it is easier to implement with respect to horizontal formation configuration, so this section will focus on the control and implementation of horizontal formation configuration commonly used in low altitude penetration.

In order to ensure the complete and neat dynamic characteristics of the horizontal formation configuration, a formation regulatory mechanism is required. That is, in the formation flight process, a reference point is needed, which coordinates the missile members in the different relative positions of the reference point, and then the desired formation is generated. The formation configuration control generally can be divided into two steps: firstly, determine the correct positions of missile members in the formation through the information obtained by integrated sensors and support network. After that make the missile members reach the specified

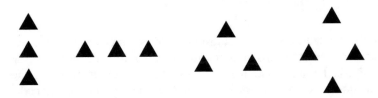

(a)Column (b) Horizontal (c) Wedge-shaped (d) Diamond-shaped

Fig. 5.4 The basic types of horizontal formation configuration, **a** column, **b** horizontal, **c** wedge-shaped, **d** diamond-shaped

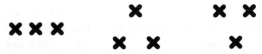

(a) Low altitude penetration (b) one missile as a high altitude navigator
(c) Multi missiles as high altitude navigators

Fig. 5.5 The basic types of vertical formation configuration, **a** low altitude penetration, **b** one missile as a high altitude navigator, **c** multi missiles as high altitude navigators

formation position at the specified time according to the formation control instructions generated by the formation decision and management.

There are three kinds of methods to select the formation reference point [3], as shown in Fig. 5.6:

(1) The formation geometric center as the reference point: The average position coordinate of all missile members in the formation is as the central coordinate of the formation, and the missile members determine their positions in real time according to the relative position of itself and the central coordinate.
(2) The leader as the reference point: Select a missile in the formation as a leader, other missile members determine their positions according to the position of the leader. The leader is not responsible for the holding of formation configuration, and holding mission is conducted by the followers.
(3) The adjacent missile position as the reference point: Each missile determines its position according to the adjacent missile in a determined position.

The formation configuration control methods usually include the formation control based on the behavior, formation control of "leader-follower" and formation control of virtual structure.

(1) Formation control based on behavior

In the flight process of missile formation, the behaviors of obtaining information from sensors and support network for each missile member can be divided into the following four situations: collision avoidance, obstacle avoidance, target acquisition and formation holding. The most important feature of the formation control method based on behavior is to determine the behavior of each missile member in the formation by means of the average weight of behavioral response control. However, the formation configuration based on the behavior control is lack of rigidity, and is mainly suitable for loose formation.

(2) Formation control of "leader-follower"

This formation control strategy is characterized by that based on the pre-set formation structure, the follower missile track the flight velocity, trajectory deflection angle and altitude of the leader missile, thus to achieve the purpose of formation holding. Because this kind of control structure is vulnerable to large disturbance, the robust control method, extremum search control method, adaptive control method, variable structure control and other methods are adopted to improve the

(a) the leader as the reference point (b) the formation geometric center as the reference point (c) the adjacent missile position as the reference point

Fig. 5.6 Three kinds of methods to select the formation reference point, **a** the leader as the reference point, **b** the formation geometric center as the reference point, **c** the adjacent missile position as the reference point

quality of formation configuration. The drawbacks of this formation are that the formation configuration is too rigid, and its threat avoidance ability is insufficient. So it is suitable for loose formation with adjacency scale.

(3) Formation control of virtual structure

Formation control of virtual structure generally adopts the method of virtual leader to coordinate other missile members to achieve the purpose of formation control. Although this method can avoid the large disturbance problem of "leader-follower" method, the synthesis, transmission and distribution processes of the position of virtual leader missile need to be at the expense of high communication quality and high computing ability. Moreover, the node positions of the virtual structure are fixed, and the obstacle avoidance function is often poor.

In addition to the above three methods of formation control, there are other methods of formation control, such as formation control and obstacle avoidance strategy based on the fuzzy logic and neural network. The control structures of these methods are often complex, and the selection of weights of fuzzy logic and neural network usually need several tests to determine which cost too much time, and the results are not universal [3, 4, 74].

In order to solve the "chain effect" problem caused by collision avoidance maneuvering during flight process of dense autonomous formation, the formation control method base on Model Predictive Control (MPC) is adopted here. The method generates the formation spacing instruction for formation maintaining based on online optimization index and online dynamic correction, and optimizes the distances between nodes to avoid collision. The ability to change the control strategy according to the specific situation of the missile formation is also shown in Fig. 5.7 the architecture of CGCS-of-MAF.

In this section, the controller of MAF will be designed based on MPC method and the related data of a certain type of missiles, a formation controller evaluation method based on the collision probability will be adopted compared with the classical PID formation controller.

5.3.1 Models of MAF

At present, the control system of formation configuration holding for loose formation with adjacency scale is mainly based on the formation mode of "leader-follower", i.e. the formation mode that follower missiles follow the leader missile, and the leader missile can be a real missile or virtual missile.

Take the ballistic coordinate system of node W (the velocity direction of node W is the axis x_p and perpendicular to the velocity and pointing rightwards is the axis z_p) as the relative coordinate system, the earth fixed axis system as the fixed coordinate system, the relative motion between formation nodes is shown in Fig. 5.8. It can be seen from the figure that the position relationship between the leader missile and the follower can be expressed as:

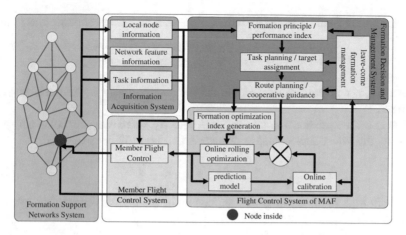

Fig. 5.7 Formation configuration control and hold system based on MPC method

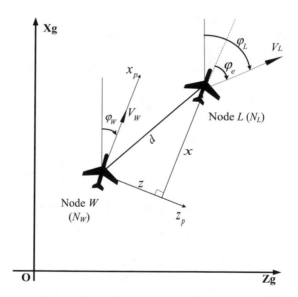

Fig. 5.8 The relative position relationship between the leader missile and the follower

$$\begin{cases} x_L = x_W + x \cos \varphi_W - z \sin \varphi_W \\ z_L = z_W + x \sin \varphi_W - z \cos \varphi_W \end{cases} \tag{5.24}$$

The relative position relationship in the direction of trajectory inclination angle of the follower between the leader missile and the follower can be obtained from the above equation. The formation maintaining method of missile formation is to make the state (x, y) reach the expected value (x_r, z_r) in the role of designed control (V_{Wc}, φ_{Wc}).

The relative position model of the leader missile and follower missile is as follows:

$$\begin{cases} x = (x_L - x_W)\cos \varphi_W + (z_L - z_W)\sin \varphi_W \\ z = -(x_L - x_W)\sin \varphi_W + (z_L - z_W)\cos \varphi_W \end{cases} \quad (5.25)$$

where (x_L, z_L) is the position of the leader, and (x_W, z_W) is the position of the follower.

Based on the relationship that absolute velocity = relative velocity + convected velocity, the kinematics equation of node L and node W is established:

$$\bar{V}_L = \bar{d} + \bar{V}_W + \varphi_W \times \bar{d} \quad (5.26)$$

Equation (5.26) is decomposed as follows in the ballistic coordinate system of node W:

$$\left.\begin{array}{l} \frac{dx}{dt} = V_L \cos(\varphi_L - \varphi_W) - V_W + \frac{d\varphi_W}{dt} \cdot z \\ \frac{dz}{dt} = V_L \sin(\varphi_L - \varphi_W) - \frac{d\varphi_W}{dt} \cdot x \end{array}\right\} \quad (5.27)$$

where x, z are respectively components of the distance d of node W to node L in the axis x_p and axis z_p of the ballistic coordinate system of node W.

Adopting the first order inertia link to model on the motion of flight velocity and trajectory deflection angle of the leader and follower, that is

$$\frac{d\varphi_W}{dt} = \frac{1}{\tau_{\varphi W}}(\varphi_{Wc} - \varphi_W), \frac{dV_W}{dt} = \frac{1}{\tau_{VW}}(V_{Wc} - V_W), \frac{d\varphi_L}{dt} = \frac{1}{\tau_{\varphi L}}(\varphi_{Lc} - \varphi_L), \frac{dV_L}{dt}$$
$$= \frac{1}{\tau_{VL}}(V_{Lc} - V_L)$$

The derivative equation of the relative position can be obtained by Eq. (5.27):

$$\begin{cases} \frac{dx}{dt} = -\frac{z}{\tau_{\varphi W}}\varphi_W + \frac{z}{\tau_{\varphi W}}\varphi_{Wc} - V_F + V_L \cos(\varphi_L - \varphi_W) \\ \frac{dz}{dt} = \frac{x}{\tau_{\varphi W}}\varphi_W - \frac{x}{\tau_{\varphi W}}\varphi_{Wc} + V_L \sin(\varphi_L - \varphi_W) \end{cases} \quad (5.28)$$

The dynamic model of missile formation holding system is composed of Eq. (5.25) and (5.28). Now the control instruction (V_{Wc}, φ_{Wc}) of flight velocity and trajectory deflection angle of the follower needs to be designed, and it will make the relative position between missiles can reach the given expected value, that is, the error between the relative position and the expected position asymptotically converges to zero.

Linearize the above Eq. (5.28) based on the assumption of small angle and small disturbance, we can get:

$$\left.\begin{array}{l} \frac{dx}{dt} = V_L - V_W + \frac{d\varphi_W}{dt} \cdot z_0 \\ \frac{dz}{dt} = V_L(\varphi_L - \varphi_W) - \frac{d\varphi_W}{dt} \cdot x_0 \end{array}\right\} \quad (5.29)$$

The kinematic model of node L and node W is:

$$\left.\begin{array}{l} \frac{dx}{dt} = V_L - V_W - \frac{z_0}{\tau_{\varphi W}} \cdot \varphi_W + \frac{z_0}{\tau_{\varphi W}} \cdot \varphi_{Wc} \\ \frac{dz}{dt} = V_{L0}\varphi_L + \left(-V_{L0} + \frac{x_0}{\tau_{\varphi W}}\right) \cdot \varphi_W - \frac{x_0}{\tau_{\varphi W}} \cdot \varphi_{Wc} \\ \frac{d\varphi_L}{dt} = \frac{1}{\tau_{\varphi L}}\left(\varphi_{Lc} - \varphi_L\right) \\ \frac{dV_L}{dt} = \frac{1}{\tau_{VL}}\left(V_{Lc} - V_L\right) \\ \frac{d\varphi_W}{dt} = \frac{1}{\tau_{\varphi W}}\left(\varphi_{Wc} - \varphi_W\right) \\ \frac{dV_W}{dt} = \frac{1}{\tau_{VW}}\left(V_{Wc} - V_W\right) \end{array}\right\} \tag{5.30}$$

5.3.2   MPC Controllers Design of MAF

It can be seen from Eq. (5.30) that the motion model of the formation of two missiles is simple and suitable for the application of model predictive control. Therefore, in this section, the MPC formation holding controller based on the state space is designed according to the characteristics of formation flying.

1. Design of MPC formation flight controller

(1) Prediction model

According to Eq. (5.30) select the state $X = [x \ z \ \varphi_W \ V_W]^T$, where x, Z is the projection of the relative distance node L to node W in the ballistic coordinate system of node W; φ_W is the trajectory deflection angle of node W; V_W is the velocity of node W; the controllable input is $u = [\varphi_{Wc} \ V_{Wc}]^T$, where φ_{Wc}, V_{Wc} are respectively the trajectory deflection angle instruction and flight velocity instruction of node W. In order to reduce the computational burden, the trajectory deflection angle and flight velocity of node L can be seen as measurable disturbance $d_L = [\varphi_L \ V_L]^T$, that is, the trajectory deflection angle and flight velocity of node L are obtained directly by support network instead of using the first order model to predict; the output $Z = [x \ z \ \Delta\varphi \ \Delta V]^T$, where $\Delta\varphi = \varphi_L - \varphi_W, \Delta V = V_L - V_W$. The discrete form of formation motion model while ignoring d_L (d_L will be added to the prediction model later, which can be regarded as the calibration of the model) is:

$$\left.\begin{array}{l} X(k+1) = AX(k) + Bu(k) \\ Z(k) = CX(k) \end{array}\right\} \tag{5.31}$$

Set $X(k|k) = X(k)$ means the state value obtained at time k, and according to $Z(k)$ and $u(k-1)$, because $u(k)$ is not yet calculated; $u(k+i|k)$ means the input value u of the next $k+i$ moment assumed at the time of k; $X(k+i|k), Z(k+i|k)$ respectively mean the predicted value of X, Z at the moment k, using $u(k+j|k)$ and $(j = 0, 1, \ldots, i-1$; P is the prediction horizon; N is the control horizon, that is $u(k+i|k) = u(k+N-1|k), N < i < P-1$, by means of the iteration (5.31), we get:

$$
\left.
\begin{aligned}
X(k+1|k) &= AX(k) + Bu(k|k) \\
X(k+2|k) &= A^2X(k) + ABu(k|k) + Bu(k+1|k) \\
&\vdots \\
X(k+p|k) &= A^PX(k) + A^{P-1}Bu(k|k) + \ldots + Bu(k+P-1|k)
\end{aligned}
\right\}
\tag{5.32}
$$

That is:

$$
X(k+j|k) = A^jX(k) + [A^{j-1} \quad A^{j-2} \quad \cdots \quad I]B
\begin{bmatrix}
u(k|k) \\
\vdots \\
u(k+j-1|k)
\end{bmatrix}, j = 1, 2, \ldots, P
\tag{5.33}
$$

Because $u(k)$ is unknown at the moment k, and $u(k-1)$ is known, set $\Delta u(k+j|k) = u(k+j|k) - u(k+j-1|k)$, then

$$
\left.
\begin{aligned}
u(k|k) &= \Delta u(k|k) + u(k-1) \\
u(k+1|k) &= \Delta u(k+1|k) + \Delta u(k|k) + u(k-1) \\
&\vdots \\
u(k+N-1) &= \Delta u(k+N-1|k) + \ldots + \Delta u(k|k) + u(k-1)
\end{aligned}
\right\}
\tag{5.34}
$$

Then for $1 \leq j \leq N$, we can get:

$$
\left.
\begin{aligned}
X(k+1|k) &= AX(k) + B[\Delta u(k|k) + u(k-1)] \\
X(k+2|k) &= A^2X(k) + (A+I)B\Delta u(k|k) + B\Delta u(k+1|k) + (A+I)Bu(k-1) \\
&\vdots \\
X(k+N|k) &= A^NX(k) + (A^{N-1} + \cdots + A + I)B\Delta u(k|k) \\
&\quad + \cdots + B\Delta u(k+N-1|k) + (A^{N-1} + \cdots + A + I)Bu(k-1)
\end{aligned}
\right\}
\tag{5.35}
$$

$$
X(k+j|k) = A^jX(k) + \left[\sum_{i=0}^{j-1} A^iB \ldots B\right]
\begin{bmatrix}
\Delta u(k|k) \\
\vdots \\
\Delta u(k+j-1|k)
\end{bmatrix}
+ \sum_{i=0}^{j-1} A^iBu(k-1)
\tag{5.36}
$$

For $N \leq j \leq P$, we get:

$$
\left.
\begin{aligned}
X(k+N+1|k) &= A^{N+1}X(k) + (A^N + \cdots + A + I)B\Delta u(k|k) \\
&\quad + \cdots + (A+I)B\Delta u(k+N-1|k) \\
&\quad + (A^N + \cdots + A + I)Bu(k-1) \\
&\;\;\vdots \\
X(k+P|k) &= A^P X(k) + (A^{P-1} + \cdots + A + I)B\Delta u(k|k) \\
&\quad + \cdots + (A^{P-N} + \cdots + A + I)B\Delta u(k+N-1|k) \\
&\quad + (A^{P-1} + \cdots + A + I)Bu(k-1)
\end{aligned}
\right\}
\tag{5.37}
$$

$$
\begin{aligned}
X(k+j|k) &= A^j X(k) \\
&\quad + \left[\sum_{i=0}^{j-1} A^i B \ldots \sum_{i=0}^{j-N} A^i B \right]
\begin{bmatrix}
\Delta u(k|k) \\
\vdots \\
\Delta u(k+N-1|k)
\end{bmatrix}
+ \sum_{i=0}^{j-1} A^i B u(k-1)
\end{aligned}
\tag{5.38}
$$

In summary, the total prediction equation of state X is:

$$
\begin{bmatrix}
X(k+1|k) \\
\vdots \\
X(k+N|k) \\
X(k+N+1|k) \\
\vdots \\
X(k+P|k)
\end{bmatrix}
=
\begin{bmatrix}
A \\
\vdots \\
A^N \\
A^{N+1} \\
\vdots \\
A^P
\end{bmatrix}
X(k)
+
\begin{bmatrix}
B \\
\vdots \\
\sum_{i=0}^{N-1} A^i B \\
\sum_{i=0}^{N} A^i B \\
\vdots \\
\sum_{i=0}^{P-1} A^i B
\end{bmatrix}
u(k-1)
$$

$$
+
\begin{bmatrix}
B & \cdots & 0 \\
AB+B & \cdots & 0 \\
\vdots & \ddots & \vdots \\
\sum_{i=0}^{N-1} A^i B & \cdots & B \\
\sum_{i=0}^{N} A^i B & \cdots & A+B \\
\vdots & \vdots & \vdots \\
\sum_{i=0}^{P-1} A^i B & \cdots & \sum_{i=0}^{P-N} A^i B
\end{bmatrix}
\begin{bmatrix}
\Delta u(k|k) \\
\vdots \\
\Delta u(k+N-1|k)
\end{bmatrix}
\tag{5.39}
$$

Finally, the prediction equation of output Z is:

$$Z(k+j|k) = CX(k+j|k), \quad j = 1, 2, \ldots, P \tag{5.40}$$

(2) Rolling optimization

Set the cost function of formation flying at each time k is:

$$J = \sum_{l=1}^{P} \left\{ \sum_{m=1}^{N} \|\omega_z(l,m) \cdot [Z_m(k+l|k) - ref_m(k+l)]\|^2 + \sum_{m=1}^{N} \|\omega_{\Delta u}(l,m) \cdot \Delta u_m(k+l|k)\|^2 \right\}$$
$$\tag{5.41}$$

where P is the prediction horizon; N is the control horizon; $(k + l|k)$ is the predicted value of the time $k + l$ calculated at time k; ref is the spacing hold instruction; ω_y is the output penalty weight matrix which reflects the efforts to maintain the distance between node W and L and efforts to match trajectory deflection angle and velocity, and is the most important weight for keeping distance; $\omega_{\Delta u}$ is the input rate penalty weight matrix, which reflects the instruction change rate and has the property of integral, so that the system can realize the no difference control. In order to simplify the solution of Quadratic programming (QP), here the weight of each horizon to (l, m) is equal, then ω is simplified as row vector $\omega_z = (\omega_x \, \omega_z \, \omega_{\Delta\varphi} \, \omega_{\Delta V})$, $\omega_{\Delta u} = (\omega_{\Delta\varphi_c} \, \omega_{\Delta V_c})$. It can be seen that in the case of no terminal constraints, the physical meaning of the weight is very clear, but in order to ensure the stability of the system, a large prediction time domain is adopted, after simulation and design, select $P = 50$; $N = 40$; $\omega_z = (200, 250, 200, 50)$; $\omega_{\Delta u} = (5, 0.1)$.

Finally, the control $u(k)$ of MPC at the moment k is the following constrained optimization problem:

$$\Delta u_{opt} = \arg \min J$$
$$u(k) = u(k-1) + (I_{2\times2}, O_{(N-1)\times(N-1)}) \cdot \Delta u_{opt}$$
$$st. \begin{cases} \varphi_{min} \leq u_1 \leq \varphi_{max} \\ V_{min} \leq u_2 \leq V_{max} \end{cases} \tag{5.42}$$

where $[\varphi_{min}, \varphi_{max}]$ and $[V_{min}, V_{max}]$ are respectively trajectory deflection angle and velocity instruction limit. Δu_{opt} is the first control of the optimal sequence, u_1, u_2 are respectively φ_c and V_c. The cost function like the type (5.41) can be reduced to the QP standard problem, which is strictly convex in the positive weight. The specific process is solved by relatively mature KWIK algorithm and is not involved here.

(3) Feedback correction

The current state $X(k)$ of the system is obtained through the support network in each computing cycle, and may be inconsistent with the state's predicted value $X(k|k - 1)$, but the state space MPC method can directly set $X(k)$ as the initial state of the system, which has a similar role in the feedback correction; and as noted previously, the trajectory deflection angle and velocity of node L can be seen as measurement interference $d_L = [\varphi_L V_L]^T$, and be transmitted to node W through network and added to the predicted value of the prediction model according to (5.30), which also plays a role in correction.

2. Prediction model with network-induced delay

When the network-induced delay cannot be ignored, the discretization of the linear motion model needs to consider the influence of time delay, considering the general form of linear time invariant continuous systems

$$
\left.\begin{array}{l}
\frac{dX(t)}{dt} = AX(t) + Bu(t) \\
Z = CX(t)
\end{array}\right\}
\tag{5.43}
$$

Its time domain solution is

$$
X(t) = e^{A(t-t_0)}X(t_0) + \int_{t_0}^{t} e^{A(t-\mu)}Bu(\mu)d\mu
\tag{5.44}
$$

When the network-induced delay τ exists, because the input u of MPC is discrete (sampling period is T_s), when the new state does not arrive, u will be equal to the input of the last moment, so

$$
\frac{dX(t)}{dt} = \begin{cases} AX(t) + Bu((k-1)T_s), & kT_s \le t < kT_s + \tau \\ AX(t) + Bu(kT_s), & kT_s + \tau \le t < (k+1)T_s \end{cases}
\tag{5.45}
$$

When $kT_s \le t < kT_s + \tau$, consider the state X transferring from kT_s to $kT_s + \tau$, then

$$
X(kT_s + \tau) = e^{A\tau}X(kT_s) + \int_{0}^{\tau} e^{A\mu}d\mu \cdot Bu((k-1)T_s)
\tag{5.46}
$$

When $kT_s + \tau \le t < (k+1)T_s$, consider the state X transferring from $kT_s + \tau$ to $(k+1)T_s$, then

$$
X((k+1)T_s) = A_1X(kT_s) + B_1u((k-1)T_s) + B_2u(kT_s)
\tag{5.47}
$$

where $A_1 = e^{AT_s}$, $B_1 = e^{A(T_s-\tau)}\int_{0}^{\tau} e^{A\mu}d\mu B$, $B_2 = \int_{0}^{T_s-\tau} e^{A\mu}d\mu B$

Avoiding ambiguity, k represents the first k sampling period, such as $Y(k) = Y(kT_s)$, set

$$
Y(k) = \begin{pmatrix} X(k) \\ u(k-1) \end{pmatrix}
\tag{5.48}
$$

Then in the case of network-induced delay, the discrete model of MPC is

$$
\begin{aligned}
Y(k+1) &= A_Y Y(k) + B_Y u(k) \\
Z(k) &= C_Y Y(k)
\end{aligned}
\tag{5.49}
$$

$$
A_Y = \begin{pmatrix} A_1 & B_1 \\ 0 & 0 \end{pmatrix}, B_Y = \begin{pmatrix} B_2 \\ I \end{pmatrix}, C_Y = (C \quad 0)
\tag{5.50}
$$

Then the prediction model can be obtained by the same method as before.

In consideration of the convenience of engineering implementation, the overload command n_{Wc}^* and velocity command V_{Wc}^* of the follower missile [1, 4] in the formation are:

$$
\begin{cases}
n_{Wc}^* = \dot{\varphi}_{Wc} V_{Wc} \\
V_{Wc}^* = V_{Wc}
\end{cases}
\tag{5.51}
$$

5.3.3   Simulation for MPC Controllers

Take the autonomous formation of two missiles as the experimental object, the initial conditions and requirements are: the velocity of the member missiles are 100 m/s, the minimum velocity is 50 m/s, and the maximum velocity is 200 m/s; the expected formation holding distance is $x = 100$ m; $z = 173.2$ m; formation spacing is $d = 200$ m. When $t = 5s$, the trajectory deflection angle of leader missile is required to turn $60°$ to the negative direction; when $t = 30$ s, turn $60°$ to the positive direction. The simulation results of PID formation controller and MPC formation controller are respectively given in Figs. 5.9 and 5.10.

As is shown in Figs. 5.9 and 5.10, during the simulation experiment of 70 s, the mean square error of formation spacing adopting PID formation controller is $\sigma_{pid} = 32.7089$ m, and the mean square error of formation spacing adopting MPC formation controller is $\sigma_{mpc} = 0.8393$ m; In addition, during the formation holding process, the error between the actual formation spacing and expected formation spacing of PID formation controller is bigger than 60 m, otherwise, the error between the actual formation spacing and expected formation spacing of MPC formation controller is less than 2.75 m. As the formation potential energy $P_E(q)$ [in Eq. (2.23)] is used to evaluate the performance of cooperative and guidance controller given in the second chapter, MPC formation controller not only has significant advantages for formation flight control in loose formation, but also for the formation flight control in dense formation which will be introduced latter.

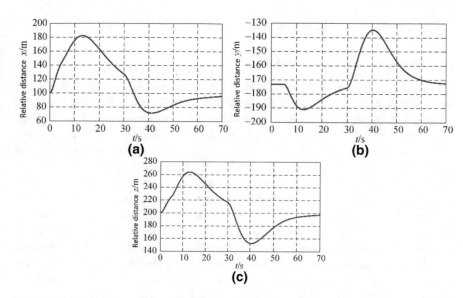

Fig. 5.9 The simulation results of PID formation controller, **a** relative distance of direction x, **b** relative distance of direction y, **c** relative distance of direction z

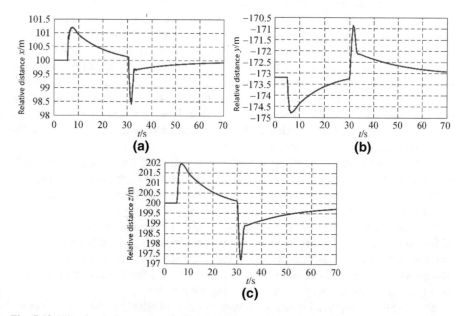

Fig. 5.10 The simulation results of MPC formation controller, **a** relative distance of direction x, **b** relative distance of direction y, **c** relative distance of direction z

5.3.4   Avoidance Obstacles Control of MAF

1. Tracking method based on behavior

This section aims to provide an engineering obstacle avoidance control method of loose formation, in the premise of maintaining a relatively stable formation, enhance the ability of the missile to avoid threat and obstacle in unknown environment. If the environment where MAF carry out tactical mission are mountainous region or flat-lands, the obstacles are usually relatively large scale natural objects (such as mountains), such obstacles are easy to extract in digital map, and generally loaded into the digital map of the flight control computer before launching. However, if the tactical mission for complex targets such as large city, port, strategic point and high military value target is carried out, in order to ensure the high penetration probability, the missile formation usually fly in low altitude or super-low altitude, and the missile members flying in low altitude are prone to collide with ground artificial obstacles (e.g., buildings, high-voltage tower and wire rod). Compared with natural objects, the scale of artificial obstacles is usually relatively small, so it is difficult to extract in the digital map, which can lead to greater risk of missile formation flight [1, 4].

The thought of control based on behavior is to stipulate several expected basic behaviors for the missile, and the missile behaviors generally include collision avoidance, obstacle avoidance, flying to target area and formation holding; react to the input information of information acquisition system and the formation support network, and output the reaction vector as the expected response of the behavior. The behavior selector synthesizes the output of each action through a certain mechanism, and takes the integrated result as the final output of the missile. The basic idea of tracking method is that a missile in the formation is designated as the leader according to certain rules, the remaining missile members are as followers, and followers track the position and flight direction of the leader with a certain distance and relative angle. The tracking method based on behavior is a kind of formation control algorithm which makes the control theory based on the behavior into tracking method, as shown in Fig. 5.11.

The tracking method based on behavior firstly divides the missile members in the formation as the leader and followers by means of role allocation principle, then the behavior of the missiles flying to the target area is converted to the behavior of the leader missile leading the entire formation to fly to the target area in accordance with the planned route; the formation holding behavior is converted to the behavior of follower missiles tracking the leader missile; the behavior of collision avoidance, obstacle avoidance is converted to that of follower missiles. The behavior selector is composed of based on behavior restrain methods, where the behaviors of collision avoidance and obstacle avoidance have a higher priority.

Suppose that the ground obstacles obeys uniform distribution in space, as shown in Fig. 5.12, M is the current position of missile, V is the missile velocity vector, r is the threat range for the obstacle, R is the distance between missile and obstacle. In order to ensure that the missile does not enter the threat area, the missile can fly along the left side of the track MA, or fly along the right side of the track MB.

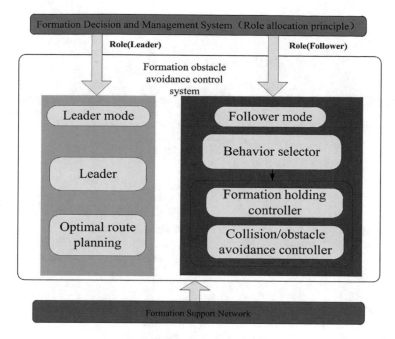

Fig. 5.11 Structure of tracking method based on behavior

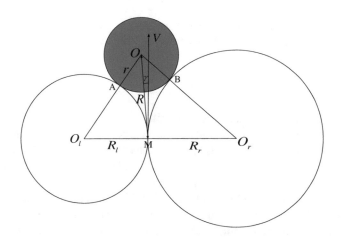

Fig. 5.12 Schematic diagram of missile obstacle avoidance method

Suppose the nearest threat source to the missile is center O, the ideal threat avoidance tracks are arcs MA and MB of circles O_l, O_r which are circumscribed to circle O and tangent to the velocity vector V, define tangent point A and B are threat avoidance navigation point.

In ΔOO_lM, based on cosine theorem we get

$$\cos(\angle OMO_l) = \sin \gamma = \frac{R^2 + R_l^2 - (R_l + r)^2}{2RR_l} \qquad (5.52)$$

The turning radius of threat avoidance of the missile is further derived

$$R_l = \frac{R^2 - r^2}{2(r + R \sin \gamma)} \qquad (5.53)$$

$$R_r = \frac{R^2 - r^2}{2(r - R \sin \gamma)} \qquad (5.54)$$

The expression of the corresponding lateral acceleration is

$$a_r = \frac{V^2}{R_r}, a_l = \frac{V^2}{R_l} \qquad (5.55)$$

where a_r, a_l respectively mean the required lateral accelerations for obstacle avoidance of the missile along the arc MB and MA.

The final obstacle avoidance overload command n_a^*

$$n_a^* = a_r \quad or \quad a_l \qquad (5.56)$$

Under the condition that the obstacle area is not large, the final obstacle avoidance velocity command and the leader missile velocity alignment instruction V_a^* is

$$V_a^* = V_L \qquad (5.57)$$

where V_L is the velocity of the leader missile.

The missile formation holding method and obstacle avoidance method are given above, and the actual formation flight process is the superposition of the above two situations. In order to search the target area efficiently, it is necessary for missiles to implement autonomous formation flight. When the missile encounters obstacles that are not marked in early route planning, the ability of temporary escape threat is often required for the missile. The formation control method based on behavior combines the formation holding process and obstacle avoidance process, and is divided into two specific steps (as shown in Fig. 5.13), respectively behavior decomposition and control implementation [1, 4].

2. Behavior decomposition

In the process of behavior decomposition, the missile members divide the mission behaviors into three parallel sub behaviors according to the current mission nature and environment information: that is, the exception handling, formation holding

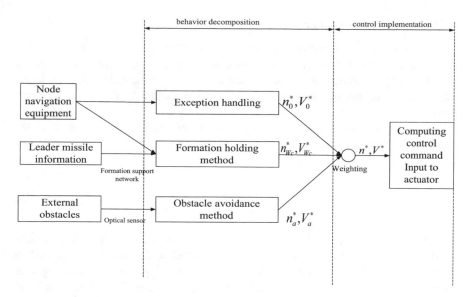

Fig. 5.13 The formation control method based on behavior

method and threat avoidance method, at the same time, the corresponding importance is given to each sub action by weighted value. The missile members obtain the information of the leader through formation support network, specifically including the velocity V_L, yaw rate ω_L and trajectory deflection angle φ_L of the leader; The missile members obtain the navigation information through their own navigation equipment, specifically including the velocity V_W, yaw rate ω_W and trajectory deflection angle φ_W, formation overload command n_{Wc}^* and formation velocity command V_{Wc}^* are generated according to Eq. (5.51). The missile members can obtain the information of the obstacle through optical sensors, including the distance R of the missile to the obstacle, the angle b between the velocity vector and the line connecting the missile and the obstacle, the obstacle avoidance overload instruction n_a^* and obstacle avoidance velocity command V_a^* are generated according to Eqs. (5.56) and (5.57). Exception handling refers that during mission of the missile members, while encountering communication link interruption and failing to hold the formation configuration, the flight control computer of the missile member will enter the state of exception handling, which must ensure that the missile carry out the mission in accordance with the one missile optimal strategy. The exception handling has the highest priority during the mission, and the overload instruction and velocity instruction that the missile member required to carry out the mission in accordance with the one missile optimal strategy are respectively n_0^* and V_0^* (e.g., $n_0^* = 0$ and $V_0^* = V_L$ mean that the missile fly along a straight line).

3. Control implementation

The control implementation process means that according to the current flight status and formation motion model, the missile member calculates the actual control instruction by weighting, and inputs to the actuator, then the rudder deflects in the yaw channel of the missile, so as to realize the motion control of the missile formation.

According to the formation holding and obstacle avoidance method, after the overload and the velocity command of each sub behavior is calculated, weighting according to the following equation

$$\begin{cases} n^* = \zeta_{n0}n_0^* + \zeta_{nW}n_{Wc}^* + \zeta_{na}n_a^* \\ V^* = \zeta_{V0}V_0^* + \zeta_{VW}V_{Wc}^* + \zeta_{Va}V_a^* \end{cases} \tag{5.58}$$

where $\zeta_{n0} + \zeta_{nW} + \zeta_{nW} + \zeta_{na} = 1$, $\zeta_{V0} + \zeta_{VW} + \zeta_{Va} = 1$, ζ_{n0} means the weight of overload instruction of exception handling process, ζ_{V0} means the weight of velocity instruction of exception handling process, ζ_{nW} means the weight of overload instruction of formation holding process, ζ_{VW} means the weight of velocity instruction of formation holding process, ζ_{na} means the weight of overload instruction of obstacle avoidance process, ζ_{Va} means the weight of velocity instruction of obstacle avoidance process.

(1) Conditions meeting the exception handling: $\zeta_{n0} = \zeta_{V0} = 1$, $\zeta_{nW} = \zeta_{VW} = \zeta_{na} = \zeta_{Va} = 0$, the formation holding and obstacle avoidance method is no longer effective, the missile carries out the mission by means of the one missile flight, or enters recovery stage along the specified navigation point.

(2) When conditions are not meeting the exception handling, the formation holding and obstacle avoidance method is effective, the weights are calculated according to whether the optical sensor has found the obstacle:

$$\left.\begin{array}{l} \begin{cases} \zeta_{nW} = \zeta_{VW} = 1 \\ \zeta_{na} = \zeta_{Va} = 0 \end{cases} \text{, Optical sensors don't detect obstacles} \\[2em] \begin{cases} \zeta_{nW} = \zeta_{VW} = 0.6 \sim 0.8 \\ \zeta_{na} = \zeta_{Va} = 0.2 \sim 0.4 \end{cases} \text{, Optical sensors detect obstacles} \end{array}\right\} \tag{5.59}$$

where $\zeta_{nW}, \zeta_{VW}, \zeta_{na}$ and ζ_{Va} can be adjusted according to the actual situation, the value of ζ_{na} and ζ_{Va} are mainly determined by the ratio of the scale of the ground obstacle to the threat range of the obstacle. The larger ratio means that the scale of the obstacle is large, so the value of ζ_{na} and ζ_{Va} must be enlarged to avoid the collision with obstacle; or the value of ζ_{na} and ζ_{Va} must be decreased, $\zeta_{nW} = 1.0 - \zeta_{na}$, $\zeta_{VW} = 1.0 - \zeta_{Va}$ must be enlarged to enhance the missile formation ability.

4. Simulation experiment

The top view of the initial position of the missile formation and the distribution of obstacles are given in Fig. 5.14, and the top view of the experimental results of the formation configuration control is as shown in Fig. 5.15. Where 0, 1, 2, 3, 4 represent missiles, 0 represents the leader missile, and the weights of formation holding and obstacle avoidance of each missile are uniform, $\zeta_{nW} = 0.7$, $\zeta_{na} = 0.3$. The ellipse with solid line represents the region of threat, and the ellipse with dotted line represents the detection zone of the missile. Only when the threat region enters the detection zone of the missile, the missile adopts the obstacle avoidance method to avoid the threat. The results of the threat avoidance experiment are as shown in Fig. 5.16, where the long solid curve means the missile flight trajectory. It can be seen that the missile formation can avoid the ground potential threats. At the same time, as the formation weights are not completely zero, in case of threat avoidance, the basic formation positions can be maintained, and prepare to form the formation again after threat avoidance.

This section introduces the formation control method based on behavior, and reduces the higher requirement of the formation support network update rate; in view of the drawback of the traditional formation control method based on behavior that the formation is not rigid, introduces the virtual structure as a reference. Through the integration of the two formation configuration control methods, taking advantages of each method, under the premise that the formation configuration is relatively stable, the introduced formation control method can enhance the ability of missile formation to avoid obstacles and threats in unknown environment.

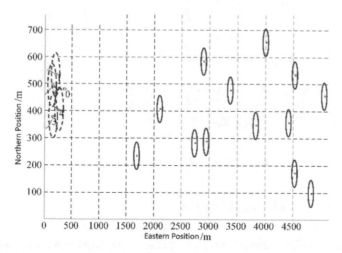

Fig. 5.14 Initial position and distribution of obstacles

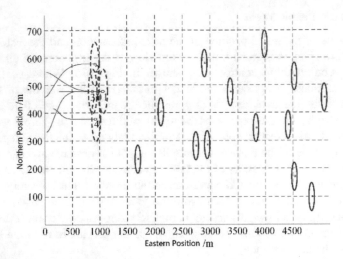

Fig. 5.15 The formation results

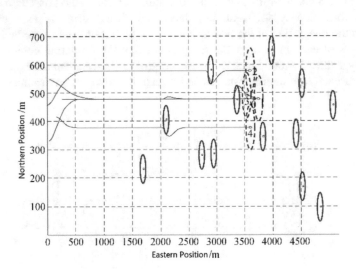

Fig. 5.16 The threat avoidance results of the formation

5.4 Dense Formation Control

According to the Definitions 2.16 and 2.17, the mathematical expectation of safety distance margin of tight formation and dense formation mean that all missile members ε_i are respectively $\sigma_i \leq \Delta\mu_{ij} < 3\sigma_i$ and $0 < \Delta\mu_{ij} \leq \sigma_i$. From the view of formation tightness, different from the loose formation, the two formation configuration make the formation spacing d_{ij} closer to its safe distance d_{si}, the probability of taking collision avoidance maneuvering for the missile members is larger, the probability of collisions between missile members is higher; From the view of formation denseness, the required free space of taking collision avoidance maneuvering around for the missile members of tight formation is existed, the free space is moderate for the missile members to take collision avoidance maneuvering around; while the denseness of the dense formation increases, that is, the adjacent degree of the formation is greater than six, which makes the missile members have no free space to take collision avoidance maneuvering around, and there is a high probability of collisions between missile members. Therefore, compared with the tight formation, the dense missile formation requires a higher performance of the cooperative guidance controller. From the definition of formation, whether tight formation or dense formation, different from the loose formation, the tightness and denseness of the formation both have a larger increase, which makes the probability of taking collision avoidance maneuvering for missile members increases, and the probability of collisions between missile members increases.

It can be seen that the main difference between tight formation and dense formation is the different denseness. From the view of formation tightness, the tight formation is not too much different from dense formation, but the dense formation makes the formation spacing d_{ij} closer to its safe distance d_{si}, and the probability of taking collision avoidance maneuvering is further increased than tight formation. Nevertheless, in the case of formation flying, there are similarities between tight formation and dense formation, that is, the probability that the missile member is always in the process of taking collision avoidance maneuvering is higher, the high probability collision avoidance maneuvering is the same characteristic of the two kinds of formation, and the difference lies only in the free space to take collision avoidance maneuvering, and the requirements of the formation distance control and collision avoidance maneuvering strategy are similar, so it is reasonable to combine the tight formation situation into the dense formation framework. In other words, the dense formation and tight formation where the safety factors of all missile members in the formation are less than 3 are included in the framework of dense formation to study.

In the framework of dense formation, because each missile member in the formation is surrounded with densely distributed neighbor missile members and there is no more free space, which makes the collision avoidance maneuvering (e.g., the avoidance of undesirable motion due to various random disturbances or some unexpected situations or faults) of missile members more difficult and complex than the loose formation, and this is also the main problem to be considered in the formation control system under the framework of dense formation. In the demand background of networking autonomous formation cooperative flight, the missile members can realize the real-time interconnection and interworking of information through the formation support network, make the control of collision avoidance maneuvering be possible, which lays a technical foundation for the realization of large scale autonomous missile formation.

Based on the collision avoidance maneuvering problem of dense formation, according to the group behavior of the animals in the nature, the Light Passing Hypothesis of the collision avoidance mechanism for dense group is given below, combining this hypothesis with MPC formation flight controller, a local prediction model of autonomous formation is established, and the Local MPC (LMPC) of dense formation based on Bionics is designed and the effectiveness of the LMPC and the Light Passing Hypothesis is verified by simulation experiments [74].

Under the framework of dense formation, the research on the whole behaviors during the flight process of the missile formation, including the problems of formation turning, formation holding, reconstruction, split and transformation, is particularly important. In this section, three kinds of formation turning and four kinds of formation change are clearly defined, and based on this, the formation configuration split method based on particle turning and formation transformation method based on the formation state are proposed. Although the two methods are not limited to the dense formation, it is still proposed through the key problem of reducing collision probability between formation members [74].

5.4.1   Analysis of Animal Groups Movement Behaviors

In the long process of natural evolution, a large number of animal groups have been bred in nature, from the group movement trajectory fossils of verifiable ancient crustacean species, to the common animal groups in daily life, such as birds, fish, ants, bees, etc. [110−113]. Among them, the study of the formation flying habits and characteristics of birds is the most valuable for the study of the theory and technology of MAF.

Many scholars have carried on thorough observation and research to the large-scale movement of birds, and collected and analyzed a large number of observational data. On the basis of these documents, this section will summarize and analyze the research conclusions of animal group behaviors which have very important reference value for the analysis and design of CGCS-of-MAF.

1. The characteristics of individual movement behavior in bird flocks

The animal behavior model of Reynolds summarizes the behavior characteristics of group animals, that is, individuals are always moving in the same direction as their neighbors and trying to stay close to them but avoiding collisions at the same time, and establishes three criteria according to the characteristics of these behaviors [114]:

(1) Velocity Matching: attempt to match velocity with nearby flock mates.
(2) Attracting: attempt to stay close to nearby flock mates.
(3) Excluding: avoid collisions with nearby flock mates.

Many scholars have modeled this kind of distance dependent behavior, such as building a virtual force model, discrete or continuous, as shown in Ref. [115].

Much of the researches in the field of multi-agent are carried out around the behavior criteria, but one point, for different groups or formations, expressions of these three criteria are generally different.

2. The characteristics of interaction between individuals in dense bird flocks

In two years, Ballerini has tracked and observed 10 large formations of birds in detail. They take pictures of the same formation of birds respectively in multiple locations through the high-speed camera, identify and calculate the position and speed of each bird in three dimensions, and then find the hidden relationship and law behind the data through the statistical methods. One of the experiments is to explore the "interaction distance" between birds, that is, the interaction range of an individual bird and adjacent birds and actually in what range an individual bird interact with such birds, the utilized data is the statistical characteristics of isotropy and anisotropy of the velocity vectors of all birds, and the results are as shown in Ref. [114]. The data processing results show that the bird's flight velocity vector is only related to the birds in the topological range in a statistical sense, that is, one bird only interacts with the nearest average 6 to 8 neighboring birds rather than interacting with all neighboring birds within a certain range, this phenomenon is also known as neighbor-follow, and the rigorous form of this conclusion has been given by Eqs. (2.21)–(2.23). The idea of building the local predictive model of dense autonomous formation is put forward based on the conclusion that "Under the circumstance of safe distance, for the most dense formation in two-dimensional plane, there are only 6 to 8 nodes around each node" [64, 74].

3. The characteristics of turning flight of bird flocks

Early in 1943, Gerard has proposed the articles on the observation and study of birds turning flight in ≪Science≫, and put forward that the birds tend to do equal radius turn rather than parallel turn when they are turning flight. Pomeroy and Heppner also analyzed the equal radius turning flight mechanism in detail through the flight observation of pigeons consisting of 12 rock pigeons, and then Ballerini team observed and analyzed the bird flocks consisting of 781 individuals and verified again the equal radius turning flight mechanism using data, as shown in Refs. [74, 113–116].

It can be seen that the turning flight of bird flocks tends to maintain their geometries in the ground reference system, rather than following the leader bird to turn. So in the process of turning flight, the bird flocks consume least energy, because the outer birds do not need to accelerate the turning to keep the relative position to the leader bird, and the original bird becomes the flanking bird.

4. The tightness distribution characteristics of bird flocks

References [74, 113–116] shows the tightness observation results of bird flocks by Ballerini team. The data shows that in bird flocks, the average distance between boundary individuals is minimum and the average distance between center individuals is maximum, there is an increasing gradient from the boundary to the center. The tightness characteristics of bird flocks will be analyzed deeply and based on that the light passing hypothesis will be proposed.

5. **Behavior characteristics of the leader and the followers of the shoal of fish**

German physiologist Von Holst obtains some interesting research results through the observation of the gregarious behavior of minnows which like clustering. In order to determine the position of nerve center which affect the gregarious tendency of minnows, Von Holst did an experiment to cut out the forebrain of a minnow, and found that it differs from other minnows in behavior: a normal minnow is usually concerned about its companions around, even in the moment of changing direction, it will look back to see whether other minnows follow it and then decide whether to change the direction; however, the minnow which has the forebrain removed will act directly based on its preference, no longer consider the companions and become "selfish". However, surprisingly, this "selfish" minnow finally become the leader of the whole shoal of fish.

To a certain extent, this experiment illustrates that if a group wants to move toward a target, then two factors are needed: the "selfish" of the leader and the "trust" of the followers [74, 117].

5.4.2   Collision Avoidance Methods of FCSM

At present, the research on collision avoidance methods in formation control is widely used in the fields of ground robot formation system, satellite formation system, ship formation system, unmanned aerial vehicle and missile formation system. All collision avoidance methods in groups are based on the control of distance between the individual in essence.

1. Collision Avoidance Methods Based on Planning or Scheduling

As the description about Loose Formation Flying Control in Sect. 5.3, route planning or scheduling planning is often used for collision avoidance control in loose formation [118].

It is precisely because the density and density of loose formation are low, not only the formation spacing is much larger than the safety distance, but even if there are occasional collision threats caused by some unexpected factors, the freedom space that missile members needed to circumvent and maneuvering is enough, so the collision avoidance problem is not prominent in loose formation, it can be resolved through the off-line planning and scheduling.

2. Collision Avoidance Methods Based on Virtual Forces

The method based on virtual force is derived from Reynolds' model of animal group behavior. It constructs a function between virtual force or potential field (repulsion, retention, attraction) and distance. In the motion of the formation, when other nodes or threats enter the detection range of a node, the node calculates the virtual force according to the distance, and then transforms it into the controlled quantity and applies it to the formation holding. It can also be applied in the collision avoidance control. Compared with the PID and other traditional methods when dealing with the distance deviation, it would be more flexible to designing a function to make a certain non-linear relationship between the distance and controlled quantity if the virtual force-based collision avoidance method is extended to the three-dimensional space. But this flexibility also makes it more difficult to design. This chapter subsequently put forward Collision Avoidance Methods according to Reynolds's Animal Group Behavior Model. But it is different from what we have applied directly on the controlled quantity. This time, we applied it on the weighing of the optimizing, which can reduce the sensitiveness between the controlled quantities and the distance related non-linear property resulted by the lack of experience of designing and the deficiency of theoretical knowledge.

3. Collision Avoidance Methods Based on height adjustment

Flight height adjustment is simple and practical. When the formation node comes across collision threats, it would maneuver to another dimension from a certain dimension. For example, in Ref. [3], when it study the decentralized UAV formation flight control methods, UAV formation flight control problem is simplified by assuming that the UAV could implement the collision avoidance maneuvers by adjusting the relative flight heights. For another example, in Ref. [3], this method is also applied in the settlement of missile formation collision avoidance. However, during the low altitude penetration of missile dense formation, in order to reduce the probability of being detected and to make better use of ground clutter reflection and earth curvature, the nodes often do not have such free spaces on the height, in other words, it have no other nodes above them (dense formation in three-dimensional). In short, flight height adjustment has the premise that the node must have such free dimension to make collision avoidance maneuver.

5.4.3   Analysis of Collision Avoidance Problem of Dense Formation

1. Dense formation collision avoidance problem

Due to combat task requirements and environment constraints, missile formation need to implement high- density low-altitude flight during a specific period or region, this time the whole formation almost has no free space in the vertical direction, and the entire formation can only maneuver in horizontal direction. In view of the independence of the longitudinal motion control loop of the missile relative to the lateral motion control loop, this book defines the formation control system in a two-dimensional plane. It can be seen from the following analysis that the definition of this two-dimensional horizontal plane is equivalent to infinity in the longitudinal direction. Compared with the traditional permissible height adjustment of the formation control method, the collision avoidance control between nodes cannot be implemented by adjusting flight height due to the limitation, so it did not reduce but increase the difficulty of the formation collision avoidance problem, but the collision avoidance mechanism could easily extended to three-dimensional space in the dense formation.

As shown in Fig. 5.17, the distance between the node ε_1 and the node ε_2 in the formation is closed to the minimum safety. It means that the node ε_1 must implement collision avoidance maneuver to ensure its own safety when the node ε_2 generates undesired disturbance. In the loose formation (Fig. 5.17a), there is a free space around the node ε_1 to avoid collision, it can take the minimum cost of their own individual to implement collision avoidance maneuver; In the case of dense formation (Fig. 5.17b), there is almost no free space around the node ε_1 to implement collision avoidance maneuver, and the collision avoidance maneuver in each direction will make the distance between the neighbor node less than the safe distance. That is to say the collision avoidance of the node itself creates an undesired disturbance to its neighbor node, which will be transmitted in the intensive formation, that is, the chain effect, which will undoubtedly increase the risk of collision of the whole formation. Therefore, in the dense formation flight, the avoidance maneuver of the node ε_1 should not only consider its individual cost, but also take into account the group cost of the entire formation, and the comprehensive balance between individual cost and group cost embodies the autonomy of networked autonomous formation system to a certain extent.

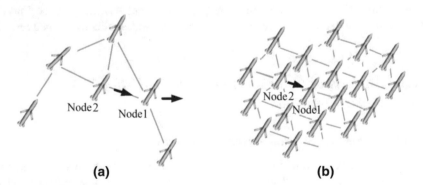

(a)   **(b)**

Fig. 5.17 The particularity of dense formation collision avoidance maneuver, **a** loose formation, **b** dense formation

2. Chain effect caused by collision avoidance maneuver

Figure 5.18 shows the chain effect diagram caused by the disturbance in the dense formation flight. The occasional maneuvering behavior of any node in the dense formation fleet will cause the collision avoidance maneuver of the neighbor node, and the disturbance caused by the collision avoidance maneuver is transmitted step by step. Because of the delay of network, the loss of the sensor, the measurement error of the various sensors and the lack of free space available for collision avoidance in the process of the dense formation collision avoidance maneuver, the relevant nodes should not implement mobility in advance according to the information of the neighboring nodes but decide their own collision avoidance motor behavior based on the real-time status information of their neighbor nodes.

The disturbance distance between the node ε_i and the node ε_{i+1} is $\Delta d(i, i+1)$, that the equivalent transfer function of member's flight control system of node ε_i is $G_i(s)$, n is the number of nodes in the chain effect. In the process to keep formation of the dense formation, the disturbance distance $\Delta d(1, 2)$ between the node ε_1 and ε_2 becomes $\Delta d(2, 3)$ between the node ε_2 and ε_3 after go through the member's flight control system $G_2(s)$ of node ε_2, then follow the step until the node ε_n, then:

$$\frac{\Delta d(n, n+1)}{\Delta d(1, 2)} = G_2(s) \cdot G_3(s) \cdots G_n(s) = G(s)^{n-1} \tag{5.60}$$

The entire system will eventually become a stable state when the number of nodes n is finite in the chain effect because of the designed transfer function $G_i(s)$ is stable. When $\Delta d(n, n+1)$ responding to $\Delta d(1, 2)$, the amplitude-frequency characteristic of $G_i(s)$ reflects the influence of the formation flight control system on the chain effect of the whole formation system. According to Eq. (5.60), if $|G(jw)| < 1$, the initial disturbance will not be amplified step by step. Taking the PID formation controller discussed in Sect. 5.3.3 as an example (with an explicit system matrix, which is easy to analyze and can be used as a reference for the MPC formation controller), the input can be considered as the leader's trajectory deflection angle and velocity $u = (\varphi_{cL}, V_{cL})^T$. The output $Z = (x, z, \Delta\varphi, \Delta V)^T$ stand for the relative distance x, z, the deviation of trajectory deflection angle and the velocity difference

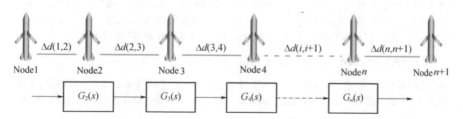

Fig. 5.18 Chain effect diagram caused by the disturbance in the dense formation

between the leader and follower. The state $X = (x, z, \varphi_W, V_W, \varphi_L, V_L, \Delta\varphi_L, \Delta V_L,$
$\Delta\varphi_{Wc}, \Delta V_{Wc})^T$ is the relative distance x, z, follower's trajectory deflection angle and
velocity, the leader's trajectory deflection angle and velocity, the instruction of
follower's trajectory deflection angle and velocity. The matrix in system (5.43) is:

$$A = \begin{pmatrix} 0 & 0 & -26.67 & -1 & 0 & 1 & 26.67 & 0 \\ 0 & 0 & -130 & 0 & 150 & 0 & -20 & 0 \\ 0 & 0 & -0.067 & 0 & 0 & 0 & 0.067 & 0 \\ 0 & 0 & 0 & -0.2 & 0 & 0 & 0 & 0.2 \\ 0 & 0 & 0 & 0 & -0.067 & 0 & 0 & 0 \\ 0 & 0 & 0 & 0 & 0 & -2.0 & 0 & 0 \\ 0 & 0.01 & -14.67 & 0 & 16.67 & 0 & -5.33 & 0 \\ 9 & 0 & -160 & -45 & 0 & 45 & 160 & -6 \end{pmatrix},$$

$$B = \begin{pmatrix} 0 & 0 \\ 0 & 0 \\ 0 & 0 \\ 0 & 0 \\ 0.067 & 0 \\ 0 & 0.2 \\ 3.33 & 6 \\ 0 & 0 \end{pmatrix}, C = \begin{pmatrix} 1 & 0 & 0 & 0 & 0 & 0 & 0 & 0 \\ 0 & 1 & 0 & 0 & 0 & 0 & 0 & 0 \\ 0 & 0 & 0 & -1 & 0 & 1 & 0 & 0 \\ 0 & 0 & -1 & 0 & 1 & 0 & 0 & 0 \end{pmatrix}, D = O_{4\times 2}$$

$$(5.61)$$

Figure 5.19 shows the leader system's amplitude-frequency characteristic curves
based on the PID formation controller. A rough understanding of the bullet's trajectory deflection angle changes between $10^{-4} \sim 1$ Hz can be get from position of
zero dB line in the figure, the relative distance deviation between the formation
nodes can be easily amplified, but for the changes above 1 Hz are not easy to be
amplified; The trajectory deflection angle and velocity difference between nodes
which were caused by the relative distance deviation, velocity variation of leader or
the variation of the trajectory deflection angle which were result from the velocity
variation of the leader can be filtered by the formation system [74].

3. **The Suppression Method of Chain Effect**

As can be seen from the above analysis, there are at least three ways to reduce the
impact of the disturbance of the individual nodes on the entire dense formation:

1. Make the $\Delta d(1, 2)$ as small as possible, so that the individual nodes are less
   likely to produce undesired maneuvers.

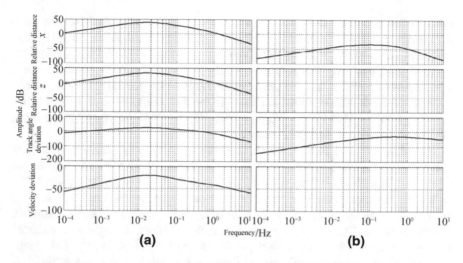

Fig. 5.19 The leader system's amplitude-frequency characteristic curves, **a** instruction of trajectory deflection angle, **b** instruction of velocity

2. For any ω, formation flight control system amplitude-frequency characteristics $|G(jw)| < 1$.
3. Make n as small as possible, so as to reduce the range of nodes affected by the chain effect.

In the above three ways, the way (1) is caused by the external environment and the uncertainties of node itself, sudden failure, the error generated from the formation keeping when the formation carries out maneuver. It is difficult to avoid in the actual formation flight environment. Way (2) could be balanced with the performance of distance keeping when designing the formation flight control system. Here we will try to use the way (3) to propose an effective chain effect suppression method suitable for dense formation flight control system.

5.4.4 Light Passing Hypothesis for Collision Avoidance Mechanism

1. Light Passing Hypothesis

From the analysis of the density distribution of birds in Sect. 5.4.1, it is known that dense birds' distribution have a characteristic that the density increases from the center to the boundary. From this we achieve the idea that when the density adjustment process of the birds is in steady state, the collision threat for all of birds from the center to the boundary should be basically the same, that is to say that all birds' feelings reach a balanced state. On the contrary, when the feelings of the birds have not yet reached equilibrium, that is, the density from the center of the birds to the boundary does not form a suitable incremental gradient, the birds should move toward the direction where the collision threat it perceived is small and eventually form a stable distribution in which the density is increasing from the center to the boundary. Such distribution indicates that the birds at the border feel a small threat of collision and the center of the birds on the contrary. In other words, the level of the feeling of threat is related to the distance to the free space outside the bird's borders. That is the intrinsic relationship between the density of the birds, the threat of collision, and the free space.

How does the bird feel the threat that corresponds to the distance to the free space near the borders of the birds? Here we give a hypothesis that the collision threat from other birds is judged by the degree of light and shade it perceived. The light is at the maximum brightness at the outer boundary of the birds and the brightness gradually decays in the process to reach the central of birds through the middle members. The bird group absorb, shelter, or block the light as a complex geometric and the effect is the light passing of the individual and the light penetration of all individuals constitutes the light passing distribution of the bird group, which makes light as a means for individuals to feel their status in large groups. With natural selection, a single bird in the bird group tend to make movement in the direction where is easier for light passing, it reflects a relatively small group density and is closer to the free space away from the boundaries of the group, thus to make it more possible to avoid threats. The collision avoidance mechanism of dense birds reflected from the light passing hypothesis is not the inherent characteristic of single bird but the acquired characteristic of the networked whole group. Light, as an input, acts on the bird group network with a certain light passing distribution. Different outputs are obtained on different individuals so that a single bird naturally has the ability to feel the overall collision threat gradient.

Light passing performance in difference locations of the bird group shows the difficulty or possibility to obtain the free space. Simply put, the essence of the light passing hypothesis is that the individual in the dense bird group moves toward the area of the free space outside the boundary of the group while avoiding the collision between the individuals, and the boundary of the group is the area in which exists many free space. Then, the disturbance caused by the collision in dense formation is passed to the free space outside the boundaries of the bird group at a lower cost. As shown in Fig. 5.20, ε_2 is in the dense group, there is no free space around it, and it can only avoid the collision to move towards the free space where is most possible to reach, so it will move to the node ε_3 instead of the node ε_1 [74].

2. Collision Avoidance Modeling Based on Light Passing Hypothesis

The model of the light passing hypothesis used for collision avoidance of missiles dense formation is established based on the overall cognitive mechanism of the collision threat of the birds through the light in the network. As is described above, the light passing performance can be valued by the possibility to achieve the free space. Here we take free space as an analogy of "light", the light passing coefficient C to characterize the light passing and the generating process of light passing coefficient to imitate the process of light passing.

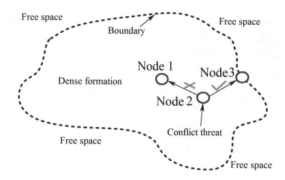

Fig. 5.20 Light passing hypothesis illustration of collision avoidance mechanism in dense formation

According to Definition 2.11, the Proximity Groups of the missile member ε_i are the collection $v(q) = \{(i,j) \in \varepsilon \times \varepsilon : \mu_{ij} < d_{imax}, i \neq j\}$ of neighbor nodes of the missile member ε_i, that is, if $\forall \varepsilon_j \in v(q)$, then, $\mu_{ij} \leq d_{imax}$. The set of neighboring group's boundary points of the missile member is $B(G, d_{imax})$. The detailed decision-making protocol would be given in the following Chap. 6 which help the nodes to autonomously determines whether or not it belongs to $B(g, d_{imax})$. Obviously, the node would has more free space in the intensive formation when $\varepsilon_i \in B(G, d_{imax})$, and the node owns almost no free space when $\varepsilon_i \notin B(G, d_{imax})$.

The opening angle of the free space of ε_i in the two-dimensional plane is set as $A(\varepsilon_i) = \angle \varepsilon_m \varepsilon_i \varepsilon_n$, where $\varepsilon_m, \varepsilon_n \in v(q)$, its distance to the bisector of $A(\varepsilon_i)$ is $l > k_{imin} \cdot d_{si}$, k_{imin} is the threat coefficient. The light passing performance of ε_i can be understood as the probability of light passing, that is, the probability of ε_i to fall into the free space, set it as $p(\varepsilon_i)$. The free space within the two-dimensional plane or "light" obey the uniform distribution, so it can be defined that: if $\varepsilon_i \in B(G, d_{imax})$, then, the coefficient $C\{\varepsilon_i\}$ of node ε_i is the transmitting probability of the node ε_i, i.e.:

$$C\{\varepsilon_i\} = p(\varepsilon_i) = \frac{A(\varepsilon_i)}{2\pi} \tag{5.62}$$

There is no light passing in neighbor node if $\varepsilon_i \notin B(G, d_{imax})$, that is, the probability of light passing $p(\varepsilon_i | \bar{\varepsilon}_j) = 0$ on the conditions of no neighbor nodes of ε_i fall into the free space, where $\varepsilon_i \in v(q)$, $\bar{\varepsilon}_j$ means ε_j allows no light to pass; Moreover, under the conditions of only one neighbor node of ε_i allow light passing, the probability of light passing $p(\varepsilon_i | \varepsilon_j)$ is

$$p(\varepsilon_i | \varepsilon_j) = \frac{A(\varepsilon_i | \varepsilon_j)}{2\pi} \tag{5.63}$$

In equation: $A(\varepsilon_i|\varepsilon_j)$ is the opening angle of the free space of ε_i that provided by the ε_j after it fall into its free space, a neighbor node must be transmitted at the same time if ε_i need to transmit light, and the probability for the two nodes to transmit light simultaneously is $P(\varepsilon_i\varepsilon_j)$

$$p(\varepsilon_i\varepsilon_j) = p(\varepsilon_i|\varepsilon_j) \cdot p(\varepsilon_j) \tag{5.64}$$

Here we take the criteria of optimism decision-making, and select the neighbor node ε_{imax} whose light passing coefficient C is largest to be the node that could transmit light like ε_i, that is,

$$\varepsilon_{imax} = \arg\max[C\{\varepsilon_j|\varepsilon_j \in n_{ai}(q)\}] \tag{5.65}$$

And if $\varepsilon_i \notin B(G, d_{imax})$, then we could define the light passing coefficient of node ε_i as the possibility of transmit light simultaneously with the neighbor node whose light passing coefficient C is largest, that is,

$$C\{\varepsilon_i\} = p(\varepsilon_i\varepsilon_{imax}) = p(\varepsilon_i|\varepsilon_{imax}) \cdot p(\varepsilon_{imax}) = \frac{A(\varepsilon_i|\varepsilon_{imax})}{2\pi} \cdot C\{\varepsilon_{imax}\} \tag{5.66}$$

Above all, the light passing coefficient of node ε_i based on the optimism decision criterion is defined as follows:

Definition 5.3 Transparency Generation Protocol, light passing coefficient characterizes the transparency of the node ε_i in the formation is

$$C\{\varepsilon_i\} = \begin{cases} \frac{A(\varepsilon_i|\varepsilon_{imax})}{2\pi} \cdot C\{\varepsilon_{imax}\} & , \ \varepsilon_i \notin B(G, d_{imax}) \\ \frac{A(\varepsilon_i)}{2\pi} & , \ \varepsilon_i \in B(G, d_{imax}) \end{cases} \tag{5.67}$$

It can be seen from Eq. (5.67) that the formula defined by the light passing coefficient (light passing generation protocol generation protocol) is a recursive formula based on optimism decision-making. The initial value is derived from the nodes owning free space and in the collection $B(G, d_{imax})$ with the condition of a certain d_{imax}. The light passing coefficient of nodes with no free space can only be obtained through the transmitting of other nodes' light passing coefficient in the network. It's a network support protocol that simulates the light attenuation in the network of light passing hypothesis.

Figure 5.21 shows the simulation result of light passing generation protocol for a formation with $n = 64$. For convenience, the light passing of the node is normalized on the entire formation area and the color value represents the light passing coefficient. It can be seen from the color distribution in the figure that it is effective to use light passing coefficient to characterize the network's light passing performance. Therefore, the light passing collision avoidance mechanism is applied to the local model predictive control (LMPC) method on the basis of MPC controller in the follow chapter.

Fig. 5.21 Simulation result of light passing generation protocol for a formation with $n = 64$

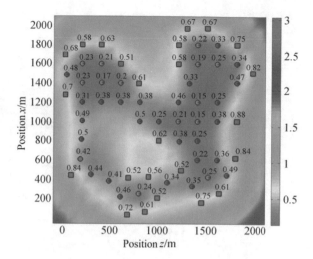

5.4.5   LMPC Controllers of Dense Formation

It is necessary for dense formation flight controllers to own the ability of collision avoidance between nodes. According to the enlightenment of the animal population behavior in Sect. 5.4.4, this section will first improve and expand the MPC-based formation controller designed in Sect. 5.3.2. The main differences are improvements and extensions of prediction model of the MPC formation controller, the cost weight and the reference pitch instruction in Sect. 5.3.2 based on three bionics laws or hypothesis, i.e. near-neighbor follow-up phenomenon, Reynolds' animal group behavior model and light passing hypothesis. As shown in Fig. 5.22 [74].

The LMPC, which has been improved by the first two bionics laws, is used independently as a formation flight controller with a collision avoidance function. The light passing collision avoidance function can be turned on or off as an extended option according to the situation. The three main improvements are also corresponded to the design ideas of architecture of the missile cooperative, guidance and control system in Chap. 2, the relationship is shown in Fig. 5.23

1. **Design of LMPC formation flight controller**

For the density of formation, the node ε_i only has relationship with about 6 to 8 neighborhood nodes $\varepsilon_j \in v(q)$ around it, thus the number is limited and smaller, and the node $\varepsilon_j \in v(q)$ has the most direct threat of collision to node ε_i. Moreover, with the enlightenment of bird's behavior of near-neighbor following-up, it is feasible to extend the formation control system of node ε_i to the local motion prediction model which contains all the neighbors (ε_i and $\varepsilon_j \in v(q)$) from the motion prediction model contains only two-node (ε_i and pilot point) given above. Thus, the formation local prediction model MPC is called the local model predictive control (LMPC) for autonomous formation.

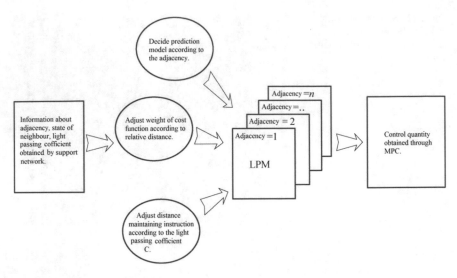

Fig. 5.22 The relationship of LMPC and MPC

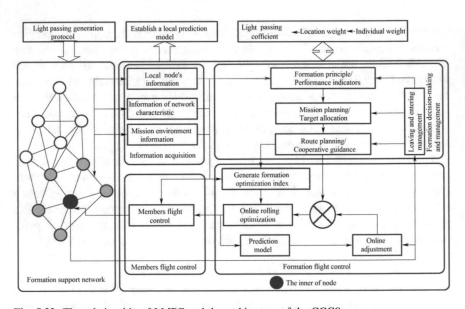

Fig. 5.23 The relationship of LMPC and the architecture of the CGCS

For the LMPC of the node ε_i, the state $X_i = (x_1\ x_2 \cdots x_j \cdots x_n\ z_1\ z_2 \cdots z_j$ $\cdots z_n\ \varphi_i\ V_i)^T$ are selected, where the x_j and z_j is the projection from the formation spacing d_{ij} of node ε_i and the neighbor node ε_j in the trajectory coordinate system; φ_i is the trajectory deflection angle of the node ε_i; V_i is the velocity of the node ε_i; n_{ai} is the number of $\varepsilon_j \in v(q)$, that is adjacency.

The controllable inputs are $u_i = (\varphi_{ci}\ V_{ci})^T$, where φ_{ci}, V_{ci} the trajectory deflection angle commands and velocity commands for the node are ε_i. The measurable interferences $d_{Li} = (\varphi_1\ \varphi_2 \cdots \varphi_j \cdots \varphi_n\ V_1\ V_2 \cdots V_j \cdots V_n)^T$ are obtained by the node ε_i through the support network, in which φ_j, V_j are the trajectory deflection angle and velocity of the neighbor node ε_j. The outputs are $Z_i = (x_1\ x_2 \cdots x_j \cdots x_n\ z_1\ z_2 \cdots z_j \cdots z_n\ \Delta\varphi_i\ \Delta V_i)^T$, where $\Delta\varphi_i = \varphi_L - \varphi_i$, $\Delta V_i = V_L - V_i$ and L represents the serial number of the pilot node in $\varepsilon_j \in v(q)$, the node can be an actual neighbor node or a virtual pilot node, and the output Z_i reflects the characteristics of the node ε_i: (1) relative positional relationship with all neighbor nodes or pilot nodes. (2) the deviation of trajectory deflection angle and velocity with pilot node. The process of modeling relative kinematics relation and flight control system are the same as the two-node modeling described above. The local motion discrete prediction model with n_{ai} as the adjacency is:

$$
\begin{cases}
X_i(k+1) = A_i X_i(k) + B_{ui} u_i(k) + B_{di} d_{Li}(k) \\
Z_i(k) = C_i(k) X_i(k) + D_i d_{Li}(k)
\end{cases}
\tag{5.68}
$$

where the coefficient matrix are as follows

$$
A_i = \begin{pmatrix}
& & \vdots & A_1 \\
O_{(2n_{ai}+2)\times 2n_{ai}} & & \vdots & A_2 \\
& & \vdots & A_3
\end{pmatrix}_{(2n_{ai}+2)\times(2n_{ai}+2)},
$$

$$
A_1 = \begin{pmatrix}
-\frac{z_{r1}}{\tau_{\varphi i}}, -1 \\
\vdots \\
-\frac{z_{rj}}{\tau_{\varphi i}}, -1 \\
\vdots \\
-\frac{z_{rn}}{\tau_{\varphi i}}, -1
\end{pmatrix}_{n_{ai}\times 2}, \quad
A_2 = \begin{pmatrix}
\frac{x_{r1}}{\tau_{\varphi i}} - V_{r1}, 0 \\
\vdots \\
\frac{x_{rj}}{\tau_{\varphi i}} - V_{rj}, 0 \\
\vdots \\
\frac{x_{rn}}{\tau_{\varphi i}} - V_{rn}, 0
\end{pmatrix}_{n\times 2},
$$

$$
A_3 = \begin{pmatrix}
-\frac{1}{\tau_{\varphi i}} & 0 \\
0 & -\frac{1}{\tau_{Vi}}
\end{pmatrix}_{2\times 2},
$$

$$B_{ui} = \begin{pmatrix} B_{u1} \\ B_{u2} \\ B_{u3} \end{pmatrix}_{(2n_{ai}+2)\times 2}, B_{u1} = \begin{pmatrix} \frac{z_{r1}}{\tau_{\varphi_i}}, 0 \\ \vdots \\ \frac{z_{rj}}{\tau_{\varphi_i}}, 0 \\ \vdots \\ \frac{z_{rn}}{\tau_{\varphi_i}}, 0 \end{pmatrix}_{n_{ai}\times 2},$$

$$B_{u2} = \begin{pmatrix} -\frac{x_{r1}}{\tau_{\varphi_i}}, 0 \\ \vdots \\ -\frac{x_{rj}}{\tau_{\varphi_i}}, 0 \\ \vdots \\ -\frac{x_{rn}}{\tau_{\varphi_i}}, 0 \end{pmatrix}_{n_{ai}\times 2}, B_{u3} = \begin{pmatrix} \frac{1}{\tau_{\varphi_i}} & 0 \\ 0 & \frac{1}{\tau_{V_i}} \end{pmatrix}_{2\times 2},$$

$$B_{di} = \begin{pmatrix} O_{n_{ai}\times n_{ai}} & I_{n_{ai}\times n_{ai}} \\ B_{d1} & O_{n_{ai}\times n_{ai}} \\ \hline O_{2\times 2n_{ai}} \end{pmatrix}_{(2n_{ai}+2)\times 2n_{ai}}, B_{d1} = \begin{pmatrix} V_{r1} & & & \\ & \ddots & & O \\ & & V_{rj} & \\ & O & & \ddots \\ & & & & V_{rn_{ci}} \end{pmatrix}_{n_{ai}\times n_{ai}}$$

$$C_i = \begin{pmatrix} 1 & & & \\ & \ddots & & O \\ & & \ddots & \\ O & & & -1 \\ & & & & -1 \end{pmatrix}_{(2n_{ai}+2)\times(2n_{ai}+2)}, D_i = \begin{pmatrix} 0 & & & \\ & \ddots & & O \\ & & \ddots & \\ & 1 & \cdots & 0 \\ & & \cdots & 1 & 0 \end{pmatrix}_{(2n_{ai}+2)\times(2n_{ai}+2)}$$

where $\begin{cases} D_i(2n_{ai}+1, 2+L) = 1 \\ D_i(2n_{ai}+2, 2+n_{ai}+L) = 1 \\ else\ D_i = 0 \end{cases}$, and the subscript r represents the balanced point during linearizing, which is taken as a reference value.

The cost function of online optimization J_i is:

$$J_i = \sum_{l=1}^{P_i} \left\{ \sum_{m=1}^{N_i} \| \omega_{zi}(l,m) \cdot [Z_{im}(k+l|k) - ref_{im}(k+l)] \|^2 + \sum_{m=1}^{N_i} \| \omega_{ui}(l,m) \cdot u_{im}(k+l|k) \|^2 \right.$$
$$\left. + \sum_{m=1}^{N_i} \| \omega_{\Delta ui}(l,m) \cdot \Delta u_{im}(k+l|k) \|^2 \right\}$$

(5.69)

In equation, P_i is the prediction horizon, N_i is the control horizon, $(k+l|k)$ stands for to predict value of time $k+l$ at time k, *ref* is the spacing hold instruction, and ω_i is cost weight which is divided into three groups, ω_{zi} is the output punishment weight matrix, the most important weight to keep spacing between nodes, which reflects the distance retention force between the node ε_j and each neighbor node and also reflects the matching force of the trajectory deflection angle and velocity of the node ε_j and the pilot node. ω_{ui} is the input punishment weight matrix, and it reflects the intensity of the trajectory deflection angle instruction and the velocity instruction of node ε_j. $\omega_{\Delta ui}$ is the input derivative weighting matrix which reflects the change rate of the instruction. In order to simplify the solution of quadratic programming (QP), we set the equal weight of (l, m) in each time period, thus the ω_i is degenerated into a row vector $\omega_{zi} = [\omega_{i1} \ \omega_{i2} \cdots \omega_{ij} \cdots \omega_{i(2n_i+2)}]$, $\omega_{ui} = (\omega_{i\varphi_c} \ \omega_{iV_c})$, $\omega_{\Delta ui} = (\omega_{i\Delta\varphi_c} \ \omega_{i\Delta V_c})$.

The weight value is decided by the Reynolds' animal behavior model. And take the method of dividing an animal's behavior into the exclusion area, the holding area and the attraction area according to the distance to other animals as a reference, the weight ω_{ij} is designed as the function $\omega_{ij} = f(d_{ij})$ related to the formation spacing d_{ij} between node ε_i and each neighbor node ε_j, the threat distance $d_{imin} = k_{imin} \cdot d_{si}$ and the adjacent distance $d_{imax} = k_{imax} \cdot d_{si}$ are the parameters of $f(\cdot)$ related to the exclusion area, the holding area and the attraction area around the node ε_i. The function $f(\cdot)$ is discrete in the real process. After simulation debugging, a set of values of weight ω_i are listed in Table 5.1.

In fact, the pilot point ε_L is "selfish" and it has no obligation to follow the following node. For the following node ε_i and node ε_L, there is a minimum value for

Table 5.1 The designing results of weight value of LMPC formation flight controller

		d_{ij}/d_{imin}	$(-\infty, 0.2)$	$[0.2, 0.4)$	$[0.4, 0.6)$	$[0.6, 0.8)$	$[0.8, 1)$	$[1, \infty)$
ω_i	ω_{zi}	ω_{ij}	800	500	340	320	100	0
		$\omega_{i(j+n_{ai})}$	750	550	360	340	100	0
		$\omega_{i(2n_{ai}+1)}$	0	0	1	5	10	0
		$\omega_{i(2n_{ai}+2)}$	0	0	0	0	1	0
		$\omega_{iL} \ \omega_{i(L+n_{ai})}$	$\omega_{iL} = \max(100, \omega_{ij}), \omega_{i(L+n_{ai})} = \max[100, \omega_{i(j+n_{ai})}],$					
	ω_{ui}	$(0.1, 0.1)$						
	$\omega_{\Delta ui}$	$(5, 0.1)$						

the weight between them, which is equivalent to the maximum weight of the other nodes. That is to say that if the node ε_i shows "trust" to ε_L, it would guarantees the matching effect. On the one hand, if ε_L take care of the followers often, the follower would responds to such "care" as well, thus resulting in the instability of the formation. On the other hand, if the weight between the followers and ε_L does not set the lower limit, then the subordinate relationship in the formation would not be obvious, and sometimes it would drag the whole followers. In particular, if it is combined with the nearest neighbor to follow up, the relationship of local interaction is shown in Fig. 5.24.

After deciding the weight value, the node ε_i would solving a QP problem online under the constraint of a linear inequality to get the optimal controlled quantity u_i of LMPC as an instruction to work on the inner ring of member flight control system.

$$\Delta u_{iopt} = \arg \min J_i$$

$$u_i(k) = u_i(k-1) + \left(I_{2\times2}, O_{(N_i-1)\times(N_i-1)}\right) \cdot \Delta u_{iopt} \quad st \cdot \begin{cases} \varphi_{min} \leq u_{i1} \leq \varphi_{max} \\ V_{min} \leq u_{i2} \leq V_{max} \end{cases}$$

$$(5.70)$$

where $[\varphi_{min}, \varphi_{max}]$ and $[V_{min}, V_{max}]$ are the trajectory deflection angle and velocity instruction limitation. Δu_{iopt} is the first controlled quantity of the optimal sequence, u_{i1}, u_{i2} are φ_{ci} and V_{ci}.

2. **The LMPC flight control based on the light passing hypothesis**

LMPC formation flight controller has collision avoidance function, but the function is for individual interests or interests of the range around neighbor nodes within local prediction model. When the node ε_i encountered a threat of collision from ε_d, meanwhile, the formation spacing d_{id} is small, the weight of ε_d would be larger than the surrounding neighbors of ε_i according to the characteristics of weight function $f(\cdot)$, thus ε_i would try to keep distance with ε_d. In other words, ε_i would become selfish to ignore the weight of the surrounding nodes. LMPC formation controller and the light passing coefficient C would be combined together to help ε_i get the

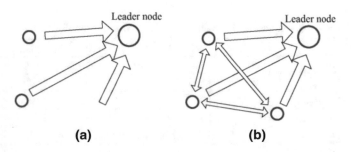

(a)   (b)

Fig. 5.24 The difference of interaction between traditional follow-up and near-neighbor follow-up, **a** traditional follow-up, **b** near-neighbor follow-up

ability to take into account the overall interests of the formation, and a light passing avoidance algorithm that helps ε_i to follow the light passing mechanism to avoid collision would be given [74].

The principle of light passing avoidance is to determine whether the degree of collision threat reaches the threshold. After reach the threshold, the distance instruction of the nodes that owns the free space would expand the threshold to the direction of the bisector of its free space opening angle; and for the nodes without free space, it should follow the distance instruction of the neighbor node that owns the largest light passing coefficient C and to restore the original instruction until the collision threat level falls below the threshold.

In addition, the node would still maintains the relative position relationship with other neighbor nodes when a neighbor node loss packet in the network communication process for the LMPC is based on the local motion model, so it has a certain robustness to the network packet loss.

	Light passing avoidance algorithm
01 :	$update\ \{\varepsilon_i\}$
02 :	$if\ d_{id} < d_{imin}$
03 :	$temp \leftarrow ref\ \{\varepsilon_i\}$
04 :	$if\ \varepsilon_i \in B(G, d_{imax})$
05 :	$ref\{\varepsilon_i\} \leftarrow ref\{\varepsilon_i\} \oplus d_{imin}$
06 :	$else\ \varepsilon_i \notin B(G, d_{imax})$
07 :	$\varepsilon_{imax} = \arg\max C\{\varepsilon_j \mid \varepsilon_j \in v(q_i) - v(q_d)\}$
08 :	$ref\{\varepsilon_i\} \leftarrow ref\{\varepsilon_{imax}\}$
09 :	end
10 :	$else\ d_{id} \geq d_{imin}$
11 :	$ref\{\varepsilon_i\} \leftarrow temp$
12 :	end

Note: *update* is the information update function, the subscript d indicates the threat node, the *ref* is the maintained reference spacing; $\oplus$ is addition in the direction of the bisector of $A(\varepsilon_i)$, $v(q_i) - v(q_d)$ is the subtract collection of $v(q_i)$ and $v(q_d)$.

5.4.6 Collision Avoidance Simulation of LMPC Controllers of Dense Formation

In the following section, we will set up a simulation experiment environment for dense formation flight and carry out comparative simulation analysis through the LMPC formation flight controller that with or without light passing function to avoid the collision. Then we will inspect the collision avoidance performance of the designed LMPC formation flight controller and analyze the effect of light passing avoidance algorithm on the collision avoidance of dense formation flight. At last we will verify the validity of the light passing hypothesis.

1. Setting the environment of simulation experiment

A simulation experiment scenario is proposed in which a wedge-shaped dense formation consisting of 21 nodes is disturbed by the faulty node. For convenience, the initial speed of all nodes are set to be consistent, i.e. for all nodes $V_{i0} = 0.44Ma(i = 1, \ldots, 21)$, the speed adjustable range is limited in the range 0.30–0.6 Ma , as shown in Table 5.2. The coordinates of the initial position distribution are shown in Fig. 5.25. The expected distance of the node is $\mu_{eij} = 220$ m, the safe distance is $d_{si} = 200$ m, and the target position of the node ε_1 is virtual leader bullet.

Assuming that the node ε_{20} is out of control for the failure at $t = 5$ s, its trajectory deflection is offset from the left by 36°. The velocity increases to $V_{20} = 0.53$ Ma and move to the direction of the geometric center of the formation; the fault condition continues 8.5 s after the control system is restored and the LMPC controller of node ε_{20} restart work then.

Figure 5.26 shows the simulation results of the LMPC formation flight controller with no light passing collision avoidance function. At the time $t = 6.5$ s, the node ε_{20} is out of control and flew towards the center of the formation (Fig. 5.26a), and at time $t = 10.1$ s, the neighbor node ε_{14} and ε_{19} first feels the collision threat and tries to keep a distance from the node ε_{20} z (Fig. 5.26b); we can see from Fig. 5.26c–f that the non-expected mobility of node ε_{20} affects almost all other nodes with only 6 nodes avoided. However, there is no collision between nodes because of the design of LMPC collision avoidance. At time $t = 13.5$ s, the control system of node ε_{20} recovers to work and start to correct relative distance from the reference position, then follow the virtual leader bullet; From Fig. 5.26d–f, we can see that all nodes that are disturbed are gradually recovered to the target position with the releasing of the threat. At time $t = 30$ s, the designated wedge-shaped formation is restored.

Table 5.2 Main parameters of nodes in the wedge-shaped formation

Node distance (m)	Safe distance (m)	Initial velocity (m/s)	Maximum velocity (m/s)	Minimum velocity (m/s)
220	200	150	200	100

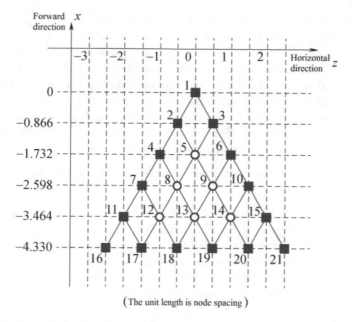

Fig. 5.25 Position distribution of wedge-shaped dense formation

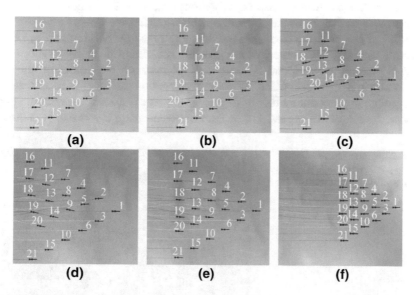

Fig. 5.26 Simulation of LMPC formation controllers without light passing collision avoidance, **a** $t = 6.5$ s; **b** $t = 10.1$ s; **c** $t = 14$ s, **d** $t = 18$ s; **e** $t = 22$ s; **f** $t = 30$ s

The validity of LMPC formation flight controller collision avoidance is verified by simulation experiments. Meanwhile, we also found that, without taking measures to measure the overall cost of the collision and did not consider the overall interests of the formation, the LMPC formation flight controller can only carry out collision avoidance behavior blindly for the individual nodes but cannot automatically acknowledge the adding overall collision threat of the formation. And the chain effect spread to all the 15 nodes in the direction of disturbance. What's more, if the scale of the dense formation does not stop increasing, the chain effect of the dense formation caused by the disturbance would be intolerable.

2. **Simulation of LMPC Formation Controllers With Light passing Collision Avoidance**

The same simulation experiment is carried out for the LMPC formation flight controller with light passing avoidance function. The light passing results of the formation network are shown in Fig. 5.27. According to the design of the light passing avoidance algorithm, after the fault node ε_{20} rushes to the node ε_{14}, the node ε_{14} should carry out avoidance maneuver towards the node ε_{10}.

The simulation results of the LMPC formation flight controller with light passing avoidance function are shown in Fig. 5.28. The node ε_{20} is out of control at $t = 6.5$ s, then move towards the node ε_{14} (Fig. 5.28a); the node ε_{14} first senses a collision threat at $t = 9$ s, and the difference to the function without light passing avoidance is that the node ε_{14} does not blindly carry out avoidance maneuver towards its left but to the direction of the node ε_{10} with the largest transmittance coefficient C in the allowable azimuth range (Fig. 5.28b).

At about $t = 11$ s, the node ε_{19} has already felt the collision threat (Fig. 5.28c), we can find that the movement of the node ε_{19} is different from the condition with no light passing as well, and because it is the node on the boundary, the node decelerates to shifted backwards by a threat perception distance in the direction of the opening angle bisecting line of its free space; At $t = 14.2$ s, the control system of node ε_{20} returns to normal but at the time when node ε_{14} is recovering to its

Fig. 5.27 Simulation results of light passing generation protocol of node formation

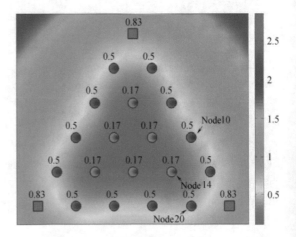

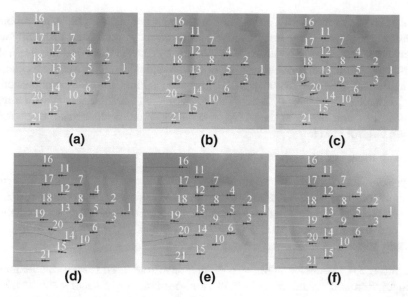

Fig. 5.28 Simulation of LMPC formation controllers with light passing collision avoidance, **a** $t = 6.5$ s; **b** $t = 10$ s; **c** $t = 11.5$ s; **d** $t = 18.9$ s; **e** $t = 24$ s; **f** $t = 26$ s

origin position, an unexpected phenomenon happened, that is, the node ε_{20} threat the node ε_{14} again in the process of resuming, what's more, the node ε_{14} independently replaced the distance command again and move towards the node ε_{10} (Fig. 5.28d); with the normalization of node ε_{20}, the node $\varepsilon_{14}, \varepsilon_{10}, \varepsilon_{19}$ and the affected node ε_{15} would be gradually turn to a stable state (Fig. 5.28e–f); Finally, the whole formation was basically restored to the original formation at $t = 30$ s.

It can be seen from the simulation that only five nodes are affected in the collision avoidance process but they did not collide with others. And they also recover to the original formation at about $t = 30$ s, but less than 10 nodes have been affected compared to the strategy without light passing avoidance. What's more, the chain effect is applied to the overall free space of the formation instead of the internal space. Experiments show that the definition of light passing coefficient and the light passing avoidance algorithm are able to reflect the light passing hypothesis and its effectiveness in dense formation flight avoidance. The light passing coefficient, as a means to reflect the overall cost of the collision avoidance process, plays an important role in improving the overall cost perception of the cooperative guidance and control system and achieving the autonomous effective coordination of the formation.

5.4.7 Formation Changing Methods

During the missiles autonomous formation carrying out missions, it needs to change the formation timely according to the needs of combat task, combat situation and combat environment. It is difficult to generate the formation that could satisfy the combat requirements because the multi-node path search is an NP-hard problem. Therefore, it is reasonable and effective to make a database for the frequently used formations, and this method is used widely by many Refs. [3, 96, 119]. For example, the common formations used in the Battlefield Code of Army Mechanized Assault Team of US are four standard formations: column and horizontal formation, wedge-shaped formation, diamond-shaped formation [120]. Balch used a similar formation of four nodes in the study of behavior-based multi-terrestrial robot formation control [121].

1. Four forms of formation change

At present, the research on the formation change is common in the ground robot system. But there are few studies on the form and concept of the aircraft formation change. Related concepts and definitions of the formation holding, reconstruction, split and transformations would be given later, and we would analyze the relationship and subtle differences among them.

Definition 5.4 Formation split: is to divide the whole large team of the geometric configuration and organizational structure into a number of smaller sub-formation according to specific needs. It is convenient for sub-formations to implement the specific missions of upper layer. In the process of splitting, the small formation basically retains its geometric configuration in the original large formation.

Definition 5.5 Formation reconfiguration: refers to the process of reorganizing the geometrical configuration and organizational structure of the formation after the nodes of the formation have left the formation due to occasional factors or the new nodes joined in the formation. Formation reconfiguration is the process caused by random factors. Generally, small adjusting range and less number of nodes would be involved in the process. Just like the simulation experiment in Ref. [74] that the wedge-shaped formation with three missiles changed into a single column with two missiles for one missile's broken.

Definition 5.6 Formation holding: is keeping the basic geometrical configuration of formation in the process of movement, that is to say, be able to restore after deformation within a certain limit or the relationship among the nodes is abnormal. The formation holding is a problem been studied a lot at this stage, and the PID-based formation flight controller and the MPC-based formation flight controller given in this section before are belonging to the category of formation holding controller. The direct effect of those controllers is to make the node tends to be consistent with its reference position. If there is a deviation between in the node and its reference position, the formation holding controller would correct it through a certain control method. For example, the formation would maintain the geometric configuration during the maneuvering process or restore the original formation after obstacle avoidance maneuver.

Definition 5.7 Formation transformations: is the process that the formation would actively switch from one formation to another for a certain need. The process of formation transformation often involves a large number of nodes, the geometric configuration and organizational structure also has a greater change after transformation.

The above definitions of the four kinds of formation transformation behavior are mostly realized directly or indirectly through the formation holding control. Missile formation is different from the ground robot formation, especially the flight missile formation that its velocity adjustment is often limited in a limited range and cannot be zero, besides that, the acceleration capacity of trajectory deflection angle is limited, which have increased the difficulty for formation transformation. At present, the missile formation transformations that been researched are almost loose formation, that is, the collision avoidance problem is not prominent and need not be dealt with because the formation spacing is relatively large; as for formation transformation, the same method is applied to transform a formation directly to another, that is to make formation transformation by changing the reference position of the formation holding controller at one time.

It would be relatively simple when the scale of the formation is not large and the node that participates in the process of formation transforming is not much. However, these factors obscure the collision threats and unstable hidden dangers that exist in the process of formation transformation, which is a matter of high attention when the tight formation and dense formation carry out formation transforming. The following is a comparison analyze of the common direct transformation methods and the indirect transformation methods based on the formation state presented in this section.

2. **Direct formation transformation methods**

In the process of formation transformation of the dense formation, the formation spacing is close to the safe distance, if the formation flight controller has no collision avoidance strategy, it would be unavoidable to intersect with the spacing where distributes part of missile members (i.e. the collision) during the moving to the target position because the limited mobility of missile members.

For this reason, the above-mentioned LMPC formation flight controller with collision avoidance function is used to implement the direct transformation of the formation. The direct transformation process is illustrated by an example of a tight formation transformation of seven missiles. Assume that a hexagon tight formation in the horizontal need to be transformed into a longitudinal formation, and the formation flying at a constant height in the process, the main parameters are shown in Table 5.3. The initial position and target position are shown in Fig. 5.29, where the node ε_6 is pilot node, and the simulation results are shown in Fig. 5.30. The node model is easier to observe after been scaled up [74].

Table 5.3 Main parameters of nodes in hexagon formation

Node distance (m)	Safe distance (m)	Initial velocity (m/s)	Maximum velocity (m/s)	Minimum velocity (m/s)
210	200	100	200	50

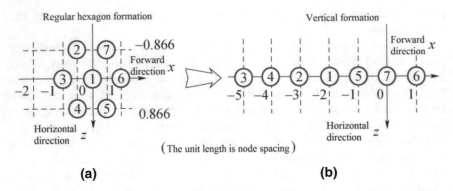

Fig. 5.29 The node's position in initial formation and target formation, **a** initial formation; **b** target formation

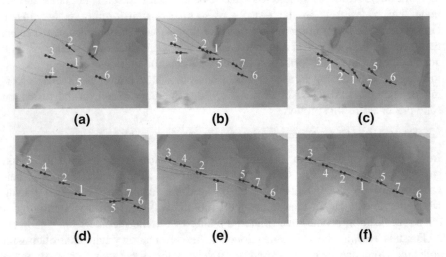

Fig. 5.30 Simulation results of direct formation transformation methods, **a** $t = 8.5$ s; **b** $t = 12.5$ s; **c** $t = 18$ s; **d** $t = 25$ s; **e** $t = 30$ s; **f** $t = 35$ s

The formation begin to transform at $t = 6$ s, all the nodes move towards the new reference position simultaneously (Fig. 5.30a); the nodes ε_1 and ε_3 cannot keep way from ε_6 so fast because of the speed limit, meanwhile, the node $\varepsilon_4, \varepsilon_5, \varepsilon_2$ and ε_7 is moving toward the formation axis, when the distance between them is less than the safe distance, the node ε_1 and ε_3 began to deviate from the formation axis to avoid collision, when the node ε_5 is too closer with the node ε_1 during the movement to the reference position, it would began to accelerate to go through, in that way, the node ε_1 and node ε_2 staggered precisely (Fig. 5.30b); at $t = 18$ s, the node $\varepsilon_6, \varepsilon_1$ and ε_3 has not yet keep enough distance, the oscillation of node ε_4 and ε_2 is small while the oscillation of node $\varepsilon_1, \varepsilon_5$ and ε_7 is larger due to the previous motion (Fig. 5.30c); at $t = 25$ s, nodes $\varepsilon_6, \varepsilon_1$ and ε_3 almost reached the desired position towards the moving direction, the node $\varepsilon_4, \varepsilon_2$ and ε_1 tends to be stable then (Fig. 5.30d); the ballot of node ε_5 and ε_7 arrived the desired position after several rounds of oscillation, at $t = 53.5$ s whole formation tends to be stable (Fig. 5.30e–f).

Figure 5.31 shows the quadrature component x, y in the velocity coordinate system and distance d of the node ε_1 relative to the other nodes during the whole process of formation transforming. From the figure we can see that the distances between node ε_1 and $\varepsilon_2, \varepsilon_5$ are closer at 12 s, and this match with Fig. 5.29, what's more, the nearest distance is less than 50 m between the nodes and didn't cause collision.

According to the simulation experiment, it can be seen that the formation flight controller with collision avoidance function could avoid collision when the missile distance is less than the safe distance. However, although there is no collision in the direct transformation process, some nodes may be in conflict with other nodes in the process of achieving the desired value and it hidden some unstable factors in the process. Therefore, a transformation method based on the formation state is given, that is, decompose the direct transformation in which are ease to cause conflicts between the nodes into several sub-transforms which are not easy to cause conflicts to weaken the unstable factors of the transformation.

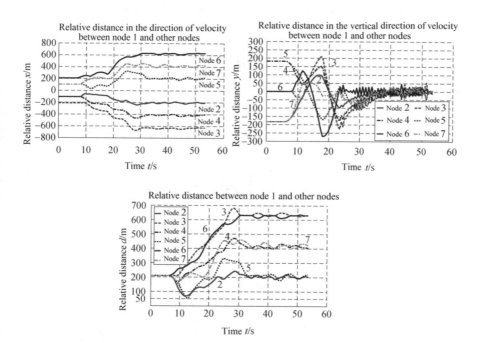

Fig. 5.31 The relative distance between ε_1 and other nodes in the process of direct formation transformation

3. **Method of Formation Transformation based on Formation State**

The formation can be described by the relative position matrix F of each follower node and leader node. In the process of the formation transforming from the initial F_1 to the target F_r, if there is a little scale of sub-form F_i that make the collision probability between each node less than a certain threshold in the direct transforming from F_{i-1} to F_i, then we call the F = {F1, F2 … Fr} is a formation state transformation from F_1 to F_r.

The state transition flow is shown in Fig. 5.32. When the formation is in a certain form, the sub-transform of the next state would be triggered if the formation index is less than the preset value.

4. **Simulation experiments of formation transformation based on the formation state**

We will carry out simulation experiments in the following to transform the hexagon formation to longitudinal tight formation based on the transformation of the formation state. The simulation conditions are the same as before, and the transformation of the formation has four states as shown in Fig. 5.33. F2 ensures appropriate space around the nodes ε_6, ε_1 and ε_3 for the rest of the nodes to enter, F3 make the nearby nodes to enter in, it is more stable than making the node ε_2, ε_7, ε_4 and ε_5 enter in at the same time.

Fig. 5.32 The state transition flow based on the formation state

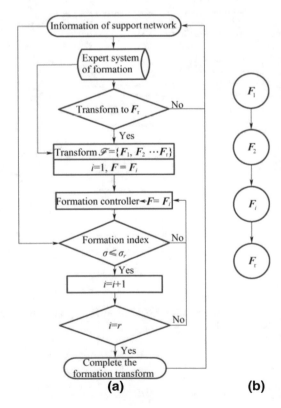

(a)   (b)

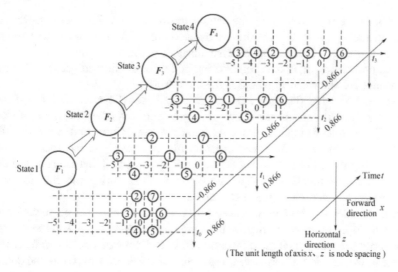

Fig. 5.33 The transforming state from hexagon formation to line formation

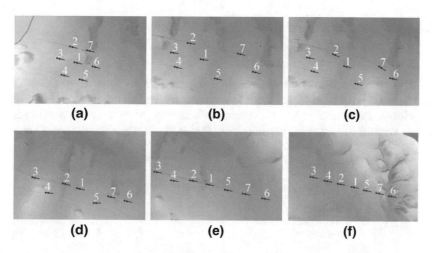

Fig. 5.34 Simulation of transformation base on the formation state, **a** $t = 4.5$ s; **b** $t = 22$ s; **c** $t = 26$ s; **d** $t = 41$ s; **e** $t = 52$ s; **f** $t = 65$ s

The simulation results are shown in Fig. 5.34:

The initial formation is the hexagon formation (Fig. 5.34a); $t = 6$ s, the formation began to switch, all the nodes first move to the F2 (Fig. 5.34b); $t = 26$ s, the nodes reached F2, the formation and then move to the next state F3 (Fig. 5.34c); when the nodes reach F3 the whole formation would finally transferred to the target formation F4 (Fig. 5.34d); $t = 65$ s, the formation becomes stable finally (Fig. 5.34e, f).

It can be seen clearly from the whole transformation that the transformation based on the shape state transition is much more stable than the direct transformation. We only observe the relative distance d (include the relative distance between node $\varepsilon_3, \varepsilon_2, \varepsilon_5, \varepsilon_6$ and neighboring node) for the limited space of the book. The results are shown in Figs. 5.35, 5.36 and 5.37.

From the relative distance between nodes, it can be seen that the distance between each node and other nodes is larger than the safety distance of 200 m during the whole formation transformation process. Compared with the direct transformation method, the transformation method based on the formation state transition is proved to be effective. However, it is also found that the state transition method consume much time than the direct method during the transformation, which reflect the contradiction of fast and stability.

From another point of view, the transformation method can be regarded as a method to form a formation, that is, the formation member transforms to the target formation's position topological structure from a difference one. Therefore, in the process of the forming of formation, especially of the dense formation, the collision between nodes is the problem that should be pay attention to.

5.4.8 Formation Turning Flight and Formation Split Methods

As a whole, the formation's characteristics are closely related to its internal membership's movement characteristics, control methods, organizational forms and other formation attributes. For the formation's turning (or curve movement), there is at least three forms, such as particle turn, rigid body turn and the deformation turn,

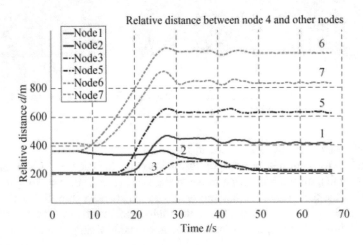

Fig. 5.35 The distance between ε_4 and other nodes

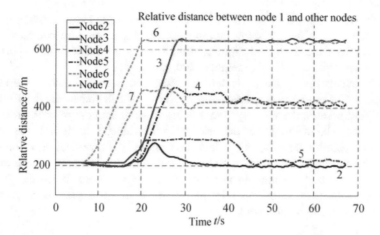

Fig. 5.36 The distance between ε_1 and other nodes

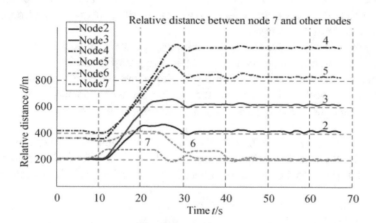

Fig. 5.37 The distance between ε_7 and other nodes

different forms reflect different network relationships and control logic in the formation [74], the reasonable use of the turning form would bring the formation benefits in some occasions the of movement.

1. **The formation's turning form**

(1) Rigid-body turning

Definition 5.8 Formation's Rigid-body-turning: is a turning form during which the missile member's geometric position in the ballistic coordinate system cannot be changed. When it has real/virtual collars (collectively called as pilot points), the missile members would keeping their position in the ballistic coordinate system of the real/virtual collars.

As shown in Fig. 5.38, the whole formation is like a rigid body moving in space in the ground coordinate system, this form of turning is the most common, which originated from the formation flight controller design to maintain the relative reference distance of the missile members and pilot points.

As it reveals during the previous formation kinematics modeling, the relative distance is the most direct relationship of the missile members in the formation maintaining system; the velocity vector is the most direct control of the formation flight control; the magnitude of the velocity in the velocity direction is the primary controlled quantity. The trajectory deflection angle in the vertical direction of the velocity is the main controlled quantity of the velocity. Therefore, the turning is often the rigid body turning in the formation flight process, if we didn't make additional adjustments for the relative reference distance (that is, the spacing command of the formation controller).

(2) Particle turning

Definition 5.9 Formation particle turning: is a turning form of the whole formation during which the missile member's geometric position in the ground coordinate system would not be changed.

As shown in Fig. 5.39, the whole formation is like a particle moving in space from the ground coordinate system, and the bird group is turning like this form.

It makes missile members turning at the same radius to avoid the speed exceeding or overload in case that the turning radius is too large when outside members making equal angular velocity turning in the process of rigid-body-turning.

The particle turning has essentially changed the internal structure of the formation. On the one hand, the spacing command of the formation controller is adjusted according to the trajectory deflection angle of the whole formation (or pilot

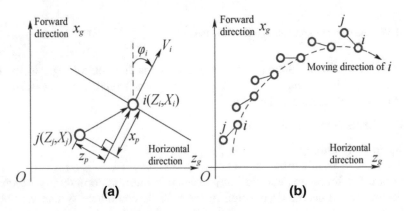

Fig. 5.38 The illustration of formation rigid-body-turning, **a** keeping the reference position unchanged in the ballistic coordinate system, **b** the turning characteristics in the ground coordinate system

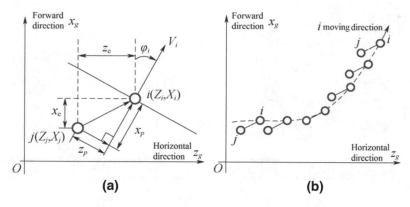

Fig. 5.39 The illustration of formation particle turning, **a** keeping the reference position unchanged in the ground coordinate system, **b** the turning characteristics in the ground coordinate system

point), and the reference distance of the ground coordinate system is defined in real time by the conversion matrix from ground coordinates to the ballistic coordinate system, the spacing command d_r in the two-dimensional plane of the particle turning is:

$$
d_r = \begin{pmatrix} x \\ z \end{pmatrix} = \begin{pmatrix} \sin\varphi & \cos\varphi \\ \cos\varphi & -\sin\varphi \end{pmatrix} \begin{pmatrix} Z_j & -Z_i \\ X_j & -X_i \end{pmatrix} = \begin{pmatrix} \sin\varphi & \cos\varphi \\ \cos\varphi & -\sin\varphi \end{pmatrix} \begin{pmatrix} x_e \\ z_e \end{pmatrix}
$$

(5.71)

The symbol description is shown in Fig. 5.39. On the other hand, the following-up or affiliation relationship between the missile members in the formation should be adjusted according to actual needs.

(3) Deformation turning

Definition 5.10 Deformation turning: refers to a form of turning during which the overall geometry of the formation is changed. As shown in Fig. 5.40, this form of turning is passive at most time, the instruction of the formation controller isn't adjusted deliberately but generated from the missile members' following-up inertia, similar to the aforementioned chain effect, but this decentralized, proximity following approach lower the requirements for formation communication systems in a sense.

2. A formation split method based on particle turning

For flight missile formation, the formation split is usually used for formation classification in the period of final guidance in the work area. If each subgroup makes curvilinear motion as a rigid body in the process of formation split, there is a risk of collision between adjacent subgroups due to the "flick" phenomenon of the rigid-body-turning. As shown in Fig. 5.41.

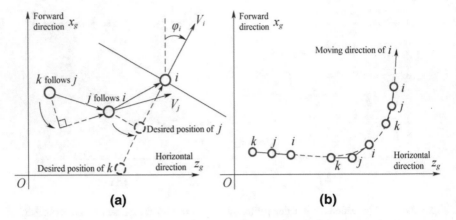

Fig. 5.40 Illustration of formation's deformation turning, **a** deformation caused by following-up, **b** the turning characteristics in the ground coordinate system

Fig. 5.41 The problems encountered in the dense formation splitting process

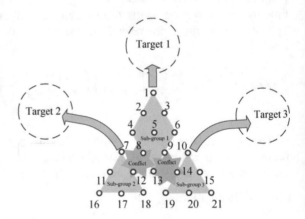

The conflict can be avoid if the sub-group switch to the particle turning to continue the curvilinear motion, this is the idea of formation split based on the particle turning.

3. **Simulation experiment of formation splitting based on particle turning**

We take the wedge-shaped formation as an example in the following section, the validity of the method are verified by simulation, formation parameters, relative position and serial number are in same with Sect. 5.4.6.

First, the formation moving to the target area and turning as a rigid body, and it would flying to the three sub-targets when approaching the target area,

The subgroups are divided into subgroups No. 1 $\{\varepsilon_i(i = 1 \sim 6, 8, 9, 13)\}$, subgroups No. 2 $\{\varepsilon_i(i = 7, 11, 12, 16 \sim 18)\}$, subgroups No. 3 $\{\varepsilon_i(i = 10, 14, 15, 19 \sim 21)\}$ similar to the scene that Fig. 5.41 shows. The pilot node is the virtual node that coincides with the node ε_1, and the formation making curvilinear motion as a rigid body; after the trigger of the splitting command, the pilot node of sub-groups 1, 2, 3 is the node $\varepsilon_1, \varepsilon_7$ and ε_{10}, the formation is switched to the particle turning. The simulation results of the whole process are shown in Fig. 5.42.

As shown in Fig. 5.42, in the turning process during t = 5.6–20.5 s , the formation is turning as a rigid body, we can see clearly that the rigid body motion of the entire formation includes not only translational but also rotation in the ground coordinate system, (Fig. 5.42a); $t = 32.5$ s, the formation began to separate to groups, we can see from Fig. 5.42b that the three subgroups basically maintained the formation at the beginning of the formation splitting in the ground coordinate system and fly to three directions led by node $\varepsilon_1, \varepsilon_7$ and $\varepsilon_{10}; t = 39.8$ s, three subgroups are almost separated (Fig. 5.42c), and compared with the internal geometric relationship of the three subgroups in Fig. 5.42b, the three subgroups like three large particles, and the motion includes only translation while without rotation, so it has no "flick" phenomenon in subgroups led by node ε_7 and ε_{10} in the process of grouping and no conflict with the sub-group led by the node ε_1 as well.

The simulation results show that, in some cases, the feasibility and application value of the formation split method based on particle turning are effective in reducing the complexity and difficulty to dealing with the collision conflict among nodes in dense formation split. It should be pointed out that the position of the sub-group pilot node in the direction of sub-group movement changes after the particle turning, and it is not conducive to the upper task of the formation in some time because the node by the front of the formation could get access to external environmental information earlier, in this time, this the pilot node should be redistribution. However, if the pilot node owns special equipment, then it needs to switch to rigid body turn or design other sub-group control method.

Collision avoidance is one of the core problems in the dense formation flight control. This section analyzes the particularity of the collision avoidance problem of dense formation, after that, we learn from the behavior of dense groups in nature, and then obtains the relevant research results of dense formation flight control:

(1) Designed the LMPC formation flight controller according to the interaction characteristics of the individual groups of birds, and then designed the LMPC weight according to the individual movement characteristics of the birds group to reduce the collision probability in the formation maintaining process.

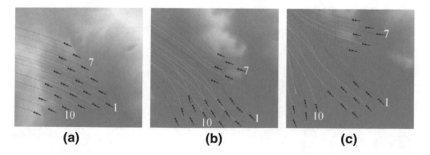

| (a) | (b) | (c) |

Fig. 5.42 Formation split simulation based on the particle turning, **a** $t = 20.5$ s; **b** $t = 35.5$ s; **c** $t = 39.8$ s

(2) The light passing coefficient model of the dense formation and the translucency hypothesis of bird collision avoidance is put forward based on the characteristics of the density distribution of birds, and the avoidance algorithm is established upon the LMPC formation flight controller. During the flight of the whole formation, the individual members of the missile use their own topological state in the formation to form the acquired characteristics which are determined by the network structure. The transmit of acquired feature, along with the existing congenital features of all the missile members in the network reveal the overall interests of the formation, and endow the missile members with the cognitive ability for the overall interests of the formation. This reflects the networked characteristics and autonomous characteristics of cooperative guidance and control of missiles autonomous formation.

(3) Gave a clear definition of the four transformations of formation, and proposes a transformation method based on formation state to reduce the collision probability in the process of transformation.

(4) Learned from the turning characteristics of the birds, and defined three kinds of turning forms, proposed the formation split method based on particle turning to reduce the collision probability in the process of formation splitting.

In summary, the network of cooperative guidance and control of missiles autonomous formation control system is not only the communication network, but also the support network; the autonomy of autonomous formation is not only reflected in the decision-making concerns with individual interests of the missile members, moreover, it is reflected in the decision-making that concerns with overall interests of the formation.

Chapter 6
Member Flight Control System (MFCS)

6.1 Configuration Design of MFCS

6.1.1 Compositions of MFCS

The Member Flight Control System (MFCS) is the guidance control system for the flight attitude and trajectory of each missile member, ε_i. The typical system structure is a three-loop structure based on the hypothesis: a missile can be equivalently modeled as the 6-DOF rigid body. The performance of MFCS is one of the important factors in determining the precision of control and guidance $\sigma_i(t)$, as well as the effect factor to quality of realizing MAF.

The Flight Control System of MAF (FCSM) give the optimized reference trajectory, formation configuration and desired attitude control instruction, which are the input of MFCS. Combining information provided by Support Networks System of MAF, Information Acquisition System of MAF gives the actually current position, formation configuration and attitude. By comparing the desired instruction information and actually current ones, the expected formation spacing, formation keeping control instruction can be obtained. Finally, through the actuator loop and controlled rudders, the instruction will be transformed into the missile flight track and attitude. The structure of the MFCS of cruise missiles is shown in Fig. 6.1 [1, 75].

6.1.2 Modeling and Design Methods for MFCS

Whether missiles can be safely and steadily formed in formation flight, is largely dependent on the information interaction and processing between missiles. Via the navigation system and various forms of sensor equipment, the members of the missile formation can be able to gather the target location information and the flight environmental information, while these information exchange is mainly through the

© National Defense Industry Press, Beijing and Springer Nature Singapore Pte Ltd. 2019
S. Wu, *Cooperative Guidance & Control of Missiles Autonomous Formation*,
https://doi.org/10.1007/978-981-13-0953-3_6

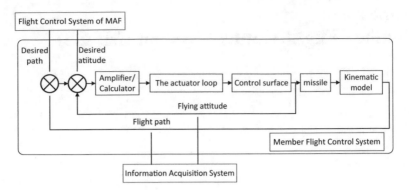

Fig. 6.1   Structure diagram of MFCS

Support Networks of MAF. However, most of the existing research on autonomous formation flight is generally based on assumption of ideal wireless communication, i.e. there is no packet loss and time-delay, which in practical, due to the wireless mobile self-organizing network itself inherent characteristics and health status, will affect the stability of autonomous formation guidance and control system.

The following describes a missiles MFCS's model based on network control system. The actual network bandwidth and load capacity are limited, meanwhile considering to amount of equipment, as masses of data is transmitted through network, consequently, which always lead to frequent information conflict and retransmission phenomenon, so that the information cannot be transmitted in the process without delay. In a closed-loop control, via communication network, the controller, actuator, controlled plant and sensors in the network control system are connected to form a loop, whose structure is shown in Fig. 6.2.

How to eliminate the adverse effect for system performance as the result of information transmission delay, communication constraints, packet loss and asynchronous is one of the main problem for network control system. The relevant research at this stage can be divided into the following two categories:

(1) Designing a specific communication protocols, which will be not affected by the negative factors, such a transmission delay and packet loss; for example, adopting a variety of congestion control algorithms.

Fig. 6.2   Structure of closed-loop networked control system

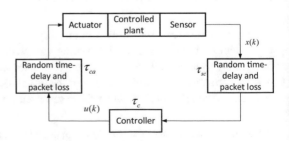

(2) Designing a reasonable control strategy to solve the mentioned problems considering the objective existence of the network conditions as a given environment.

Currently, the main network control system modeling are summarized below [54–63, 122–133]: Ref. [54] studies modeling and controller design method for the system which include the short time delay, multi-packet, multi-input multi-output, long delay and packet loss. Reference [122] gives that the modeling methods of satellite networks based on random model and queuing theory. Reference [123] focuses on the short delay (single packet, multi-packet, packet loss), Long delay (clock-driven, event-driven), multi-input multi-output long-delay system modeling. Reference [124] completes modeling and analysis for the network control system which has time-varying transmission delay. It explores the robust stability problem based on discrete Lyapunov stability theory and matrix inequality theory. About the packet loss problem, the stability equivalence theorem between this kind of switched system model and pulse system is given, which shows that the problem of network control system can be transformed into a problem of pulse system. Reference [123] makes the network control system with long delay and packet loss be built into a Markov jump system, and the network control system with over-one-period network delay is built into a discrete time model with structural uncertainty. In [125], the multi-model theory is combined with the network control system, and the delay is mapped into the index of the construction model; the model is a Markov chain.

In this section, a stochastic robust design method for networked control system based on multi-structure Markov jump model is proposed, combined with stochastic robust analysis and design (SRAD) and background of MAF support network. Furthermore, it gives analysis of Member flight control system modeling and stability, taking into account the stability of the closed-loop system and adjust time, response time, peak time, delay time, overshoot and other dynamic performance requirements. It is a robust flight controller which can comprehensively measure system stability and dynamic performance indicators.

For the actual distributed network control system, the time delay can be divided into two kinds according to the control function: one is the network communication transmission delay in the forward channel between controllers and actuators, as well as the transmission delay in the feedback channel between sensors and the controllers. The other is to send the real time measured signal data to actuators, which means actuators integrated control and directly drive the controlled plant. Without an independent controller, there is no transmission delay for the control data in the forward channel between actuators with control function inside, only the delay in the feedback channel exists between sensors and actuators. The former is called a "network control system with dual-channel delay", while the latter is called a "distributed network control system with feedback channel delay". Here we mainly focus on the latter one.

6.1.3 Network Control System Based on the Markov Jump Models

Assume that the state equation of the controlled plant is:

$$\begin{cases} \dot{x} = Ax(t) + Bu(t) \\ y(t) = Cx(t) \end{cases} \tag{6.1}$$

Here, $x(t) \in R^n, u(t) \in R^k, y(t) \in R^m$ represent respectively the state of the controlled plant and control input and output vector, A, B, C are constant matrixes of the corresponding dimension.

Assume that sensor nodes are time-driven to perform periodic sampling with the period T on the controlled plant, subsequently transmit the collected data from controlled plant to the controller receiving buffer through the communication network. Controller nodes use event-driven mode, as long as the sensor information reaches the controller side, the controller starts to calculate the corresponding controlled variable, and then immediately sends result to receiving buffer of actuator side. The actuator nodes uses the time-driven mode with the same period T as sensor nodes for periodic synchronization sampling. The actuator read the control signal from receive buffer every period time T, which ensure the use of latest control signal. Under such control mode, an example timing diagram for information transmission in closed-loop networked control systems is shown in Fig. 6.3.

In the figure, $u_{x(k-n)}$ represents the control signal calculated according to sensor information $x(k - n)$ at time $(k - n)T$, where $k, n \in Z$. The time at which the controller calculates is marked as t_1, t_2, t_3 and t_4. It is assumed that the control signal calculated at time kT has resulted in packet loss due to network reasons. From the figure, we can see that the control signal used by the actuator at the time kT is $u_{x(k-2)}$. As a result of the packet loss and the existing network long delay, there is no new updated control signal arriving in this cycle from time kT to time $(k + 1)T$, so that at time $(k + 1)T$ the actuator nodes could only still use the control signal $u_{x(k-2)}$ as same as the last sample circle.

The transmission delay d_x at time kT is defined as: suppose that the controlled variable on the actuator side is $u_{x(k-n)}$ at time kT, let $d_k = n$. $\{d_k | 0 \leq d_k \leq d_m < \infty$,

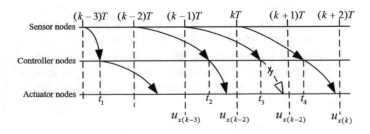

Fig. 6.3 An example sequence diagram of information transfer in system

$k = 0, 1. \ldots\}$ is a bounded random integer sequence, where d_m represents the maximum transmission delay. And the random delay sequence $\{d_k | 0 \le d_k \le d_m < \infty, k = 0, 1 \ldots\}$ can be modeled as a finite state Markov process. Set $S_d = \{0, 1 \cdots, d_m\}$ as its state set, the transition probability from the state i to j can be expressed as

$$p_{ij} = Prob\{d_{k+1} = j | d_k = i\} \tag{6.2}$$

Here, $i, j \in S_d$

It can be seen from Fig. 6.3 that the delay d_x in each step (each circle) is increased by up to one. For example, in the case of Fig. 6.3, $d_k = 2$, since the actuator always adopt the latest control signal, it is impossible to use a control signal which is older than $u_{x(k-2)}$ at the time $(k+1)T$, so the value of d_{k+1} is at most three. In the meanwhile, it can be seen that, in an ideal case, the delay can be decreased to 0 at each step. It means that time delay will increase/decrease when packet is lost/the old signal is ignored if new signal update. Thereby obtaining a delay state transition probability matrix is shown as follow:

$$P_S = \begin{bmatrix} p_{00} & p_{01} & 0 & 0 & \cdots & 0 \\ p_{10} & p_{11} & p_{12} & 0 & \cdots & 0 \\ \vdots & \vdots & \vdots & \ddots & \ddots & \vdots \\ \vdots & \vdots & \vdots & \ddots & \ddots & 0 \\ \vdots & \vdots & \vdots & \vdots & \ddots & p_{d_m-1 d_m} \\ p_{d_m 0} & p_{d_m 1} & p_{d_m 2} & \cdots & \cdots & p_{d_m d_m} \end{bmatrix} \tag{6.3}$$

Here, $0 \le p_{ij} \le 1, \sum_{j=0}^{d_m} p_{ij} = 1 \quad i, j = 0, \ldots, d_m$

Assuming that the probability of transmission delay d_k is equal to i at time kT can be expressed as $\pi_i(k) = Prob\{d_k = i\}$. We introduce a Markov state probability distribution as $\pi(k) = [\pi_0(k) \quad \pi_1(k) \cdots \pi_{d_m}(k)]$, then we can get

$$\pi(k+1) = \pi(k)P_S \tag{6.4}$$

The discrete time model of the generalized controlled plant, including the network, which is discretized by (6.1)

$$x(k+1) = \Phi x(k) + \Gamma u(k) \tag{6.5}$$

The state feedback control law is

$$u(k) = K_{d_k} x(k - d_k) \tag{6.6}$$

Here, $\Phi = e^{AT}, \Gamma = \int_0^T e^{At} dt \cdot B, d_k \in S_d$. K_{d_k} is the feedback control law gain associated with the delay state d_k.

Construct the augmented vector $\tilde{x}(k) = \left[x(k)^T \ x(k-1)^T \cdots x(k-d_m)^T \right]^T$ Eq. (6.6) can be transformed into

$$\tilde{x}(k+1) = \left(\tilde{A} + \tilde{B} K_{d_k} \tilde{C}_{d_k} \right) \tilde{x}(k) \tag{6.7}$$

$$\tilde{A} = \begin{bmatrix} \Phi & 0 & \cdots & 0 & 0 \\ I & 0 & \cdots & 0 & 0 \\ 0 & I & 0 & \cdots & 0 \\ \vdots & \ddots & \ddots & \ddots & \vdots \\ 0 & 0 & \cdots & I & 0 \end{bmatrix}, \tilde{B} = \begin{bmatrix} \Gamma \\ 0 \\ \vdots \\ 0 \end{bmatrix}, \tilde{C}_{d_k} = \begin{bmatrix} 0 & \cdots & 0 & I & 0 & \cdots & 0 \end{bmatrix}$$

where $d_k \in S_d, \tilde{x}(k) \in R^{(d_m+1)n}$, I is the unit matrix with the same dimension as Φ, $\tilde{C}_{d_k}$ consists of $d_m + 1$ sub-blocks, and the d_kth block is the array I.

6.1.4   Stability of Network Control System Based on the Markov Jump Models

In Ref. [126], the stochastic delay is modeled as a Markov process with a finite number of states. Based on this, a network control system supported by multi-structure Markov jump model is established, and the stability of system is analyzed by a proposed $V - K$ iterative method. But it only study the system stability without considering the optimal design of the system performance. In this chapter, it gives the stability analysis of the networked control system based on the multi-structure Markov jump model, which is with the background of DTDMA network protocol.

According to the conclusion of Ref. [126], (1) it is known that the system (6.5) is stochastic stable, when there is a finite value $N(x_0, s_0) > 0$ only associated with $x_0, s_0 \in S_d$ for any initial state, which satisfy that $\sum\limits_{k=0}^{\infty} E\left[\|x(k)\|^2 | x_0, s_0 \right] < N(x_0, s_0)$.

(2) The system (6.5) is mean-square stability stable, when $\lim\limits_{k \to \infty} E\left[\|x(k)\|^2 | x_0, s_0 \right] = 0$ is satisfied at any initial state $x_0, s_0 \in S_d$. (3) The system (6.5) is exponentially stable, when there is a constant attenuation rate $\alpha > 1$ for any initial state $x_0, s_0 \in S_d$, which satisfy $\lim\limits_{k \to \infty} \alpha^k E\left[\|x(k)\|^2 | x_0, s_0 \right] = 0$. We can obtain the following stability theorem of the network control system based on the multi-structure Markov jump model [79, 127]:

Theorem 6.1 *To prove that System* (6.5) *has mean square stability is equivalent to the existence of a symmetric positive definite matrix* $F_0, F_1, \cdots, F_{d_m}$ *which satisfy the following condition:*

$$\sum_{i=0}^{d_m} p_{ji} A_i^T F_i A_i < F_j, j = 0, \ldots, d_m \tag{6.8}$$

Here, $A_i = \tilde{A} + \tilde{B} K_i \tilde{C}_i$

The lower bound of the decay rate α is required to satisfy the modified inequality (6.8) after substituting F_j with βF_j, where $\beta = 1/\alpha$

Using the conclusion of Theorem 6.1, to make the system (6.7) has the decay rate $\alpha = 1/\beta$, the conditional (6.8) turns to be

$$\sum_{i=0}^{d_m} p_{ji} A_i^T F_i A_i < \beta F_j, j = 0, \ldots, d_m \tag{6.9}$$

According to the Schur complement, Eq. (6.9) can be written as follows:

$$\begin{bmatrix} \beta F_i & A_0^T F_0 & \cdots & A_{d_m}^T F_{d_m} \\ F_0 A_0 & p_{j0}^{-1} F & \cdots & 0 \\ \vdots & \vdots & \ddots & \vdots \\ F_{d_m} A_{d_m} & 0 & \cdots & p_{jd}^{-1} F_{d_m} \end{bmatrix} > 0, j = 0, \ldots, d_m \tag{6.10}$$

The above equation is a quadratic matrix inequality about variables β, K_i, F_i, where $i \in S_d$. When β, K_i are given, it turns to be a linear matrix inequality about F_i. Similarly, it is the linear matrix inequality about β, K_i when F_i is given, and LMI (Linear Matrix Inequality) method is used to solve the minimum β and corresponding K_i which make all the $d_m + 1$ inequalities in (6.10) are satisfied, so that the lower bound of the decay rate $\alpha = 1/\beta$ can be obtained as well as controller gain under various delay conditions. However, the above method needs the assigned state transition probability matrix P_S which is difficult to get in the practical application. In addition, the system is often non-linear and contains uncertain parameters, which make it harder to use in practice. For this purpose, we explore a stochastic robust analysis and design method (SRAD) to optimize the system design.

6.1.5   Robustness Design Methods for MFCS

At present, the common network control system design methods are as the following:

(1) Augmented state method. Reference [128] proposes augmented state model for deterministic discrete time network control systems aiming at networks with periodic delay characteristics. By introducing the delayed input and output variables to augmented state and establishing a new state space equation, the

system is described and solved with the original state equation simultaneously. However, due to the use of augmented state, system complexity greatly increases, and control system computing and storage space are highly required for implementing.

(2) State estimation method. Reference [123, 125] suggests a delay compensation method based on deterministic predictor. Using the past measurement signal, system's past state reconstructed by observer, combined with advance forecast system state by predictor to generate a control signal for compensating the adverse effects of the delay. However, since the performance of observer and predictor depend on the accuracy of model building, which means an accurate model for plant is required.

(3) Adopt the state estimation to reduce communication delay or improve controller performance. The cause of existing communication delay is largely due to communication overload. Reducing network load can significantly reduce the delay in the networked control system. In [131], state estimation is used to decrease the traffic which will lead to communication delay reduction. Reference [125] adopt a predictor to estimate the plant state between two successive samples and then generate a control signal to improve the performance of the networked control system.

(4) Optimal control method. In [122], a networked control system with stochastic delay characteristics is described by a linear stochastic system model, and the influence of communication delay in a networked control system is transformed into a linear quadratic Gaussian (LQG) problem. Simulation results show that this method is better than the method based on deterministic predictor, though it requires a lot of storage space to retain past information. In addition, this method assumes that the sum of communication delays is less than one sampling period, which may require a relatively large sampling period in practical applications and may not apply in systems that require fast response.

In summary, the current study considers system random delay, which is exclusive of the noise and (external) nonlinear interference that cannot be ignored in the real system. So that a network control system control method that can overcome the random delay, system noise and external perturbation is worthy of much more study.

This section will base on the optimal controller structure mentioned in (4) above, and gives a stability analysis and controller design of network control system with multi-structure Markov jump model, solved by the Stochastic Robust Analysis and Design (SRAD) method [1]. With the optimal quadratic controller as the reference structure, genetic algorithm is used to optimize the design, considering fully of the closed-loop system stability affected by uncertain parameters and its dynamic performance including response time, peak time, delay time and overshoot etc. The robust flight controller can be obtained for taking into account both stability and dynamic performance of the system, whose specific design method can be found in [1, 76, 77].

1. System Parameters Uncertainty and Features Point Selection

According to the mathematical model of missile overload control, the uncertain vectors such as mass, inertia, thrust of the missile, aerodynamic parameters, network delay and packet loss are defined as $v = (v_1 \ v_2 \ldots v_{29})^T$. The uncertain vectors can be divided into 22 missile parameters as $(v_1 \ v_2 \ldots v_{22})^T$, and seven parameters about network delay and packet loss represented as $(v_{23} \ldots v_{29})^T$. In general, each component of $(v_1 \ v_2 \ldots v_{22})^T$ can be assumed being subject to a normal distribution with a mean of 1.0 and a standard deviation of 0.2. Each component in $(v_{23} \ldots v_{29})^T$ is subject to uniform distribution. Considering engineering realization of system, the PID controller structure is selected for stochastic robust design. According to the missile flight profile, we usually select a feature point for design in the level flight segment with fixed height [1, 76].

2. Stochastic Robust Analysis and Design (SRAD) of Networked Flight Controllers

(1) Considering the absence of delay in model (6.5) and (6.6), select the quadratic index function as

$$J = \frac{1}{2} \sum_{k=0}^{\infty} x^T(k)Qx(k) + u^T(k)Ru(k)$$

(2) Select the stochastic robust controller based on PID structure, it can get that $u(k) = -k_{pc}Kx(k)$. The designed vector is $\Omega = [Q_i \quad R_i \quad k_{pc}]$, where Q_i and R_i is respectively the diagonal element of Q, R, and k_{pc} is the robust control law.

(3) For each specific value Ω_i of the design vector Ω, the linear control law gain K can be solved according to the controlled plant model (6.5) which ignore the delay, $K_i = K$, $i = 0, 1, \ldots, d_m$.

(4) Let $\beta = 1$, and use step (3) to obtain $K_i (i = 0, 1, \ldots, d_m)$, and then solve F_j $(j = 0, 1, \ldots, d_m)$ which satisfy Eq. (6.10).

(5) Getting the better solution of β and K_i, i.e. correcting K_i. To be specific, we solve the optimization problem of Eq. (6.10) with the minimum β by F_j $(j = 0, 1, \ldots, d_m)$ resulted from step (4), and then obtain the corresponding β and K_i.

(6) When the missile model is perturbed according to the set distribution law, select the uncertain parameter vector as $v = (v_1 \quad v_2 \quad \ldots \quad v_{27})^T$.

(7) When the initial network delay state and the transition probability matrix elements are perturbed, i.e. the matrix (Φ, Γ) in the controlled plant (6.5), the initial delay state distribution $\pi(0)$ and the delay state transition probability matrix P_S make a change according to the set distribution law. We can set a group of uncertain parameter vector $v = (v_1 \quad v_2 \quad \ldots \quad v_{29})^T$, and the initial delay state distribution is $\pi(0) = [\pi_1 \quad \pi_2 \quad \pi_3] + [\zeta_1 \quad \zeta_2 \quad \zeta_3]$, while the

delay state transition matrix is $P_S = P_{S0} + \begin{bmatrix} \Delta_{11} & \Delta_{12} & 0 \\ \Delta_{21} & \Delta_{22} & \Delta_{23} \\ \Delta_{31} & \Delta_{32} & \Delta_{33} \end{bmatrix}$; where ζ_i follow uniformly distributed in the interval of $[-0.2, 0.2]$, $i = 1, 2, 3$ and $\sum_{i=1}^{3} \zeta_i = 0$. The non-zero elements in Δ_{ij} obey uniformly distributed over the interval $[-0.1, 0.1]$, $i, j = 1, 2, 3$ and $\sum_{i=1}^{3} \Delta_{ij} = 0$.

(8) When the model parameters and the network delay are perturbed, i.e. the state matrix in the controlled plant system (6.5) is perturbed, we can get the simulation results of stability and performance index including response stability, setting time, rising time, delay time, overshoot and control signal clipping. And then the weighted sum of the probability which is below the index requirement can be calculated as $J(\Omega) = \sum_{j=1}^{M} w_j \hat{p}_j^2$, where probability $\hat{p}_j$ is obtained by Monte Carlo simulation. Adopt Genetic Algorithm to solve an optimal vector $\Omega = \Omega_{opt}$ for satisfying $J(\Omega_{opt}) = Min(J(\Omega))$.

(9) According to Ω_{opt} and the controlled plant model (6.5) to solve a robust control law K, complete the design. Need to further explain here:

(1) For obtaining numerical solution, a sufficient small positive number ξ can be used to replace zero element in the transition probability matrix P_S. In the meantime, the nonzero element in the same row subtracted with ξ need to be ensure to satisfy the conditional expression (6.4).

(2) According to Theorem 6.1, system (6.7) under the mean square stability conditions, with any initial state $x_0, s_0 \in S_d$, $\lim_{k \to \infty} \|x(k)\| = 0$ is true for the probability of one, i.e. so-called absolutely stability. The stability of the system under the perturbation of various parameters can be determined by whether or not the Eq. (6.10) can be established, which need to solve K_i in the case of $\beta = 1$ in the step (5). A variety of robust performance will be determined by repeated Monte Carlo estimation method simulation based on the disturbed parameters including $\pi(0)$ and P_S.

6.2   The Example of Robustness Design for MFCS

This section completes the design of control law with longitudinal attitude controller and overload controller for a certain type of missile, and analyzes the difference of two control methods in rapidity through the comparison between the simulation results; and then designs a networked robust controller, the validity of the design method is identified by comparing with the conventional design method of controller.

6.2.1 Overlord Controller Design of the Missile

1. The model of overlord control

The control system of the missile should not only consider the non-linear factors such as the kinematics coupling, the aerodynamic coupling and the control coupling of the missile, but also make the state variable of the control system model easy to measure, so that the robustness design and implementation of the guidance control system are more simple and feasible, here using the missile overload control mathematical model [1].

(1) Nonlinear motion model of longitudinal channel

When the benchmark motion is selected as the horizontal non-slip flight, $\phi = 0, \beta = 0$, then the vertical motion state vector is selected as $[\omega_z \quad h \quad n_y \quad \varepsilon \quad \eta \quad \delta_z]$, and the control vector is $[\delta_{zc}]$, the state vector components are in turn the component on the z axis of the missile coordinate system of the rotational angular velocity, the missile coordinate system relative to the ground coordinates, the height, the normal overload on the y axis of missile coordinate system, $\varepsilon = g \int_0^t \left[n_y(\tau) - n_y^* \right] dt$, $\eta = \int_0^t [h(\tau) - h^*] dt$ and the elevator angle, where ε, η are the integral compensations which are used to eliminate the steady-state error of the command response, for the command n_y^*, h^*. The vertical model of missile longitudinal overload control is:

$$\dot{\omega}_z = \frac{a_{11}}{a_{14}} n_y + a_{12} \omega_z + \left(a_{13} - \frac{a_{11}a_{15}}{a_{14}} \right) \delta_z$$

$$\dot{h} = \varepsilon$$

$$\dot{n}_y = a_{14} \omega_z - a_1 n_y + \left(a_1 a_{15} - a_2 a_{14} - \frac{a_{15}}{T_\delta} \right) \delta_z + \frac{k_\delta}{T_\delta} a_{15} \delta_{zc}$$

$$\dot{\varepsilon} = g \left(n_y - n_y^* \right)$$

$$\dot{\eta} = h - h^*$$

$$\dot{\delta}_z = -\frac{1}{T_\delta} \delta_z + \frac{k_\delta}{T_\delta} \delta_{zc}$$

And linearize it as follows:

$$\dot{x} = A_p x + B_p u + \Gamma_p E_p$$

where A_p, B_p, Γ_p the matrix of the corresponding dimension, and the element is comes from the aerodynamic data interpolation of a certain type of flying missile.

(2) Horizontal lateral overload control mathematical model

Select the lateral vector of the lateral motion as $[\,n_z\quad \omega_y\quad \xi\quad \delta_y\quad \omega_x\quad \phi\quad \lambda\quad \delta_x\,]$, and the control vector is $[\,\delta_{xc}\quad \delta_{yc}\,]$, where δ_x is aileron deflection, δ_y is rudder angle, and the command is n_z^*, ϕ^*, and the integral compensations are used to eliminate the steady-state error of the command response, so let $\xi = \int_0^t \left[n_z(\tau) - n_z^*\right]$ and $\lambda = \int_0^t [\phi(\tau) - \phi^*]d\tau$. Then the motion state equation of yaw channel is:

$$\dot{n}_z = a_3 n_z + a_{16}\omega_y + \left(a_4 a_{16} - a_3 a_{17} - \frac{a_{17}}{T_\delta}\right)\delta_y + \frac{k_\delta}{T_\delta}a_{17}\delta_{yc}$$

$$\dot{\omega}_y = \frac{a_8}{a_{16}}n_z + a_9\omega_y + \left(a_{10} - \frac{a_8 a_{17}}{a_{16}}\right)\delta_y$$

$$\dot{\xi} = n_z - n_z^*$$

$$\dot{\delta}_y = -\frac{1}{T_\delta}\delta_y + \frac{k_\delta}{T_\delta}\delta_{yc}$$

And the motion equation of the roll channel is:

$$\dot{\omega}_x = a_6\omega_x + a_7\delta_x$$

$$\dot{\phi} = \omega_x$$

$$\dot{\lambda} = \phi - \phi^*$$

$$\dot{\delta}_x = -\frac{1}{T_\delta}\delta_x + \frac{k_\delta}{T_\delta}\delta_{xc}$$

where the parameters are as follows [1]:

$$a_1 = \frac{1}{mV}\left(qSC_y^\alpha + P\right), \; a_2 = \frac{1}{mV}qSC_y^{\delta_z}, \; a_3 = \frac{1}{mV}\left(qSC_z^\beta - P\right), \; a_4 = \frac{1}{mV}\left(qSC_z^{\delta_y}\right),$$

$$a_5 = \frac{qSL}{J_x}m_x^\beta, \; a_6 = \frac{qSL^2}{J_x V}m_x^{\bar{\omega}x}, \; a_7 = \frac{qSL}{J_x}m_x^{\delta_x}, \; a_8 = \frac{qSL}{J_y}m_y^\beta, \; a_9 = \frac{qSL^2}{J_y V}m_y^{\bar{\omega}_y},$$

$$a_{10} = \frac{qSL}{J_y}m_y^{\delta_y}, \; a_{11} = \frac{qSL}{J_z}m_z^\alpha, \; a_{12} = \frac{qSL^2}{J_z V}m_z^{\bar{\omega}_z}, \; a_{13} = \frac{qSL}{J_z}m_z^{\delta_z},$$

$$a_{14} = \frac{qS}{mg}\left(C_{x0} + C_{xi} + C_y^\alpha\right), \; a_{15} = \frac{qS}{mg}C_y^{\delta_z}, \; a_{16} = \frac{qS}{mg}\left(C_Z^\beta - C_{x0} - C_{xi}\right), \; a_{17} = \frac{qS}{mg}C_z^{\delta_y},$$

$$a_{18} = T_\delta, \; a_{19} = k_\delta$$

2. Overload controller design

The structure of the controller is:

$$\delta_{zc} = k_{\omega_z}\omega_z + k_h(h - h^*) + k_{n_y}\left(n_y - n_y^*\right) + + k_{\varepsilon}g \int_0^t \left[n_y(\tau) - n_y^*\right]d\tau$$

$$+ k_{\eta} \int_0^t [h(\tau) - h^*]d\tau + k_{\delta_z}\delta_z$$

$$\delta_{yc} = k_{n_z}\left(n_z - n_z^*\right) + k_{\omega_y} + k_{\xi} \int_0^t \left[n_z(\tau) - n_z^*\right]d\tau + k_{\delta_y}\delta_y$$

$$\delta_{xc} = k_{\omega_x}\omega_x + k_{\phi}(\phi - \phi^*) + k_{\mu} \int_0^t [\phi(\tau) - \phi^*]d\tau + k_{\delta_x}\delta_x$$

The three-channel controller parameters K_z, K_y, K_x based on LQR structure are designed by SRAD method as follows:

$$K_z = \begin{bmatrix} k_{\omega_z} & k_h & k_{n_y} & k_{\varepsilon} & k_{\eta} & k_{\delta_z} \end{bmatrix} = \begin{bmatrix} -21.081 & -0.126 & -2.961 & 0.116 & -0.403 & -0.002 \end{bmatrix}$$
$$K_y = \begin{bmatrix} k_{n_z} & k_{\omega_y} & k_{\xi} & k_{\delta_y} \end{bmatrix} = \begin{bmatrix} 1.937 & -14.185 & 23.622 & 0.0674 \end{bmatrix}$$
$$K_x = \begin{bmatrix} k_{\omega_x} & k_{\phi} & k_{\lambda} & k_{\delta_x} \end{bmatrix} = \begin{bmatrix} -2.170 & -6.502 & -0.000 & 0.021 \end{bmatrix}$$

3. Overload controller performance analysis

Take the longitudinal channel as an example to investigate the response of the missile to the height command ($H^* = 50$ m). The simulation results are shown in Fig. 6.4.

It can be seen through the simulation results, the flight height converged into the steady state in about 5 s with the use of overload control scheme, while it need for 8 s or so into the steady state with the use of attitude control scheme, so the overload controller is faster. In the attitude control scheme the flying state of the missile is controlled by changing the attitude angle, that is, the deflection angle is generated by the rudder surface so that the aerodynamic force and the aerodynamic moment acting on the missile are changed to change the attitude movement of the missile.

In other words, the effect of the steering gear deflection is aerodynamic and aerodynamic moment, the second effect is the attitude change. Need to pay attention to, the sooner it responses, the faster rudder rotates, and it will be put forward higher requirements for the performance of the steering gear. In this case, the stochastic robust design method is used, and the steering surface deflection is used as a performance requirement in the cost function, then the appropriate feedback

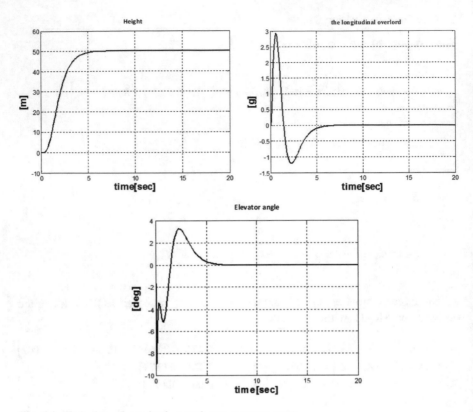

Fig. 6.4 Pitch channel overload control state response curve

coefficient is designed under the constraint of the steering gear. The design method of the flight control system for the missile members in the cruise missiles autonomous formation can be found in [1, 75, 76].

6.2.2 Networked Flight Controller Design of the Missile

This section designs a member flight control system based on the longitudinal overload model of a certain type of winged missile.

1. Network delay setting

Assume that the sampling period of the sensor is $T = 0.05$ s, and the minimum transmission delay of the network is one cycle and the maximum transmission delay is three cycles, namely $S_d = \{1, 2, 3\}$. Uncertainty perturbation models for other network parameters are as follows:

The initial delay state is:

$$\pi(0) = [0.5 \quad 0.3 \quad 0.2] + [\zeta_1 \quad \zeta_2 \quad \zeta_3] \qquad (6.11)$$

The delay state transition matrix is:

$$P_S = \begin{bmatrix} 0.49 & 0.49 & 0.02 \\ 0.3 & 0.6 & 0.1 \\ 0.3 & 0.6 & 0.1 \end{bmatrix} + \begin{bmatrix} \Delta_{11} & \Delta_{12} & 0 \\ \Delta_{21} & \Delta_{22} & \Delta_{23} \\ \Delta_{31} & \Delta_{32} & \Delta_{33} \end{bmatrix} \qquad (6.12)$$

where ζ_i in the interval $[-0.2, \ 0.2]$ is on the uniform distribution and $\sum_{i=1}^{3} \zeta_i = 0 (i = 1, 2, 3)$. Nonzero elements in Δ_{ij} are uniformly distributed over the interval $[-0.1, \ 0.1]$ and $\sum_{j=1}^{3} \Delta_{ij} = 0 (i, j = 1, 2, 3)$. Network delay distribution is shown in Fig. 6.5.

2. Stochastic Robust Analysis of Networked Flight Controllers

Firstly, the simulation results using the classic LQR controller are shown in Fig. 6.6. It can be seen from this figure that although the flight height state can converge at this time, but the elevator is jittery and the normal overload is outside the allowable range, the conventional controller performance is greatly affected under this network delay condition.

The parameters of the networked flight controller obtained by SRAD method are given below. The missile longitudinal overload control model is selected and the mass, inertia, thrust and the aerodynamic parameters of the missile and the uncertainty vector of the network delay and loss are set according to (6.11), (6.12).

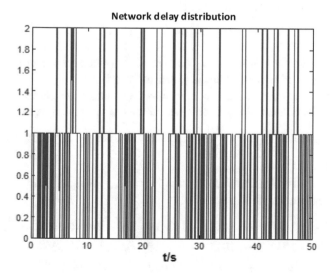

Fig. 6.5 Network delay distribution

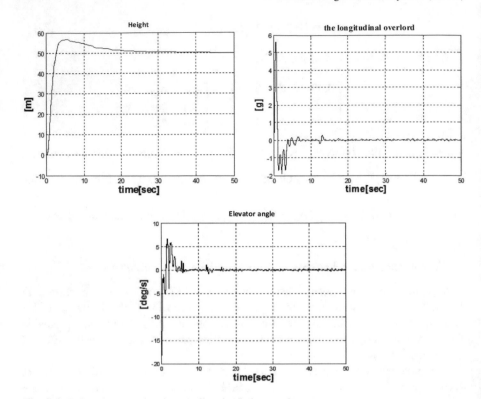

Fig. 6.6 Delayed conventional controller simulation results

The full state measurable overload control linearization mathematical model [1] of the missile pitch channel is given as:

$$\dot{x} = A_p x + B_p u + \Gamma_p E_p \qquad (6.13)$$

where the state vector is $\begin{bmatrix} \omega_z & h & n_y & \varepsilon & \eta & \delta_z \end{bmatrix}$, the control vector is $u = \delta_{zc}$, and the specific form of the controller is given in Sect. 6.2.1.

Taking the longitudinal channel design as an example, we choose the weighted square of the 18 non-satisfying design requirement probabilities as stochastic robust cost function, namely $J(\Omega) = \sum_{j=1}^{M} w_j \hat{p}_j^2$ and $M = 18$, design requirements and weight values see Ref. [1].

The designed stochastic robust gain is $K_i = \begin{bmatrix} k_{\omega_z} & k_h & k_{n_y} & k_\varepsilon & k_\eta & k_{\delta_z} \end{bmatrix}$, and the specific parameters are as follows:

$$K_1 = \begin{bmatrix} -12.1489 & -0.1753 & -0.9314 & 0.0410 & -0.5404 & -0.0037 \end{bmatrix}$$
$$K_2 = \begin{bmatrix} -12.1067 & -0.1811 & -0.9418 & 0.0431 & -0.5397 & -0.0044 \end{bmatrix}$$
$$K_3 = \begin{bmatrix} -10.2047 & -0.3492 & -1.5650 & 0.2061 & -0.5057 & -0.0105 \end{bmatrix}$$

The simulation results of stochastic robust controllers under delay conditions are shown in Fig. 6.7. Compared with the simulation results in Fig. 6.6, it can be seen that the state response is stable when the stochastic controller is used, the elevator is not violently jitter, and the control quality is improved obviously. However, there are some steady-state errors and the increase of response time, which indicates that the delay still has some effect on the control system. The effect of network delay on the performance of the control system is further analyzed in the following nonlinear closed-loop system simulation experiment.

3. The nonlinear closed-loop system simulation experiment

The experimental trajectory of the simulation experiment is set as follows: the missile starts climbing from a height of 1500 m, and then keeps at a height of 2000 m, after 100 s starts diving down, and finally settles down to 5 m before 250 s. The Mach number Ma = 0.65–0.8, and the maximum allowable overload of 3g using virtual target navigation.

Limited to space, here only part of simulation results is given. Figures 6.8, 6.9, 6.10, 6.11, 6.12 and 6.13 show a nonlinear closed-loop system simulation experiment using a stochastic robust controller. The simulation results show that in the

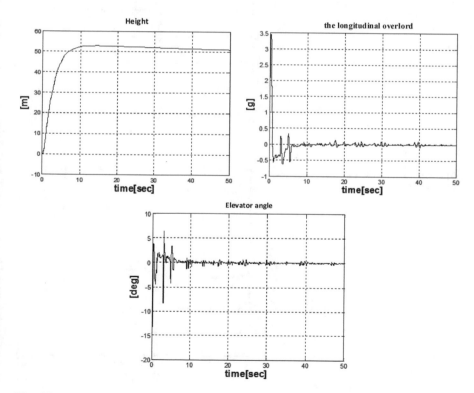

Fig. 6.7 The simulation results of stochastic robust controllers under delay conditions

Fig. 6.8 Longitudinal
trajectory curve

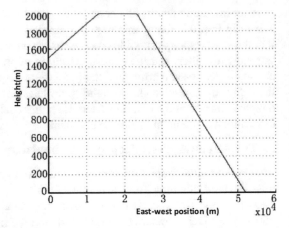

Fig. 6.9 Longitudinal
overload response curve

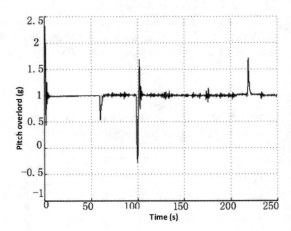

Fig. 6.10 Elevator angle
response curve

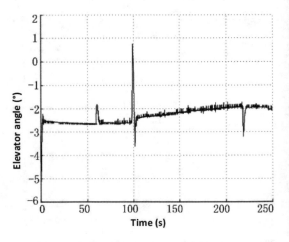

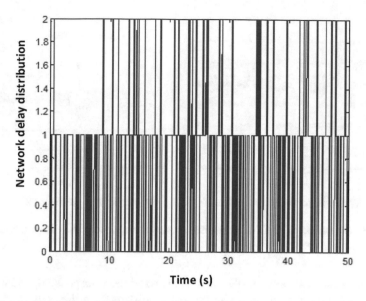

Fig. 6.11 The initial delay distribution of the network

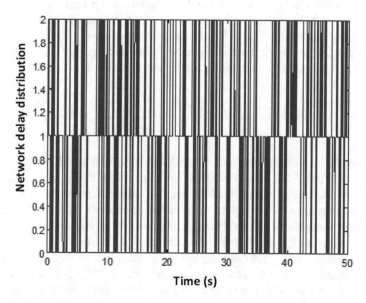

Fig. 6.12 Delay distribution after the change of network state

case of delay and packet loss, the stochastic robust controller and the traditional controller designed based on experience both can make the missile fly safely and accurately according to the program trajectory and meet limit requirements of the rudder angle and overload. However, the vertical overload response curve obtained

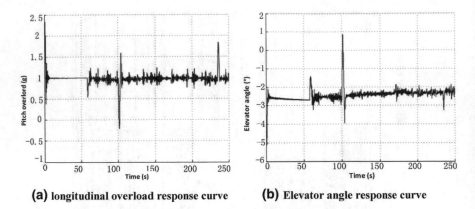

(a) longitudinal overload response curve   **(b)** Elevator angle response curve

Fig. 6.13 The simulation result of the stochastic robust controller after the change of the network parameters

by the stochastic robust controller is smoother and more responsive, which shows that the stochastic robust controller has better robustness to the network transmission delay and packet loss. Similarly, comparing the graphs in Figs. 6.11 and 6.12, it shows that the stochastic robust flight controller can be used to track the input more quickly and the system performance is better than that of quadratic controller with the weighted matrix Q and R set based on experience.

If the initial delay distribution of the network is shown in Fig. 6.11, the values of the initial delay state and the transfer matrix are changed as follows:

$$\pi(0) = [0.1, 0.5, 0.4], P_s = \begin{bmatrix} 0.3 & 0.7 & 0 \\ 0.3 & 0.5 & 0.2 \\ 0.3 & 0.4 & 0.3 \end{bmatrix} \tag{6.14}$$

The simulation result of network delay distribution is shown in Fig. 6.12. It can be seen from Fig. 6.12, the probability of long delay and packet loss of data transmission increases. Figure 6.13 shows the response curve of the stochastic robust controller after the change of the network parameters. The simulation results show that the controller designed by LQR will cause the system to be unstable if the empirical values of Q and R arrays are still used. However, the same stochastic robust controller is used, although the system performance is slightly worse due to network factors, it can still meet the system requirements. It is shown that the stochastic robust controller has the robustness to suppress the delay of the uncertain network.

The members of the missiles formation execute cooperative flight based on the support of network, and the performance of the member flight control system directly affects the ability of the formation, that is, by directly affecting the sensing and control error component $d_{ti}(t)$ which is closely related to the safety distance $d_{si}(t)$, thus affecting the intensive formation ability of the missile formation.

Therefore, this chapter establishes a networked member flight control system model considering the characteristics and health status (information packet loss, delay and robustness, etc.) inherent in the support network itself, and applies the stochastic robust analysis and design method (SRAD) to design a the networked stochastic robust flight controller of the missile members. At last, the stability of the member flight control system is analyzed by simulation experiment and the rationality and the validity of the stochastic robustness analysis and design method (SRAD) are identified.

Chapter 7
Support Networks System (SNS) of MAF

7.1 Foreword

According to the analysis of the SNS of MAF in Chaps. 1 and 2, the SNS of MAF reflects the swarm intelligence in a networked way. It can enable every member to assess the influence of its future behavior to the whole formation based only on the present and past information acquired from local interaction in a short time. In formation flight, the maneuver of collision avoidance of local nodes could lead collision in other parts of the formation. The SNS should cognize the characteristic of the formation and assess the influence of chain effect to the formation. To accomplish this assessment, the SNS should enable a single node to cognize its status in the networks and its relationship between other nodes and offer collective cost information to every node in autonomous flight formation control. This is one of the characteristics the SNS different from normal communication networks.

Based on the characteristic of networked autonomous formation, we proposed the SNS and networks support protocols which are the extension and supplement for the traditional communication networks and networks communication protocols. The SNS which is the tie between each member and the virtual part attached to every member playing a vital role in the coordination and cooperation of the formation.

The effect of cooperative operations depends on the ability of information exchange between missile weapon systems to a large extent. The SNS of MAF need to complete the sharing of information under high dynamic conditions with large capacity and high update rate. For the medium-sized missile formation, the communication rate of the SNS should be above the level of 1 Mbps at least, the update rate should be above 1 Hz at least and dynamic autonomous networking capabilities are needed to quickly establish a dynamic network. Due to the uncertainty of the missile formation dynamics and the scale of formation, the network structure will change continuously. Therefore, the networking protocol should be able to adapt to this change, and it is not sensitive to the network structure to quickly entry

© National Defense Industry Press, Beijing and Springer Nature Singapore Pte Ltd. 2019
S. Wu, *Cooperative Guidance & Control of Missiles Autonomous Formation*,
https://doi.org/10.1007/978-981-13-0953-3_7

the network at the same time. The SNS of MAF need to consider the phenomenon of local network interruption and multiple nodes out of the network (i.e.) because of distance and environmental interference and other factors in the actual flight environment. On the other hand, multiple nodes can also integrate into a larger network based on the communication protocol (i.e. join the team) to achieve a stable and reliable operation of the system.

7.2 Support Networks System (SNS)

7.2.1 Analysis of Support Networks

The SNS of MAF is the communication links and networks that transmit and exchange formation information between the various missile members according to a certain communication protocol using an automated wireless transceiver device. Ad Hoc is a wireless network communication system with full potential for development. It is composed of a group of wireless mobile nodes with self-organizing, self-adapting and self-healing ability, so that it can be used without any fixed communication facilities supporting the MAF. In this section, we mainly refer to the structure and protocol of wireless ad hoc network (Ad Hoc), and give the design method of SNS and network support protocols which meet the requirements of MAF.

1. **The common problems of communication network at present**

MAF share their own flight status and decision-making information through the SNS. The high-speed sharing and high rate sharing of information is needed for high-dynamic network. The current communication networks have the following problems:

(1) Lack of dynamic networking capabilities. Autonomy and dynamics are the basic attributes of the SNS of MAF. For the autonomy of the SNS, the corresponding distributed algorithm is needed to quickly establish a dynamic network to determine the status of each node in the network. Due to the dynamics of the network, because of its structure and scale will be constantly changing, the network protocol need adapt to this change.

(2) Node leaving and joining management capacity is insufficient. In the missile formation process, there is always exists a network of clustering and integration. Because of distance and environmental interference and other factors caused by local network interruption in the actual flight environment, the formation of multiple nodes leave the network. On the other hand, multiple nodes can also be based on the communication protocol for integration, joining the network to integrate a larger network. The existing communication network does not consider the special operation mode of the network structure, so the

node's leaving and joining response time is longer, and it is difficult to achieve stable and reliable operation of the system.

(3) The reliability and safety of the communication network is insufficient. For the SNS of MAF, it is needed to consider fault-tolerant capacity of the network after some nodes failure. The reliability of the communication network requires that the network protocol is fault tolerant when the network faces unexpected events. It can recover quickly in burst events include loss of nodes in the network, changes in topology, loss of links and loss of synchronization and other functions of the network temporarily failed, which will not cause the SNS collapse. Further requirements for reliability and safety are in the various harsh electromagnetic confrontation environment, which can cause the network failure, such as key nodes in the network is lost, most of the function is lost, can still maintain the existence of the network and for the necessarily safety of the important information.

2. **Analysis on the Demand of SNS of MAF**

1. Networking requirements of SNS

Autonomous and dynamic characteristics are the basic attribute of the SNS of MAF, which put forward some requirements for the organization and the network structure.

(1) The node has the capability of self-networking. Multiple nodes can form a network autonomously in the process of high-speed movement according to a certain agreement, without human intervention and in accordance with the need to autonomously form the cluster head.

(2) The network has the dynamic characteristics, which allows the structure and scale of the network to change at any time, and the nodes can join and leave autonomously.

(3) The number of network nodes is not fixed. The number of nodes should meet the size of the missile formation requirements.

(4) It can achieve hierarchical or peer-to-peer network structure, and nodes can be hierarchical managed.

(5) Multiple networks can merge into a larger network based on a certain agreement.

(6) The network can be decomposed into several smaller networks according to certain agreement.

2. Data transmission requirements

The missile members in the formation share their own flight status and decision-making information through the SNS. High-speed and high-rate sharing for information is needed for SNS under high dynamic conditions, which need to pay attention to the performance of transmission capacity, transmission efficiency and transmission delay and et al.

(1) Each node in the network can share information about the other nodes.
(2) The amount of data that the node needs to send each time should meet the requirements of the cooperative guidance and control system (at least one hundred bytes for each node in medium-sized missile formation).
(3) Each node needs to update the information of the other nodes stored in it in every second (the network update rate for medium-sized formation should be at least above the Hertz level).
(4) Data is highly real-time and transmission delay is small, which should meet the requirements of maximum maneuverability of the missile.

3. Reliability and security requirements

The environment of missile formation flight is changeable, and there are reduction caused by the defensive situation, so the network should have the ability to face the fault, as well as the regeneration capacity in the extreme circumstances, which proposed higher demands of robustness for SNS.

(1) The network function and performance should be sufficient robustness on the node changes including both normal and abnormal situations.
(2) Changes in the network structure or adjustment should not have an impact on the main performance of the network.
(3) Under certain conditions, the data transmission error rate should meet the formation requirements, and information should be basically reliable.
(4) The loss of part of the network function should be able to quickly recover in the allowed time, and the lost link should be quickly rebuilded in the allowed time.
(5) Loss of the main node will not cause the crash of entire network.
(6) A sudden loss of a large number of nodes will not produce network crash.
(7) The ability to sustain the network while supporting a serious error in a network under certain strong interference conditions and to recover quickly after interference is needed.
(8) The network has the security to meet the requirements of the formation tasks.

7.2.2 Compositions of SNS

1. The functional architecture of the system

In this section, we first propose several basic elements of the SNS, which is the key technologies and the main problem of SNS of MAF. Secondly, according to the requirements of the missile formation and the analysis of the previous requirements, it is necessary to construct a set of simulation and verification system which supports simulation, testing and verification. In addition to the integrated simulation of MAF, it can also design the indicators of key technologies and performance of SNS.

The basic elements of the SNS are: transmission equipment, network support protocol and data information standards. Transmission equipment includes a variety of wireless transceiver equipment, providing a physical basis for information transmission. The network support protocol ensures interconnection between multiple transmission devices. The data information standard defines the information content and format of the transmission.

SNS of MAF is mainly composed of channel transmission, network control, message processing, cooperative operation and other functions. The hierarchical structure described in the functional view is shown in Fig. 7.1.

In according to the relationship between the peer layers from low to high, SNS can be divided into the functional levels of waveform processing and transmission channels, communication networks, interactive information between the missiles, combat application processing and other functional levels. Each level supports its upper level, and ultimately complete cooperative operations of SNS.

(1) Function of waveform processing and channel transmission

Safe and reliable signal transmission among missile members of the whole network is the basic function of SNS. This layer includes RF signal processing, digital signal processing, codec and other functions, requiring the ability of anti-interference and anti-multipath fading. Anti-jamming capability is the main consideration of the transmission waveform, which need use the modulation and coding technology, interleaving technology and spread spectrum technology. They can effectively improve the system's anti-jamming capability. In addition, according to the missile tactics and combat methods, considering the impact of the operating frequency on the link loss and the formation of the maximum range of the missile for the design reference, omni-directional antenna is used to ensure the full connectivity between the members of MAF.

(2) Function of network control

SNS is a wireless transmission network of the missiles distributed in a certain battle area. The networking control function of this layer needs research special networking protocols applicable to the SNS according to the requirements of the structure of the missile members and the update rate of information. The networking protocol solves the problems of fast networking, the node joining/leaving network, high message update rate requirements and other issues.

In terms of the access control of the members of SNS, the network node addressing mode should ensure that each node gets a unique platform identification number in the whole network. The minimum independent units participating in the operation of the network are the participating nodes. Different nodes in the network can have different identities (such as leader, follower, leading nodes and so on) according to different tasks. The sending capacity of different nodes is different. The access time is ensured of the missile members with different transmission needs.

In the topology control of SNS, the network structure and the identity of the network node may change, which can be the result of formation cooperation and

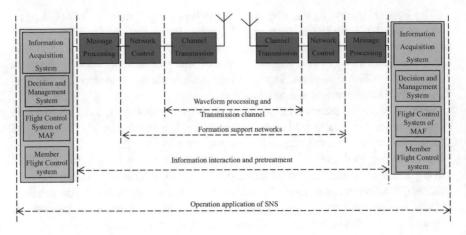

Fig. 7.1 Functional architecture of the SNS

control. It may be driven by other events (such as ground control instructions, missile members leave for some reason). According to guidance of leader or autonomous cooperation, the topology of SNS can be formed under the corresponding network control protocol, which can be used to form a master-slave network or multi-node peer-to-peer ad hoc network.

(3) Function of message handling

For the wireless transmission characteristics of the SNS, it is necessary to pre-process the information such as autonomous formation flight and cooperative attack transmitted on the wireless link before entering the wireless link in order to format the message transmission. Thus improving the efficiency of information transmission, achieving information consistency and information sharing, meeting the requirements of automatic identification and rapid processing of the tactical task message between missile members. It needs to study the types of information interaction, information usage, sharing requirements, and compile the formatted message set of the flying missile to support the efficient, reliable and standardized information transmission and exchange between the missile members.

(4) Function of network cooperative combat application processing

The formation support for networked cooperative operations includes the offline support function before the launch of the missile and the real-time support function after the launch, as shown in Fig. 7.2.

The off-line support system includes the formation task interpreter and the program route planning function, which is used to form the initial setting and task binding associated with the formation of the SNS.

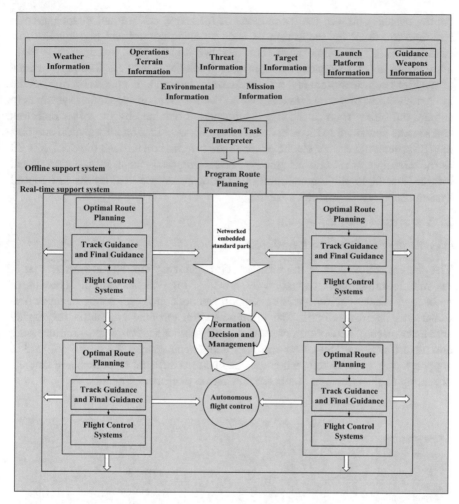

Fig. 7.2 Function of formation support network cooperative combat application

Real-time support system includes formation decision-making and management, formation flight control, member flight control and information acquisition systems. Following the basic principles of cooperativity, the formation decision-making and management system is responsible for the trade-off between individual cost and group cost, the conflict mediation of the formation, the management of joining and leaving team, implementation of the task planning and target assignment, and cooperative route planning and cooperative guidance and control. It is the central system of the MAF to optimize the instructions of formation and formation guidance to ensure the required tasks, following the self-organizing rules and principles of cooperativity. Based on the optimization indicators and formation requirements of the formation decision-making and management system, the formation flight control system optimize the instructions of formation control and maintenance in real-time to ensure the realization of node collision control and high-quality formation control and maintenance. The member flight control system is the flight attitude and trajectory control system for each missile member ε_i. The performance of the flight control system is the main factor to determine the precision $\sigma_i(t)$ of the guidance control of the missile member ε_i. The information acquisition system is to acquire the information of the target information obtained by the target detectors and sensors contained in the missile members, its own motion information and task load information, network characteristics state information through the SNS, as well as the binding combat task categories, tactical programs, target information, threat information, flight environment and other information acquired by the fire control system and target instruction system.

2. **The structure of the system**

(1) The physical framework of the SNS

The physical framework of the wireless self-organizing support network system of the missile autonomous formation is shown in Fig. 7.3. The system consists of embedded wireless self-organizing support network protocol, communication terminal and interface circuit. The communication terminal completes the digital signal processing, networking control, security authentication, message processing and other functions. The power amplifier and antenna part is the functional load to support the network system, which can be installed different amplifiers and antenna according to the structural characteristics in the platform integration.

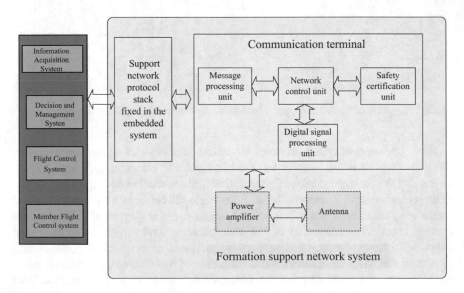

Fig. 7.3 Physical framework for supporting network systems

Digital signal processing unit completes the data link framing, waveform processing, broadband modulation and other functions. Network control module completes the network formation, network control, communication resource allocation, network user addressing, business data transmission and other functions. The security authentication module provides authentication of data integrity, source legitimacy, and so on. The message processing module completes the information conversion and preprocessing of the communication module with acquisition system, the formation decision and management system, the formation flight control system and the member flight control system.

(2) The design framework of the support network system

The design of SNS includes support network itself and the design of a variety of support network aided simulation test system. According to the three basic elements of the SNS, the design of the SNS can be equivalent to the design of the network support protocol and the transmission equipment, which conforms to the characteristics of the transmitted data. For the final SNS, the corresponding simulation and test module are also needed. Therefore, the design framework for the entire SNS is shown in Fig. 7.4. The design framework of the whole system is composed of five parts: wireless communication terminal, upper layer protocol, channel model, scene model and simulation environment. The upper protocol design is the core part, including the network layer, link layer and media access control sub-layer design. The other four parts complete the final verification and test of the supporting network system together with the above layer protocol.

The modules are described below. Upper layer protocol and wireless communication terminal together form a SNS. The wireless communication terminal, which is a key equipment in physical layer, provides services for the upper agreement. Therefore, the service capability of the communication terminal has a strong constraint on the design of the upper layer protocol.

Fig. 7.4 Design framework for SNS

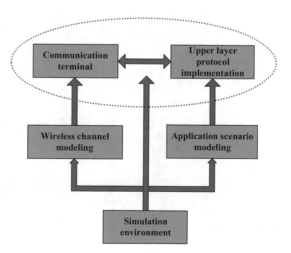

Protocol design will follow the idea of hierarchical design. With reference to the actual needs, the protocol system is divided into four levels, namely, application layer, network layer, link layer and physical layer. The application layer is defined by the user for the user process. The physical layer uses real or virtual physical devices. It is focused on the design of the network layer, the link layer, including the media access control sublayer. The design goal is to provide services that meet the application layer requirements and the corresponding quality of service under the support of physical devices, that is, the constraints that the physical layer can provide. The protocol hierarchy and the relationship between each other are shown in Fig. 7.5.

The physical layer provides the basic physical transmission capacity to complete the data codec work, the realization of the data split frame and framing. MAC layer is the main function to control the access of the shared channel with network nodes and provide access protocol. The link layer provides a well-defined service interface to the network layer, providing error control and flow control.

Wireless channel modeling and application scenario modeling are mainly to provide a relatively realistic test environment for supporting the design verification of SNS. By supporting the network system in the environment, the performance of the support network can be basically verified. The wireless channel model provides a transmission channel from the superior to the poor, which can test the adaptability of the network system to the channel conditions. Application scenario modeling provides a variety of typical application environments and various emergencies for the operation of the network system, which can test the reliability of network systems and other performance indicators.

The simulation environment provides a supportive environment for the simulation, verification and testing of the entire communication network system. The entire communication network system, including various model builds, is verified [74].

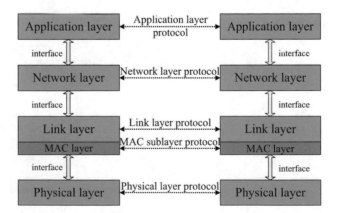

Fig. 7.5 Protocol hierarchy

7.3   Protocols for Support Networks System of MAF

The network of MAF is a whole of a number of members through a form of contact in the time and space, in the architecture, and in the task function. In a sense, the network is integration. SNS is the carrier of network. The protocols are the specific embodiment of the SNS. In this book, the SNS and protocols for SNS is the expansion and supplement for traditional communication networks and network communications protocol based on the characteristics of MAF. The purpose of the protocols is "unified" or "consensus". The protocols for SNS are intended to emphasize the integration and supporting of the SNS. The network not only reach an agreement in the interactive individual information, but also to reach an agreement in the cognition of overall characteristics and the cognition of characteristics of task objectives, which is a strong support for the coordination of MAF. It includes network communication protocols, network characteristics cognition protocols and task characteristics cognition protocols.

As the content of the network support protocol is very rich, it cannot reach every aspect of a matter here. As a consequence, this chapter will first define the concept of network support protocol, and design the wireless self-organization network communication protocol based on the specific needs of the MAF. The wireless self-organization network communication protocols are based on self-organized TDMA (STDMA), which are the basis of interconnection between nodes, and is also a manifestation of the autonomy of the formation. And based on the hierarchical network model, we design the cluster head decision-making protocol with redundancy. Finally, as for the network feature cognition protocol, we design a distributed boundary decision-making protocol for the cognition of the network boundary, which is basis of light transmission protocol in the fifth chapter, belonging to network characteristics cognition protocol. The specific research of task characteristics cognition protocol refers to the contents of the fifth chapter.

7.3.1   Concept of Support Protocols for SNS

In the networked system of MAF, the completion of the task depends on the information. And the existence and acquisition of information is in a variety of ways, which are pre-launch information offline, and online detection or generated in real-time. The information is sometimes straightforward, sometimes obscure, sometimes single, and sometimes miscellaneous, redundant, disorderly. The network support protocol acts as the sender of the information, the aggregator of the information, and even the producer of the information, providing a transparent "information pool" for the overall upper application, so that the individual information can be acquired in time. Figure 7.6 shows the schematic diagram of the network support protocol.

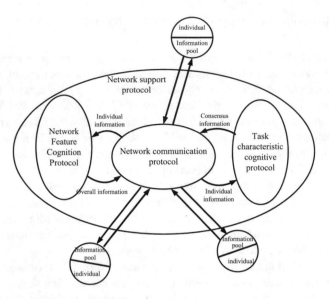

Fig. 7.6 Network support protocol diagram

Definition 7.1 The meaning of the network support protocol is to organize and maintain the overall level and structure of the formation, to support and protect the interaction and collaboration of the formation, to interact and guarantee the organization and maintenance of the overall level and structure of the formation. The protocols are a series of agreements which supports the individual in the formation, based on the hierarchical structure in the formation, combining with the overall characteristics of the parameters, relying on interaction with other individuals to efficiently complete the specific functions and specific assigned tasks. The extension of the network support protocol includes, but is not limited to, network communication protocols, network feature cognition protocols, and task feature cognition protocols, where each protocol can be subdivided into more specific sub-protocols.

Under normal circumstances, the network communication protocols ensure that the interoperability between individuals to convey the individual information (individual position, attitude, speed, etc.) at the same time. The network feature cognition protocols provide information for the overall network. The task characteristics cognition protocols provide information for mission objectives. The concept of network communication protocols is consistent with the usual concept. And the definition of network feature cognition protocols and task characteristic cognition protocols is as follows.

Definition 7.2 Network characteristics cognition protocols refer to a group of protocols that have a unified understanding of the overall characteristics, parameters, and status of the individual in the network through the SNS. For example, the online estimation protocol for health status (connectivity, packet loss rate, delay rate, update rate, etc.). In this chapter, the individual decision protocol will be presented to determine the boundary of the network in a distributed way and light transmission protocol in Chap. 5.

Definition 7.3 Task characteristics cognition protocols refer to a cluster of protocols that collectively identify, reorganize, classify, and refine the tasks, environments, or target information acquired by the individual through the network, so that the overall upper application has a unified understanding of the task objectives. Such as a data fusion protocol for threats or target information, a target calibration protocol that uniquely identifies the target.

In summary, the network communication protocol allows individuals to understand each other and to ensure the interaction of members. Network characteristics cognition protocol enable the individual to have a unified understanding of the overall network. The task characteristics cognition protocols enable the MAF with a unified understanding of the outside environment. They play a role in supporting the overall (or network) upper application at different levels.

The full set of physical hardware and software of SNS for MAF is a kind of data link between the missiles, is a wireless self-organizing network in the form of network. In this chapter, the network communication protocol in SNS mainly draws on the related research results of these two aspects [74].

7.3.2 Communication Protocols Design of Support Networks

1. Networking analysis

Autonomy and dynamics are the basic attributes of the SNS. Obviously, the network is similar to a typical wireless mobile ad hoc network. The wireless mobile ad hoc network can be defined as an autonomous wireless multi-hop network with no fixed infrastructure, no fixed router, all nodes are mobile, and can be dynamically maintained in contact with other nodes. Therefore, we can put forward the requirements of communication network system protocol refer to the design ideas of wireless mobile ad hoc network, combining with the background and requirements of the MAF.

The structure of SNS is based on the wireless self-organizing network. The network structure of the wireless self-organizing network is divided into two types: the planar structure and the hierarchical structure (Figs. 7.7 and 7.8). The flat structure is a kind of peer-to-peer network. There is no grade among nodes. The advantage is that the structure is simple and the network is robust (not affected by the damage of a special node). The disadvantage is that the scalability is poor and the network size is limited. The size of the formation increases, the routing maintenance overhead increases exponentially, consuming a lot of bandwidth, which is suitable for small networks. The hierarchical structure of the network is

Fig. 7.7 Peer-to-peer
network

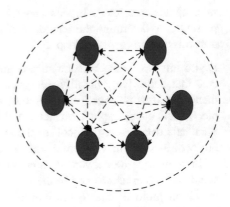

Fig. 7.8 Layered network

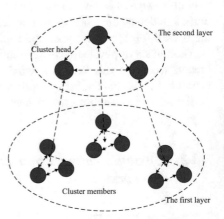

usually divided into clusters, according to the cluster decision (or election) protocol for each cluster to determine a cluster head, which responsible for the maintenance of routing and other network management. The scalability is good, but the cluster head undermine the robustness of the network to a certain extent.

The problem of network structure choice of the SNS should be considered in two aspects. On one hand, based the characteristics of the mission of MAF, according to the five types of MAF scale in second chapter, we can see that the scale of formation is limited. The radius of the formation is usually not too large, so the relay requirements are not high. However, the missile members need to interact with each other and the information update rate is high. The communication network operation requires high robustness. Based on this feature, this chapter will focus on the plane structure of the peer-to-peer network model of the communication network, which is stable, fast to become a network. Each node is in the same status and is suitable for MAF's requirements. On the other hand, the scalability of MAF is considered, based on the hierarchical structure of the clustering network model, we will also propose a cluster decision-making protocol to enhance the robustness of the SNS.

For the self-organization of the network, it is necessary to propose the corre-sponding distributed algorithm to establish the network quickly and determine the status of each node in the network. Due to the dynamics of the network, its structure and scale will change randomly, so the design of the protocols should adapt to this change, that is, should not be sensitive to the network structure.

For the network structure, it is required to achieve a peer-to-peer and hierarchical network structure. The peer network can be seen as a planar structure with the same status of each node (see Fig. 7.7). The advantage of this structure is that there is no central node, which ensures that the network has no effect on network security after some nodes are lost. All nodes of the network share a channel, sending information within the sending time, and other nodes accept the information, so that the whole network becomes an "information pool" to achieve the purpose of data sharing. Since the channel is a broadcast channel which takes a simple form of broadcast routing. Each site is within the communication distance, so it does not have to consider multi-hop routing, and the routing protocol can be done more simple. As shown in Fig. 7.8, this kind of network structure requires the support of the physical layer. If there is only one physical channel, it can only complete one peer-to-peer network, which will be powerless for the hierarchical structure. If there are multiple channels, it can be easily achieved hierarchical structure.

2. **Network structure design**

As mentioned earlier, in view of the limited size of MAF, according to the char-acteristics that the radius of the distribution will not be too large, the relay requirements are not too high, and the high requirements of information update rate and stability of the SNS, the network is set to peer-to-peer network mode. Because the peer-to-peer network structure is stable, the networking speed is fast, and the status of each node is the same, it is suitable for the mission execution of MAF.

Figure 7.9 shows a basic network plane structure, which includes a first common node, an n-th ordinary node, a first management node, a second management node, and a relay node, where the first common node, the nth The first node and the relay node constitute a network, and the second management node belongs to another network. The main interfaces between the nodes are J1, J2, where J1 is the data interface and J2 is the signaling interface. The relay node is located between the n-th ordinary node and the first management node, which serves as a data relay.

Fig. 7.9 Basic network plane structure

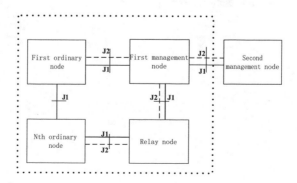

And the first management node and the second management node are connected through J1 and J2.

We set the network to the peer-to-peer network mode, and temporarily do not consider the situation of multiple networks running together, which can remove the second management node and relay nodes (Fig. 7.10). The simplified network includes a first common node, an n-th ordinary node and a management node, where the main interface between the nodes is still J1 and J2. The first ordinary node is connected to the n-th ordinary node 2 through J1, and only data information is exchanged. Ordinary node and management node are connected through J1 and J2, which transmit data information and instruction information.

Each node forms a network, relying on the form of broadcast to exchange information, the topology is shown in Fig. 7.11. In accordance with the principle of ID minimum, the network specifies the first node as the management node, which is responsible for the management of the entire network to ensure network security in all cases. In addition to send and receive data, the remaining nodes also inherit the management node according to the ID ascending order principle, so that the new management node is generated immediately after the management node is lost.

Fig. 7.10 Equivalent basic network plane structure

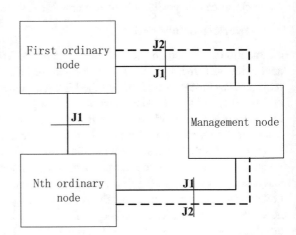

Fig. 7.11 Network topology

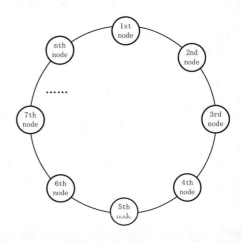

Fig. 7.12 Network slot
setting diagram

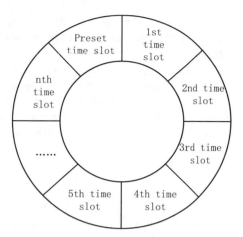

The network time slots are set up as shown in Fig. 7.12. All time slots form a
closed loop. Each node occupies the corresponding time slot according to its ID
number. Each node sends its own information according to its sort in the slot until it
finish sending Information. After a round of transmission, we leave a preset length
of the message slot for some nodes to send the highest priority information.

For the division of the wireless ad hoc network protocol level, referring to the
OSI model, the network is divided into four levels according to the special
requirements of the missile autonomous formation application, namely the application layer, the network layer, the link layer and the physical layer. It is based on
the fact that the network protocol can meet the characteristics of sudden, real-time
and short transmission period of the communication network. At the same time,
from the viewpoint of survivability of the network, the protocol should have strong
system reconstruction capability. The application layer is the user process, which is
managed by the user (i.e. the formation control system on the missile). The physical
layer adopts the actual or virtual physical device. Therefore, the design focuses on
the network layer and the link layer. The network layer satisfies the function of the
node networking. The link layer satisfies the takeover function of the node to the
public channel.

The basic service of the link layer is to transfer the data from network layer of
the source machine to the destination network layer. Its functions include providing
a well-defined service interface to the network layer, dealing with transmission
errors, and adjusting the data flow. In the link layer specifies the network service
mode and framing method.

For supporting network systems, requiring higher data reliability and shorter
latency, the use of connectionless, unacknowledged services is a better choice, with
minimal latency and simultaneous data reliability issues that can be addressed by
errors control. But for some important data, the mechanism of confirming can be
used, which access the benefits of confirming and not confirming services comprehensively. The missile data frame includes a frame header, a control domain,

a payload field, and a frame tail. Therefore, this protocol uses a fixed-length frame structure. The structure is as simple as possible, which is easy to synchronize of the frame.

The most important part of the link layer is the MAC protocol, which is used to coordinate multiple users to share a channel. The purpose is to make the network achieve maximum throughput under a certain business load. The core issue is how to assign a single broadcast channel to the in a number of competitive users. At present, the main methods of channel allocation are dynamic allocation and static allocation.

(1) static channel allocation. Each time the user occupies a fixed channel in a certain way in the life cycle, such as Time Division Multiple Access (TDMA), which divides the time into a periodic frame. And each frame is subdivided into several time slots for transmitting signals to the network. Under conditions that satisfy timing and synchronization, each node can receive signals from other nodes in each time slot without interruption. Similar to static frequency division multiplexing (FDM) and static code division multiplexing (CDM).

TDMA makes each user have their own fixed virtual channel, and the user will not produce interference. But the static mode is an efficient distribution mechanism only in the situation of fixed users, and each user has a heavy traffic burden. Obviously, when a user traffic is small, so the channel resource utilization rate allocated to the user is very low, and the user cannot assign its channel resources to other users, resulting in inefficient channel utilization.

(2) dynamic channel allocation. The dynamic allocation of channels allows the user no longer occupying the channel resources, and releasing resources occupied by idle users, to make the network utilization higher under certain conditions. But the program brings the problem of multi-user conflict. Therefore, how to avoid the user's conflict caused by the use of the channel causing the efficiency reduction become the main problem of the design.

For the algorithm that dynamically allocates a multiple access channel, there exist multi-user coordination, non-coordination, and partially coordinated channel access algorithms.

The coordinated (non-conflicting) protocol solves the problem of conflict. Under high load conditions, the protocol has a high channel utilization. But for low load conditions, the efficiency becomes lower. Typical protocols include Bitmap Protocol and Binary Countdown Protocol.

For a non-coordinated (conflicting) dynamic channel allocation algorithm, the ALOHA protocol and the Carrier Detection Multiple Access Protocol (CSMA) are typically used. The basic idea of ALOHA protocol is that users do not have to consider the status of other users to send data, sending whenever it wants to send. After a conflict, it waits for a recurrence. The channel utilization of this method is 18.4%. The improved slotted ALOHA protocol increases channel utilization to 36.8%. Obviously, there are inefficient problems of the protocol for more users and networks of large business. CSMA protocol means the user to listen to the channel

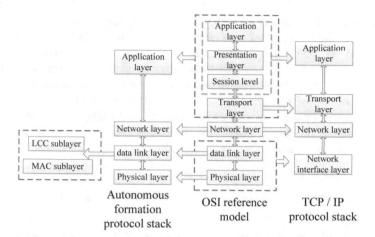

Fig. 7.13 Network communication protocol stack level design for self-organizing formation

before sending data, only to send when the channel is idle. The protocol reduces the likelihood of collisions and improves channel utilization.

Partial coordination (limited competition) protocol combines the two to obtain the advantages of both to obtain optimal or suboptimal channel utilization under various load conditions.

According to the actual demand, with reference to the ISO International Organization for Standardization proposed OSI reference model and the most widely used TCP/IP protocol stack, the autonomous formation network communication protocol stack is divided into four levels, namely the application layer, network layer, data link Layer and physical layer, as shown in Fig. 7.13.

The main function of the application layer is to process the data submitted by the network layer, to obtain useful information, to develop the appropriate operation according to the information or to send the data to the network layer. Application layer protocol is mainly defined by the upper application interface specifications, such as access to the number of neighbors and access to long machine status parameters. The main purpose of the network layer is to realize the transparent transmission of data between the two nodes. In the case of multi-hop, the network layer using the fusion mechanism of neighbor discovery ZRP regional routing protocol. Data Link Layer Logical Link Control (LCC) sublayer protocol is mainly responsible for framing and error control, ensuring the update rate under the premise of using a combination of network coding error retransmission method. The physical layers are based on spread spectrum technology. The above details refer to [2, 4, 65]. Now, we discuss the key technology of network communications protocol stack for SNS, which is the sublayer protocol design of MAC protocol. Also known as channel access protocol, MAC protocols are a variety of network protocols, especially in the network protocol of Ad Hoc network protocol. The core problem is to solve the multi-user how to use the limited channel resources to balance the network quality of service (QoS) dynamically, with high utilization and high reliability. In order to meet the application requirements of real-time traffic transmission, the MAC protocol of tactical data link generally adopt TDMA access mode. And its unique burst communication mode has good anti-intercept and anti-jamming capability, using time-dimensional channel, which is flexible to become a network. In order to meet the needs of the field environment, it is widely used in a variety of tactical data link. This section, we will mainly refer to the United States and NATO active data link Link-16 and Link-22. We will design the protocol based on fixed Time slot and reservation slot switching of the self-organized time division multiple access protocol.

3. **Advantages of networking protocols**

The support network protocol for the missile formation in this section has the following advantages [74]:

(1) It has multi-node autonomous networking capability. In accordance with the principle of the smallest ID selected management node, it has no human intervention. The nodes automatically join the network in the process of movement according to the node network method.
(2) It has the ability of fast networking capability, which reduce the collision problem under single channel condition and improve the ability of multi-node fast networking, so that multi-node network can be completed in a short time to achieve the stable operation of the network.
(3) It has the ability to manage the node to autonomously leave and join the network. The structure of the network can be changed, and the node can join and leave at any time. At the same time, the conflict mechanism is used to resolve the conflict.
(4) The node access response time is less than 100 ms level.
(5) It can meet the wireless transmission rate of more than ten Mbps level and the total number of nodes can meet the requirements of a certain size of the missile autonomous formation.

7.3.3   Media Access Control (MAC) of STDMA

The self-organized time division multiple access (STDMA) technique divides the time axis into several information frames. Each information frame is divided into several time slots for users to send data, instructions and other information. Each user periodically broadcasts information packets that subscribe to the current time slot and the future time slot, and constructs the slot status table through these messages. The information contained in this table is the basis for all users to dynamically subscribe to the time slots in order to realize the spontaneous management of the channel slots by each user. Since the slot resources in this case are managed by the user, it does not need the participation and control of the ground station, so it is called self-organizing time division multiple access mode. STDMA is a complex technology, and this chapter mainly on the following two aspects:

1. Slot allocation

In STDMA system, the sharing of information between nodes is carried out in the form of time slot, which is the most fundamental guarantee for the normal operation of data link. Therefore, the efficiency of slot allocation directly affects the efficiency and stability of the whole data link, which makes the time slot allocation algorithm STDMA technology become a research hotspot and the main point to realize TDMA. The algorithm includes fixed, reservation, competition, dynamic method and so on. In the specific implementation, often according to the specific circumstances, we comprehensive use of a variety of algorithms to deal with, as far as possible to improve channel utilization.

2. Self-organizing network management

STDMA self-organization characteristics on the one hand reflect in the allocation of time slots, on the other hand reflect in the network adjustment and management. Node network, node loss, node active leaving network and node replacement and other functions are autonomously completed in the absence of external control. Therefore, the protocol design of network management is of great significance to realize the self-organization of STDMA.

7.3.4 Time Slot Allocation Protocols Based on STDMA

In this section, the composite STDMA slot allocation protocol based on time slot fixed allocation and reservation is dynamically switched according to the network size between the fixed assignment and the reservation assignment. After each mode of operation is selected, based on the current service the internal dynamic adjustment of the frame structure and time slot division adapt to the network of high dynamic, improving the node information update rate, channel utilization and system capacity. Mode switching and the allocation of each slot are done by the node autonomously, easy to achieve self-organization and management of the network, reflecting the autonomy of the network.

1. Time slot structure

Time slot division structure is shown in Fig. 7.14, each frame is divided into n time frames, and each time frame is divided into m time slots. The values of m and n can be adjusted by the number of nodes in the network and the specific needs. In principle, each node in each frame must acquire at least one fixed time slot and avoid channel access for a long time. The assignment of the reservation slot is a

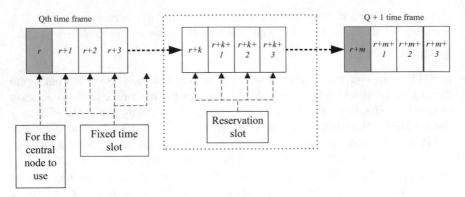

Fig. 7.14 Slot structure

dynamic and competitive process, which is allocated in turn according to the reservation request of each node.

The first time slot (the shaded area in the figure) is used by the central node, and the other time slots are used by ordinary nodes, so the central node has the most chance of sending. The difference between the central node and the other nodes in the network is only two aspects:

(1) When the new node is joining the network, the central node is responsible for accepting and assigning the ID within the network.
(2) It has the most opportunities within the network to send, which can be the core location logically in the network or nodes with high update demand.

There are two basic structures of frame:

(1) Structure 1: only contains fixed time slot (not including the virtual box);
(2) Structure 2: frame tail to increase the appointment slot (including the virtual box).

According to the upper task needs, the formation system switches between the two structures dynamically, and the two structures can also be adjusted according to the needs of the internal.

In the following example, the TDMA mechanism is used to construct the 16 nodes network slot structure. The node sends its own information in the time slot occupied. A time slot can only be occupied by a node. The slot structure is shown in Fig. 7.15. With 280 ms as a frame and three frames as a super frame, each frame is divided into seven time slots. Each time slot is $280 \text{ ms}/7 = 40 \text{ ms}$. Each node sends data in time slot. All nodes send at least one time in a super frame. The three time slots of "slot 1", "slot 8" and "slot 15" are occupied by the central node to ensure that each center node can transmit data and improve the information update rate of the center node. "Slot 7", "slot 14" and "slot 21" are not occupied by nodes in the network, but are reserved as slots when the network is full [74].

First frame:	Time slot1	Time slot2	Time slot3	Time slot4	Time slot5	Time slot6	Time slot7
Second frame:	Time slot8	Time slot9	Time slot10	Time slot11	Time slot12	Time slot13	Time slot14
Third frame:	Time slot15	Time slot16	Time slot17	Time slot18	Time slot19	Time slot20	Time slot21

Fig. 7.15 Slot structure diagram

2. Fixed slot allocation

Fixed slot allocation ensures the time slot occupied by the same node in each transmission cycle, which mainly to ensure the node's basic transmission opportunities. The advantage is that there is no transmission conflict in a separate network environment. The disadvantage is that sometimes some time slots may be idle, resulting in reduced efficiency. In order to improve the channel utilization and enhance the self-organization of the system, the inherent time slot is used as a tool to ensure that the nodes in the network have sending opportunity. On the one hand, the fixed time slot is provided for all the nodes in the network. On the other hand, the proportion of resources of fixed time slot in the entire time slot decreased, so that more time slots are provided for the efficient use of the reservation.

The time slot allocation method is as follows: each node acquires the time frame number k of the time slot by taking the remainder of the ID to the time frame number, and obtains the time slot in the time frame k by taking the remainder of the ID to the number of time slots in the time frame number. Each node can find the corresponding fixed time slot according to its own sending ID in the network. Therefore, the allocation process of the fixed time slot does not need human control, and it does not need the instruction of the central node.

The allocation of such fixed time slots is characterized by the following:

(1) Better uniformity. Uniform slot allocation can reduce the complexity of slot management.
(2) The rules are simple. The algorithm is simple and the processing time is shortened, which is helpful to improve the real-time performance of the system. The calculation is easy and the stability is high, so that the nodes can be well coordinated.

3. Reservation slot assignment

The reservation slot is a time slot that dynamically adjusts the usage right among the nodes in the network. It can be used by different nodes in different periods. It is

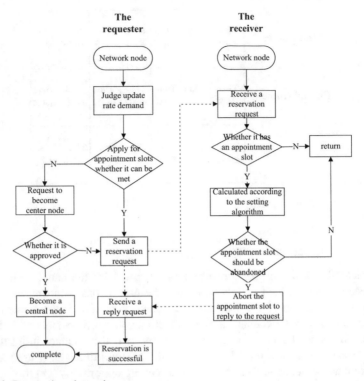

Fig. 7.16 Reservation slot assignment process

mainly used to meet the dynamic change of node traffic demand. Each node can apply for the reservation time slot directly to the data transmission, and to achieve the timely transmission of tactical information to avoid the transmission delay caused by data loss, transmission rate, transmission layer communication interruption and slow recovery, thus further improving the packet wireless network transmission real-time and emergency capability. Figure 7.16 shows the process of reservation of the slot allocation algorithm.

When the nodes in the network need to improve the update rate, it will be classified according to the needs of the classification. First we determine whether it can get the reservation time slot to meet its sending needs. If it can't be met, the node becomes the highest update rate in the network, that is, the central node. If the request is approved, replace the original as a central node, otherwise, only by applying for reservation slots to meet the demand. In order to improve the utilization of time slots, the time slots are occupied in some form by other nodes in the network at any time when the network is running. So the node obtains the reservation slot, which means that the need for other nodes in the network to give up the reservation of the reservation slot. After the other nodes in the network receive the reservation request, determine whether the node occupies the reserved time slot, if not occupied, then do not deal with or should be in accordance with the reservation algorithm to calculate whether the node should abandon the reservation slot, where the reservation algorithm can be set in a specific application. If the result of the calculation is to abandon the reservation slot, it returns the reservation request and completes the assignment of the reserved time slot. In the case where there are multiple reservation requests in the network, the nodes in the network will be processed according to the order of the timeslot numbers. The whole process does not require ground stations and central nodes to intervene.

4. Two types of switching operation principle

In the operation of fixed allocation mode, as the size of the network increases, the frame structure can be adjusted accordingly. For example, let 10 nodes are the working mode cutoff value, i.e. 10 nodes or less work in fixed allocation mode. We set it into three time frames, and the frame structure adjustment is shown in Fig. 7.17.

Each box in the graph represents a time slot, and each frame is divided into three time frames from top to bottom in the figure, and the operation mode is followed by three time frames. The number in each box represents the number of the node occupying the slot. The figure shows the slot allocation pattern when the network size is 5, 6, 7. Whenever there is a new network node, the system automatically adjusts the frame format, for the new network node to increase the fixed time slot. Although this may cause some time slots to be idle, but can meet the needs on the basis of the use of fixed allocation of advantages of small conflict, simple control, high real-time, reducing management complexity and saving processing time.

The following situation is suitable for the use of fixed allocation mode: When the network traffic is small, and there is no channel competition. Network size is larger, but the traffic is more uniform. Traffic is uneven, but relatively stable. In these cases, the slot allocation basically does not need to be adjusted, the fixed allocation has the highest efficiency.

When the traffic increases, and the distribution is inhomogeneous and dynamic, then we adopt the combination of the model of fixed and reserved, so that nodes can be dynamically adjusted according to demand to send, ensuring that the big traffic

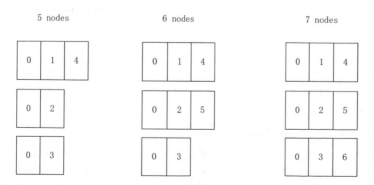

Fig. 7.17 Adjustment of fixed slot allocation mode

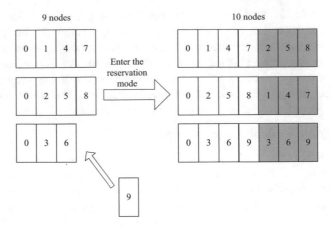

Fig. 7.18 Schematic switching diagram

node to complete the information transmission, avoiding small traffic nodes waste time slots. When the composite mode is required to be started, the frame length is expanded so that the number of time slots increases and the fixed slot allocation is invariant. The reserved slot is added to each time frame at the end and the reservation is made according to the default initial setting assigned to the nodes in the network, as shown in Fig. 7.18.

Let 10 nodes be the mode cutoff value, then when the network scale expands to 10, the network enters into the reservation mode. It can be seen from Fig. 7.18 that during the conversion process, each time frame automatically adjusts the length, increases the reservation slot (shaded part) at the end of frame, and allocates each reservation slot to the network node according to the predetermined value. It should be noted that the switching of the mode can only be done before the start of each frame so that the adjustments made by each node must be carried out during the transmission of the original fixed slot data. We have sufficient time of frame structure adjustment before the reservation slot data is transmitted, which ensures a smooth transition of the operating mechanism. Figure 7.18, for example, when the tenth node into the network, get the number 9, and then in order should be 0 (center node), 1, 4 and 7 to send, and then enter the reservation slot. So that the time of the four fixed slots is used to complete the mode adjustment, and the transmission process will not have any stagnation. When the traffic is reduced and stable, it is adjusted to no reservation slot status. The fixed slot is reserved in the composite mode for smooth switching of the mode and ensures the basic transmission of the node [74].

5. **Time slot synchronization mechanism**

We can see that since the support network of missile autonomous formation is a wireless self-organizing network, it is a multi-hop temporary autonomous system without central node, and TDMA-based time slot allocation mechanism must be synchronous access. The nodes need to keep the time slot of the precise synchronization. Therefore, it is a prerequisite to solve the problem of slot allocation in the self-organizing network without center nodes. Currently self-organizing network synchronization is divided into quasi-synchronous, external synchronization, master and slave synchronization.

(1) Quasi-synchronization

Network nodes use high-precision independent clock to achieve synchronization. As with the Cesium clock, the timing signals generated by these clocks appear at the same rate, and the error is acceptable. Quasi-synchronization system between the nodes without the timing signal transmission link and thus do not take up the channel.

(2) External synchronization method

Synchronization of nodes in the network is based on the external clock source. Using the global positioning system time scale as a time base. Such as the GPS timing.

(3) Master and slave synchronization method

Master-slave synchronization is a kind of synchronization method which is widely used at present. In the self-organizing network, the temporary master is used as the time master node to realize the synchronization of the network. The highest level clock is called the reference clock (PRC). The reference clock signal is sent to each slave node of the network through the transmission link. Each slave node locks the local clock frequency at the reference clock frequency, thus realizing the inter-node synchronization. Master and slave synchronization is easy to implement, without the problem of network instability. But because of dependence, so the reliability is low. It is desirable that the clock from the node should have a high degree of stability and can be maintained during the master node clock failure.

In the above three ways, the operation of the external synchronization mode network must rely on the global positioning system time scale signal, with high cost of the specific realization. In order to improve the network anti-destruction ability and to avoid the subordinate between the nodes and containment, military communication network generally use quasi-synchronous mode. But by the accuracy of the limit, synchronization performance can't be guaranteed. So master and slave synchronization method is often used in many applications. The self-organizing network also uses master-slave synchronization method.

Synchronization steps:

Step 1. Use the temporary head of the main node as the master node to achieve the synchronization of the network, which can be selected from the organization of the network management node.

Step 2. ID0 node to send the time scale for a period of time (according to the node itself, the selection of a reasonable time), the remaining nodes will be cleared after their own timing, restart the timer.

The biggest problem with master-slave synchronization mode is that the master node may be disconnected from the network due to movement or damage after networking. The self-organizing network given in this section is to select the management node according to the minimum ID principle. The loss of the node will not bring the problem of losing the synchronization time.

On the other hand, the failure of the node or the communication interference will affect the correct transmission of the synchronization information, resulting in local connectivity. When the synchronization mode is used in the mesh network, it may lead to the occurrence of the conflict. To solve this problem, dynamic TDMA mechanism, which after any node has sent the data frame, it immediately sends the identity to next node, then the next node to start sending. This allows the nodes to send data in the order, avoiding the stability of the system when connectivity lost.

6. **Performance analysis**

(1) Real-time

This protocol reduces the transmission of control signaling and improves the channel utilization. The management of time slot does not need to send control commands to the ground station and the central node, reducing the amount of non-user data, improving the utilization of time slot, and ensuring that nodes can obtain the channel in time access and data transfer. According to the network traffic situation, the system determines the use of fixed or reserved distribution. The allocation of the frame structure and time slot allocation can be dynamically adjusted so that each node can be obtained in accordance with the needs of the corresponding opportunity to send, so as to improve the real-time of data transmission.

(2) Update rate

In order to improve the update rate, this agreement weakens the role of the central node and the ground station, simplifies the channel access process, improves the transmission ratio of the user data packets, and improves the update rate of the effective data. Taking into account the no uniformity and dynamics of the traffic volume, the fast reservation slot allocation adjustment strategy increases the number of time slots for big traffic nodes to ensure their business needs, while reducing the number of nodes with less traffic, avoiding the channel idle, so as to optimize the entire network update rate.

(3) Time slot conflict

The fixed slot allocation of this protocol only depends on the sending ID number of the nodes in the network. Because of the uniqueness of the ID number in the network at the same time, it is ensured that the slot selection will not conflict. At the same time, when there is a reservation request, the reserved reservation slot number is not confirmed by the requesting party. After receiving the reservation request from the other nodes, the reply is followed by the order and the reservation slot is allocated in order, thus reducing the occurrence of the conflict.

7.3.5 Network Management Protocols Based on STDMA

1. Node state definition

Under normal circumstances, the state of the node can be divided into the following:

(1) Independent state: The state that the node remains before it communicates with any other node.
(2) Networking state: The state that the node has been communicated with other nodes (at least one node) to form a stable mutual transmission of data.
(3) Free state: The state that the node temporarily leaves the original network, keeping receiving other nodes' user information frame in the network, but does not send its own user information frame.
(4) Leaving state: The state that the node lost the ability to maintain communication with other nodes, while no longer continue to join the original network and permanent leave from the original network.

2. Definition of node information frame

In order to complete the networking and data transmission and reception, we define the format and content of data sent by the node:

(1) Handshake frame: Handshake frame is the information format for each node to build the initial network. When the central node initiates the initial networking, the handshake frames are sent to find the independent nodes in the network to build the initial network.
(2) Confirmation frame: It is divided into two application scenarios: In the initial networking phase, the independent node sends the frame when it receives the handshake frame to form the initial network with the central node. In the network phase, the frame is sent when the independent node arrives at the network time to join the formed network.
(3) Control frame: The control frame is the information format in which the central node acknowledges the reply to the independent node that needs to join the network. The center node sends the frame after receiving the independent node confirmation frame, as the reply frame of the confirmation frame. The frame contains the relevant information necessary for the independent node network.
(4) User information frame: It is the format of the frame that data information of network node is transmitted in real-time.

(5) Leaving frame: It is the format of the frame that the network used to inform the node leaving.
(6) Reservation frame: It is the format of the frame used by the network node to apply for the reserved slot.
(7) Reply to the reservation frame: It is the format of the frame that the network node used to forward the reservation slot.
(8) Replace the center node frame: It is the format of the frame that the network replaces the central node.

3. **Initial networking protocol**

Figure 7.19 shows the flow of a node's initial network and the process of sending and receiving data from the node.

In the initial network access process, a node is first in the listening state. Define the listening time T_s, if not received within the information in T_s, then the node broadcast actively. After the broadcast, the node continues to listen and follow the cycle. When the information received after the following four Happening:

(1) If the node receives a handshake frame from the central node, it continues to listen after sending the acknowledgment frame.
(2) If the node receives the data frame, it will detect the adjacent frame. If the interval is less than the preset value T_0, it will be detected. If the interval is greater than the preset value T_s, an incoming handshake signal will be sent randomly. And the center node of the existing network will respond to the handshake signal and assign ID to the newly added node.

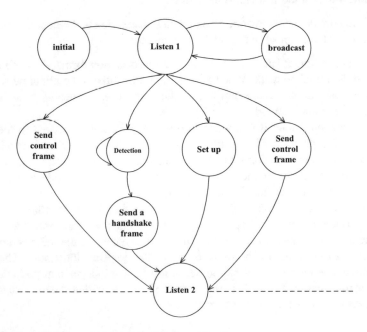

Fig. 7.19 Network diagram of the node join the network

(3) If the node receives the control frame, then the node allocates the network ID and joins the network.

(4) If the node is a central node, only the acknowledgment frame from the normal node is received, and then the central node sends the control frame, allowing the normal node to enter the network.

After the networking, the ordinary node sends and receives data in turn in accordance with the ID assigned by the central node, and the new node can always join the network.

At present, the MAC mechanism is a token-like mechanism. Each node maintains a counter and updates it in real time. When a node receives the data, if the node detects that the counter value is equal to its own ID, indicating that this node can send data at this time. Other nodes are in the state of listening in the network and only receive other node data information if their own ID are not equal to the counter value.

If the number of ordinary nodes in the initial network is greater than two, there will be more than two ordinary nodes receiving the handshake signal at the same time. If the above two or more ordinary nodes at the same time to return to confirm the signal, there will be a conflict. In the wireless environment, conflict detection can't be done like Ethernet. This is because the conflict in the wireless environment occurs on the receiver, not on the sender. When the transmission side transmits a frame, even if a collision occurs, since the signal that is being transmitted on the transmitting antenna of the transmitting side has the maximum energy, the other signal is filtered as noise, so that the receiving antenna can't judge whether the collision occurs or the noise signal. But the receiver is damaged due to signal collision.

In addition, in the communication network system design of the formation support network, the conflict avoidance mechanism is adopted. When the common node receives the handshake signal at the same time, if the channel is idle and the data is not sent immediately, we select a time slot in a predetermined number of time slots randomly to send. This mechanism effectively avoid the occurrence of conflict. If there is still a collision in the reservation time slot, the conflicting node resends the acknowledgment frame in the next time slot.

The initial networking process is shown in Fig. 7.20. When the node in the independent state starts running, it first listens to determine whether there is data transmission. If there is no data transmission, which means that there is no network at present. The node sends the handshake signal periodically as the central node, waiting for a new node to join the network. If the data is intercepted, the data type is recognized. If it is a handshake frame, which means that the network has not yet been formed. However, there is already a node for data transmission. This node is identified as a network center node. The node randomly replies an acknowledgment frame request network for a certain period of time. The central node receives the confirmation frame, the reply control frame is used for the node to join. And the control frame includes the information needed for the new node networking. The requesting network node joins the network according to the contents of the received control frame and completes the networking.

Fig. 7.20 Initial networking flowchart

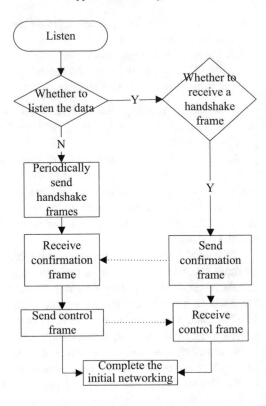

4. **Network access protocol**

As shown in Fig. 7.21, when a node in an independent state listens to the presence of a network, it can apply for joining the network. The specific process is the independent node to listen to the user information frame, it can determine the existence of a network. Then it listens to a round and records network information, including the total number of nodes, time slot allocation. And in joins randomly in the scheduled node network stage (The network provides for a certain time for the node to join after a reserved fixed time slot according to the size of the network in the maximum send ID node), sending confirmation frame. When the central node receives the acknowledgment frame, it replies to the control frame. The node applied to join and other nodes within the network according to the control frame for network adjustment, completing to join. Because the network needs to adjust the channel access mode automatically according to the network scale, the network node needs to judge the network size after receiving the control frame. If the scheduled size is reached, the reservation mechanism is started and the network slot structure is adjusted.

The data transceiver of the node after joining the network is shown in Fig. 7.22.

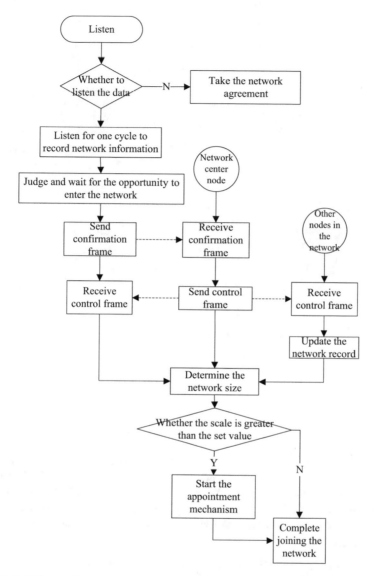

Fig. 7.21 Initial network entry process

(1) When the acknowledgment frame is received, it enters the detection link, defining the Flag. Flag = 1 when the node is in the transmission slot, otherwise Flag = 0. It sends data when Flag = 1.

(2) If the node receives the data frame, indicating that the other node is sending, defining the count variable Counter from 0 to receive data every time increment 1. When Counter is equal to the node ID, the node sends data. When each node cycle to ID is equal to the total number of nodes in the network, set the timer

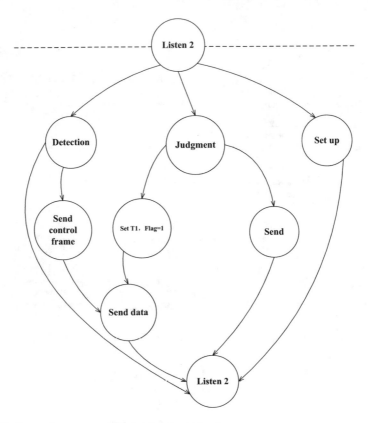

Fig. 7.22 Transceiver process after joining the network

time T_1. If it not receives the handshake frame within T_1, then it continues to send a new round of data. If it receives the handshake frame within T_1, there are new ordinary nodes to join the network. The maximum number of nodes assigned ID to the new node, and then continue to send and receive data.

(3) If the node receives a control frame, the handling is the same as initial networking protocol.

When there is a reservation mechanism, the main difference between the modes of operation when the network installed is that the corresponding adjustments made by all nodes of the whole network are different. In a network with more than 10 nodes, the addition of a new common node necessarily causes a node that originally occupies a preset slot to discard the preset slot as a fixed time slot for the newly added node. Where the transfer is entirely in accordance with the node ID, while other nodes have to include the corresponding adjustments, such as counting.

The above is the whole process of the node joining the team, so the network has been able to run steadily. For application background of the MAF, there exists the possibility of missiles out of the communication area or crash. Thus, the network can still be stable and reliable if one or more nodes leave the network.

5. Leaving protocol

Node leaving includes two cases, active exit and passive loss. After the node is lost, it is detected by the subsequent node and broadcasted. The network node will start a timer after receiving the user information frame sent by other nodes. The timer time is set to the reserved slot length between two adjacent transmissions. When the new data is received, it will close off the old timer, restart a new timer. If it has not yet received new data in the time, then the nodes of this time slot are lost and each node will be recorded internally. When the node sends data, it checks the record. If there is a node lost, it sends to the whole network of the missing notification. When the node receives the notification, each node will clear the current record.

As shown in Fig. 7.23, when the nodes in the network receive node loss notification, according to the situation before and after the loss of the network, select the appropriate treatment. If the node is lost before the non-reservation state, then the loss node must also be in the non-reservation state, there is no need to adjust the slot mode and only need to change the current node to the largest node lost ID and to give up the original time slot. The time slot of the node is logically replaced. If more than one node is lost, it is processed in turn. If the loss is before and after the reservation state, there is no need to adjust the mode, only the network node need to adjust the slot occupancy according to the state of the node. If the loss of the node causes the network size to change from the size of the reservation to the size of without reservation, it needs to adjust the slot mode. The time slot does not include the reserved time slot, only the fixed time slot, the network is restored to the non-reservation state [74].

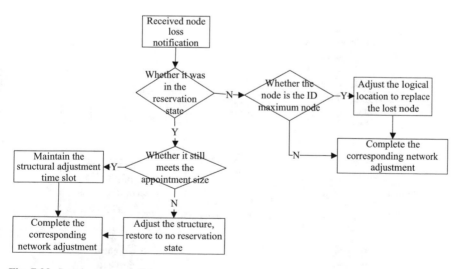

Fig. 7.23 Leaving network flow

6. Central node adjustment protocol

There are two reasons for the replacement of the central node, first because the central node is lost, and second because the objective conditions need to adjust the central node. The latter is mainly due to the particularity of the central node. Although status of the central node in the network and ordinary node is no essential difference, but it has the most chance to send. So sometimes in order to improve the data update rate of a node, or taking into account the network topology, it is need for network center nodes to dynamically adjust. For the situation of the loss of the central node, it is judged by the node. When the node receives the notification frame, each node obtains the sending ID of the lost node by querying the slot table, and judges whether the center node is lost according to the sending ID. If the central node is lost, the existing maximum sending ID node in the network selects the new central node according to the set criterion and sends the replacement center node frame to notify the sending ID of the new central node of the whole network. For the active replacement of center node, the replacement center node frame is usually sent by the ground within the reserved time slot.

The specific process shown in Fig. 7.24, after receiving the replacement center node frame, if the node is the original central node, then automatically leave the network, and then as a normal node to re-enter the network. If the node is a newly selected center node, then it adjusts the information such as the sending ID of the node, updates the time slot record, occupies the logical location of the central node. Assuming the function of the central node, the network will send the original ID in accordance with the node loss processing. If there are other nodes, the network information can be recorded and the node does not need to adjust.

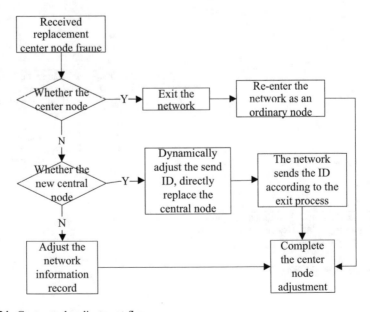

Fig. 7.24 Center node adjustment flow

7.3.6 Distributed Redundant Cluster Heads Selection Protocol

For large-scale Ad hoc networks, which are usually hierarchical structure. The primary problem of hierarchical structure is the choice of a cluster head. And cluster head to a certain extent, undermine the robustness of the network. The choice principle is to use a certain price as an indicator to find the optimal set of nodes to maximize or minimize the index. The search of the optimal solution is theoretically a NP-hard problem, so the researchers are usually designed to be heuristic, distributed clustering decision-making protocols or algorithms, as described in Ref. [134]. According to the WCA method, the load on the nodes, velocity and the degree of connectivity are weighted, and then cluster head is selected with the largest of weight of the message sent and received. Reference [135] uses the geographic location information to divide the network into several geographic blocks and then select a cluster head in each block. The LEACH method of random selection of cluster head is proposed in [136], which is equivalent to the probability of randomizing the cluster head and adjusting the probability of the number of heads of the cluster. The HEED method in [137] is the improvement of LEACH method. In [138], the distributed particle swarm optimization algorithm is used in nodes, but the size of the network is limited. These studies did not take into account of the redundancy of the cluster head. And if the formation size reached a certain degree after the use of hierarchical structure, the robustness of the cluster head should be guaranteed. For this problem in this section, we propose the redundant cluster head decision protocol.

1. Distributed Redundant Cluster Headset Decision Process

The distributed redundancy cluster decision protocol has two optional parameters: (a) redundant of cluster head c, (b) the maximum number hop between cluster head and cluster member k. After the initial networking of the formation, the color of all nodes is black and changed according to the redundancy cluster decision-making protocol, which is stabilized by the final color to decide the cluster head.

Specifically, according to WCA, each node i maintains a weight:

$$W_i = \omega_1 n_i + \omega_2 Qos_i + \omega_3 Power_i \tag{7.1}$$

where n_i is the number of neighbor of node ε_i with the same color within k hop, Qos_i is the network quality of the node ε_i, $Power_i$ is the residual energy of the node, and the weight of the three are $\omega_1, \omega_2, \omega_3$. The basic principle of weight setting is to select the nodes with more residual energy, stable network status and higher connectivity. Below is an example of redundant cluster decision-making with $c = 2$ [74]:

Step 1. The black node ε_i determines its own weight and exchanges weight with its k-hop neighbor of same color, and then sends the "red" message to the maximum weight node. At the same time, if the node ε_i received a "red" message, then it turns red and goes to step 2 immediately. If the node ε_i receives a "green" message, it turns green and goes to step 3. It repeats step 1 if the node does not change color within a certain period of time.

Step 2. Once the black node ε_i turns red, send a "green" message to the k-hop neighbor immediately.

Step 3. For a green node ε_i, when there is no black neighbor within k hop, it exchanges weight with its k-hop neighbor of same color. Then the "blue" message is sent to the node of maximum weights. At the same time, if the node ε_i receives a "white" message, then it turns white. If the node ε_i receives more than two different nodes of the "green" message, then it immediately turns yellow. If the node ε_i receives a "blue" message, it will turn blue immediately and turn to step 4. If the node ε_i do not change color within a certain period of time, it repeats step 3.

Step 4. Once the green node ε_i turns blue, it sends a "white" message to the k-hop neighbor immediately.

After stabilization, each node becomes one of four colors: red, blue, white or yellow. Obviously, the blue and red nodes are cluster head nodes, where the red nodes can be the primary cluster head, the blue node is the backup cluster head. The yellow and white nodes are member nodes, where the yellow node is covered by at least two red nodes within k hop, and the white node is covered by at least one red node and a blue node within k hop. In addition, any cluster head node (blue and red) can reach another cluster head node within two hops.

When $c = 1$, there is only one cluster head, each node will become one of three colors: red, green or yellow. The red node is the cluster head node, and the green and yellow nodes are member nodes. The process of redundant cluster decision is shown in Fig. 7.25.

2. **Simulation of distributed redundant cluster decision protocol**

500 nodes are randomly distributed in the range of 10,000 m², and the communication radius of the node is 1000 m. The simulation results of redundant cluster decision protocol are given in four groups. See Fig. 7.26 and Table 7.1.

The simulation results show that when $c = 1$, the redundant cluster decision protocol is equivalent to the cluster-head decision-making protocol in the normal sense. When $c = 2$, the protocol can double-cover the cluster decision, which can be used as a cluster decision protocol to support the network in the case of

Fig. 7.25 Redundant cluster head decision protocol flow

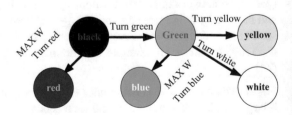

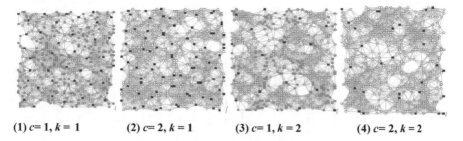

(1) $c=1, k=1$ (2) $c=2, k=1$ (3) $c=1, k=2$ (4) $c=2, k=2$

Fig. 7.26 Simulation results of redundant cluster decision protocol

multi-hop network. It can effectively reduce the impact of cluster node failure or damage on the whole network under the hierarchical network structure.

7.3.7   Distributed Border Decision Protocol

The formation as a whole network, whose nodes have their own characteristics such as the mobility of the node, location, attitude angle, flight speed, the importance of airborne equipment, but also has the characteristics of individuals relative to the overall because of the network such as the degree of scarcity of the airborne equipment in the formation, the relative position of the individual in the formation. In addition, the network itself has the characteristics of network induction delay, information update rate and so on. The acquisition of these information belongs to the category of network characteristic cognition protocol in the definition of this book, and it is expected that the network characteristic cognition protocol is task-dependent to a certain extent. That is, the type and function of the cognitive protocol required by the formation are often different depending on the specific task of the formation. In Chap. 5, the problem of collision avoidance between nodes in intensive formation is a key problem that needs to be addressed. In this chapter, we will discuss the problem of boundary decision making in node flight control. The boundary decision-making protocol is a kind of network feature cognition protocol which the node in the formation determines whether it belongs to the network boundary. For this reason, this part proposes a distributed boundary decision protocol as an example of the concept of network characteristic cognition protocol.

1. **Distributed Boundary Decision Protocol Flow**

According to Definition 2.11, the proximity groups $v(q)$ of the missile members ε_i are the collection of neighbor nodes of the missile members ε_i, and $v(q) = \{(i,j) \in \varepsilon \times \varepsilon : \mu_{ij} < d_{imax}, i \neq j\}$, that is $\forall \varepsilon_j \in v(q)$ such that $\mu_{ij} \leq d_{imax}$. In a two-dimensional plane, the set of proximity group boundary points of the missile members ε_i is $B(G, d_{imax})$, that is, the boundary of the network is a set of boundary

Table 7.1 Cluster header assignment results

$c = 1, k = 1$

Color	Red	Blue	Black	White	Green	Yellow
Nodes	40	0	0	0	250	210
Cluster head 40			Member 460			
Proportion			92%			

$c = 1, k = 2$

Color	Red	Blue	Black	White	Green	Yellow
Nodes	17	0	0	0	163	320
Cluster head 17			Member 483			
Proportion			97%			

$c = 2, k = 1$

Color	Red	Blue	Black	White	Green	Yellow
Nodes	43	38	0	0	239	180
Cluster head 81			Member 419			
Proportion			84%			

$c = 2, k = 2$

Color	Red	Blue	Black	White	Green	Yellow
Nodes	15	14	0	0	249	222
Cluster head 29			Member 471			
Proportion			94%			

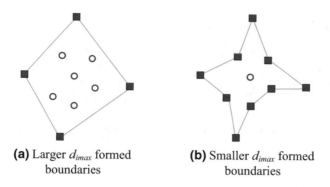

(a) Larger d_{imax} formed
boundaries

(b) Smaller d_{imax} formed
boundaries

Fig. 7.27 Relationship between adjacent distance and boundary

points $B(G, d_{imax})$ of graph G, which has the property that all edges of the points $B(G, d_{imax})$ belong to a closed polygon. So that the rest of the points are in the polygon. It is clear that the boundary of the network is related to the topology and the choice of the adjacency distance d_{imax} of the network, as shown in Fig. 7.27, where the larger adjacency d_{imax} distance makes the edge of the boundary smaller and tends to be convex, the smaller adjacency distance d_{imax} makes the boundary more edges and tends to concave boundary. The choice of adjacency distance d_{imax} depends on the requirement of the boundary application of the upper layer, such as the communication radius when studying the network communication. In the third chapter, when the light transmission is studied, the adjacency distance d_{imax} is $d_{imax} = k_{imax} \cdot d_{si}$, where k_{imax} the adjacency coefficient of the node is, d_{si} is safe distance.

Based on the above analysis and taking into account the scalability of the formation, the real-time and the Ad hoc characteristics of the formation network, the boundary decision-making protocol should be distributed and have the ability to judge the concave boundary. At the present stage, most of the research results are centralized [139, 140]. Although some of the literatures propose distributed concave boundary decision algorithms, they all assume that high network node density and node uniformity [141–143], and the network boundary studied in this section depends on the adjacent distance d_{imax}. Adjacent distance d_{imax} is a variable parameter and network node density and node uniformity are relative, so these assumptions are not applicable. In view of this problem, this part proposes a distributed boundary decision protocol based on extended message, which can be used to decide the concave boundary of the formation network well according to the adjacency distance d_{imax}.

The overall process is shown in Fig. 7.28 [74], which is divided into two processes. One is the convex boundary decision, each node ε_i according to the adjacent distance d_{imax} collects the information of the surrounding nodes, and determines whether it is a convex node in a distributed way (Fig. 7.28a), then determines the convex boundary composed of convex nodes (Fig. 7.28b). The other one is the concave boundary decision, each convex node sends extended message to a specific

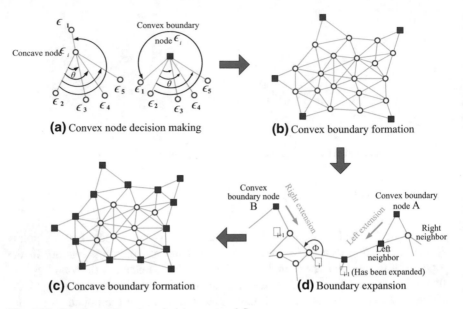

Fig. 7.28 Distributed boundary decision protocol flow

node in distributed from (Fig. 7.28d), so that adjacent convex nodes are connected to the entire concave boundary (Fig. 7.28c).

2. **Convex boundary decision**

From Eq. (2.2), $v(q) = \{(i,j) \in \varepsilon \times \varepsilon : \mu_{ij} < d_{imax}, i \neq j\}$ is a set of neighbor nodes of node ε_i whose radius is adjacent distance d_{imax}, that is $\forall \varepsilon_i \in v(q) = \{(i,j) \in \varepsilon \times \varepsilon : \mu_{ij} < d_{imax}, i \neq j\}$ such that $\mu_{ij} < d_{imax}$. As shown in Fig. 7.28a, the angle information or position information of $v(q) = \{(i,j) \in \varepsilon \times \varepsilon : \mu_{ij} < d_{imax}, i \neq j\}$ is collected by the node ε_i at first (the position information is not necessary in this protocol, but when the angle information can't be obtained, the angle can be determined by the position solution), and then arbitrarily choose a reference edge e_{ir}, find the counterclockwise angle $\theta_j = \angle e_{ir} e_{ij}$ with beginning edge e_{ir} and the ending edge e_{ij}, where $j \in \{v(q) - \varepsilon_r\}$. The final calculation of decision-making is as follows:

$$\Delta\theta_i = \min\{\theta_j \cup 2\pi \mid \pi < \theta_j \leq 2\pi\} - \max\{\theta_j \cup 0 \mid 0 \leq \theta_j \leq \pi\} \quad (7.2)$$

If $\Delta\theta_i > \pi$, the node ε_i is a convex node. The node ε_j in which the minimum value is obtained in (7.2) is called the left neighbor of the node ε_i. The node ε_j that gets the maximum value is called the right neighbor of the node ε_i. For the entire network, all convex nodes eventually form the convex boundary of the network.

3. Concave boundary decision making

The convex boundary is not a complete network boundary. In order to obtain the desired concave boundary, this protocol is based on the convex node, which sends the left (right) extension message to the left (right) neighbor, and then passes the process. For the left (right) boundary expansion, the following are left extension without special instructions, and the right extension of the same. If a node receives an extended message during the extension process, it will be decided to be the extended boundary node. When a node determines that the stop condition is satisfied, the extension is completed. Boundary expansion, which is a seemingly simple process, is complex in the actual implementation process due to the variety of connections between nodes, mainly as expanding collinearity, extending cross-over phenomena and extending the common point phenomenon.

(1) The determination of the extension point

As shown in Fig. 7.28b, the node ε_i receives the extended message from ε_{i-1}, in order to determine the next extension node that is to whom to send the extended message. The node ε_i will compare the angle $\theta_j = \angle e_{i(i-1)}e_{ij}, j \in \{v(q) - \varepsilon_r\}$ and select the node $\varepsilon_{i+1} = \arg\min(\theta_j)$, which select the smallest node for the next expansion node. The minimum angle $\Phi = \angle e_{i(i-1)}e_{i(i+1)}$ is called the expansion angle. The node ε_i then sends an extended message to the node ε_{i+1} and updates its status as a boundary node. Finally, all convex boundary nodes and extended boundary nodes form a complete network boundary.

(2) The expansion of common point phenomenon

As shown in Fig. 7.29, the phenomenon of common point include the internal boundary public point (Fig. 7.29a) and external boundary public point (Fig. 7.29b), there are two extensions that go through the public point. If the termination condition is simply designed as follows: ① A node receives the left and right extension messages in one communication cycle. ② The node receiving the extended message is the boundary node, then the expansion at the common point may stop unexpectedly. In order to solve this problem, the extension angle is introduced in the design of the termination condition, and the termination condition of the boundary extension are:

Condition 1: If the node ε_i is a boundary node of the extension angle $\Phi > \pi$, the extension stops (Node A in Fig. 7.29c).

Considering that the expansion is carried out of the left and right directions, and the different left (right) expansion directions at the common point are opposite, we have

Condition 2: If the left (right) extension of the node ε_i (such as node I in Fig. 7.29) is ε_{i+1} (as node H in Fig. 7.29c) and the node ε_i receives the right (left) extended message from ε_{i+1}, then the extension of node ε_i stops.

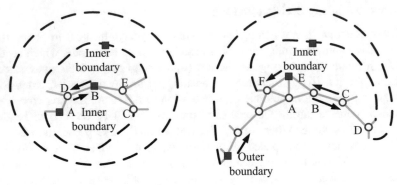

(a) Inner boundary phenomenon   **(b)** Internal and external boundaries of common point phenomenon

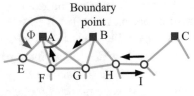

(c) Extended stop condition

Fig. 7.29 Common point phenomenon and extended stop condition

(3) The expansion of collinear phenomenon

As shown in Fig. 7.30, when a boundary node (convex or extended boundary node) calculates the next expansion point, the expansion angle of multiple nodes simultaneously obtains the minimum value, as shown in Fig. 7.30, C and B is the minimum expansion angle, but if sends message to C will get undesirable results. So when A knows the position information of C and B, t will choose the nearest distance node B for the expansion node. When A only have the angle information, it will sent extend the message to C and B at the same time and leave the decision to C and B. Because the collinear node C (B) can only determine whether it is only

Fig. 7.30 The expansion of collinear phenomenon

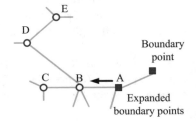

between the boundary node A and another expansion node B (C), node B will continue pass the extended message, and C is out of the expansion (not updated state).

(4) The expansion of cross phenomenon

The expansion of cross phenomenon is an undesired extension caused by an extended path in the process of expansion, intersecting its previously expanded path. As shown in Fig. 7.31a is a complex cross phenomenon, when A is extended, according to the principle of the extension point selection, it will be A–B–E—F–D (where the dotted line represents a series of extension points), i.e. BE and FD has crossed.

We can extract a basic pattern from this cross phenomenon, as shown in Fig. 7.31b. During the extension process, CA and DB intersect at a virtual point S, where C and D are not connected (otherwise C will expand D), and the number of expansion nodes between A and D is variable. Let's look at the adjacency between A and B, B and C.

$$\because CS + DS > CD > d_{imax} \quad \nexists AC + BD \leq 2d_{imax}$$

$$\therefore AS + BS < d_{imax}$$

$$\because AS + BS > AB$$

$$\therefore AB < d_{imax}$$

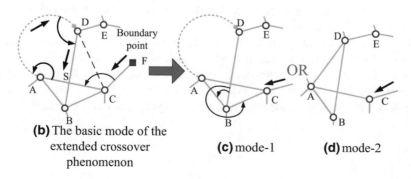

(a) Extended cross phenomenon

(b) The basic mode of the
extended crossover
phenomenon

(c) mode-1   **(d)** mode-2

Fig. 7.31 Extended crossover and its decomposition

So there is a connection between A and B. Similarly, if B and C are not connected, then A and D must have a connection (i.e. there is no other extension between A and D). Therefore, the basic cross mode can be decomposed into two sub-modes: mode-1 and mode-2, as shown in Fig. 7.31c, d. The algorithm of the two modes is designed as follows [74]

Algorithm for Determining Crossover Phenomenon in Extended Boundary Point (Left extension, and right extension is the same)
01: $\theta_L \leftarrow$ Minimum counterclockwise angle from DB to BA_i
02: $\theta_R \leftarrow$ Maximum counterclockwise angle from DB to BA_i
03: **if** $\theta_R - \theta_L > \pi$
04: Match mode-1
05: **else if** A_i and D are connected and $\theta_L < \pi$
06: Match mode-2
07: **else**
08: No crossover
09: **end**

In addition, in order to trigger the cross judgment algorithm, it is necessary to add the discarding point information in the extended message. If the next extended node of a node is the discarded point, the judgment algorithm is triggered. If the intersection occurs, we exchange extension point and the extended point to make the expansion go on.

(5) Boundary expansion process

In summary, the general flow of boundary expansion taking into account the above three phenomena is given in Fig. 7.32 as follows:

4. **Simulation of Distributed Boundary Decision Protocol**

The validity of the above distributed boundary decision protocol is verified by an example of experiment. Suppose that the condition of 300 nodes is randomly distributed in the range of 5000 m^2 and the adjacent distance of all nodes is assumed to $d_{imax} = d_{max} = 300$ m. The simulation results are shown in Fig. 7.33. In the figure, the red square point is the convex boundary node and the red dot is the extended boundary node.

The simulation results show that the distributed boundary decision protocol proposed in this section makes it easy for each node to determine its own state (whether it belongs to the boundary), especially for the distributed decision of the concave boundary of large-scale Ad hoc network and only have the angle information, which weakening the requirements of the agreement on the node.

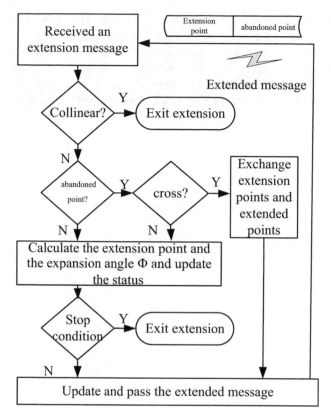

Fig. 7.32 Boundary expansion total flow

Fig. 7.33 Simulation of distributed boundary decision protocol

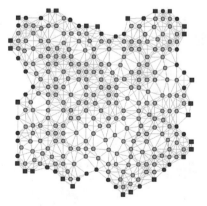

In this chapter, the concept of network support protocol is defined first. Then, the design method of wireless ad hoc network communication protocol based on STDMA is given for the specific needs of autonomous formation, and the autonomy of formation network is well reflected. A cluster decision protocol with redundancy is designed for hierarchical network model, and a distributed boundary decision protocol is proposed for network feature cognition protocol.

Chapter 8
Simulation and Verification for the CGCS

8.1 Foreword

At present, most of the reports on the field of CGCS-of-MAF are mainly focused on the theoretical methods, and more in-depth research results are also involved in the UAV's formation. For example, more research of UAV close coupled formation flight control based on the intelligent control method, the performance of formation flight control system is verified by digital simulation, and the simulation results are limited to the reference of practical engineering application. In recent years, in order to adapt to the rapid development of aircraft formation cooperative guidance and control theory, better verify formation cooperative guidance and control system architecture and design methods, and explore and discover the problems in the process of combining theory with engineering practice, the domestic and foreign research institutions have begun to pay attention to the research of the hardware in the loop (HIL) experimental verification method and the hardware test platform of the aircraft formation flying control algorithm [144, 145]. The United States and Australia have respectively developed the HIL open control platform [146–148], for the development and testing of missile flight control algorithm. In addition, the United States How et al. developed [149] testing platform for multi plane formation flight, though still failed to achieve autonomous formation flight control test completely, its formation test still depends on the ground station remote control, but there is no doubt that it has made the cooperative guidance and control technology to the engineering application of a big step. Domestic research results are reported less, the research on the cooperative guidance and control of MAF is still in the stage of theoretical tracking, and some research institutions such as colleges and universities have made some progress in the study of formation simulation [150−152].

© National Defense Industry Press, Beijing and Springer Nature Singapore Pte Ltd. 2019
S. Wu, *Cooperative Guidance & Control of Missiles Autonomous Formation*,
https://doi.org/10.1007/978-981-13-0953-3_8

8.2 Simulation and Verification for the CGCS

The HIL for the CGC-of-MAF's simulation platform is still in the exploratory stage in domestic, the IPCLab of Beihang University earlier carried out in this area related research and has built a "demonstration verification platform for CGCS of MAF (CGCS of IPCLab)", the platform has the characteristics of open architecture, flexible function and convenient use, the related experimental results of the cooperative guidance control of missile autonomous formation are completed by this platform. In this chapter, we will take the CGCS of IPCLab as an example to introduce the design and implementation of the simulation and verification experiment of cooperative guidance control system.

The CGCS of IPCLab is mainly composed of three parts: digital simulation system for the CGCS-of-MAF, hardware in the loop (HIL) simulation test system for the CGCS-of-MAF, embedded autonomous formation equivalent flight test verification system, including the embedded small UAV system, member flight control system, wireless self-organizing support network system, cooperative guidance control module and other major equipment. The CGCS of IPCLab can provide an important basic supporting role for the related research of MAF and its application and realization.

8.2.1 Function Framework for the CGCS of IPCLab

The functional framework for the CGCS of IPCLab includes three levels: digital simulation, HIL's simulation and physical equivalent verification, it is a hierarchical simulation design verification system with a relatively complete system, which can be convenient to analyze, detect and verify the theoretical method, system framework and technology integration in different stages, and improve the overall efficiency of system design, implementation and simulation verification. The function system framework for the CGCS of IPCLab shown in Fig. 8.1.

8.2.2 The Mathematical Simulation System for the CGCS of IPCLab

The digital simulation system for the CGCS of IPCLab is a set of computer system software for simulation and verification of the formation cooperative control method, it has the characteristics of high integration, good visualization, easy cutting and flexible operation, and covers the assembly area, flight area, shift area, terminal guidance area and performance evaluation of missile autonomous formation. The system can carry out independent simulation of each individual link, and

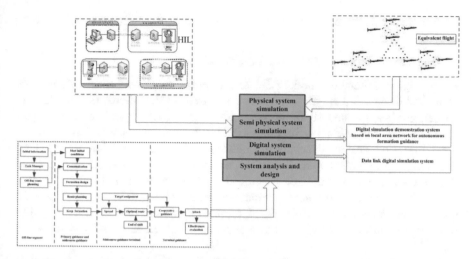

Fig. 8.1 The functional system framework for the CGCS of IPCLab

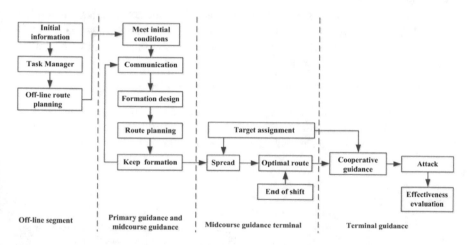

Fig. 8.2 The simulation process for the CGCS of IPCLab

can also verify the whole system simulation. The digital simulation process for the CGCS of IPCLab shown in Fig. 8.2.

The software system for the CGCS of IPCLab can work in two modes: First, the centralized digital simulation mode, that is, all formation nodes and support network models are implemented on the same terminal, easy to design, debug and verify the formation of cooperative guidance control algorithm, but limited to the terminal processing capacity, the number of nodes is limited; The other is the distributed digital simulation mode, that is, the formation node model grouping is realized on multiple terminals respectively. The support network is simulated under

TCP/IP protocol, the network communication part of the formation with a better simulation effect.

1. Digital simulation system based on Simulink

In the digital simulation model of a centralized, one part of a formation flight controller design and simulation system, is based on the Microsoft VS platform written in C++, the underlying code is more flexible, rich interface; another part based on the mathworks of Simulink/MATLAB platform construction, it convenient to visual design and improve efficiency. The main section is based on Simulink, where the single-node module includes (Figs. 8.3, 8.4, 8.5, 8.6, 8.7 and 8.8): the kinematic solution sub-module (Fig. 8.3), the force and torque resolver sub-module (Fig. 8.4), the aerodynamic parameter solving sub-module (Fig. 8.5), the air environment sub-module (Fig. 8.6), and engine sub-module (Fig. 8.7). The five sub-modules are combined into a complete single-node module as shown in Fig. 8.8.

The single node module and the formation flight controller sub-module (Fig. 8.9) are combined to form the formation node module with the ability of formation (Fig. 8.10).

Combining the multiple node formation module and communication module (Fig. 8.11) of the formation, it can obtain the digital simulation system of MAF (Fig. 8.12), which is based on the virtual reality of the integrated demonstration sub-module as shown in Fig. 8.13.

2. Distributed formation digital simulation system based on TCP/IP

The main difference between distributed simulation and centralized is that TCP/IP protocol is used to simulate the communication protocol in the local area network, and several terminals are connected through the wired network.

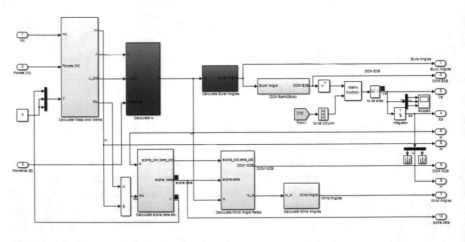

Fig. 8.3 The kinematic solution sub-module

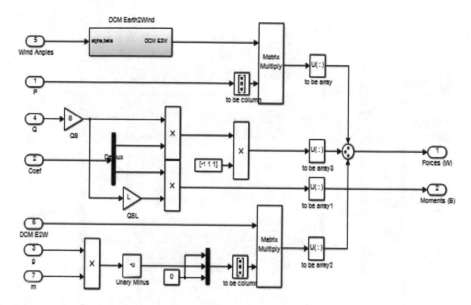

Fig. 8.4 The force and torque resolver sub-module

The simulation process of the network communication protocol is shown in Fig. 8.14. The layer protocol simulation process is shown in Figs. 8.15, 8.16 and 8.17.

Under the condition of wired network communication and TCP/IP protocol, because the quality of service of the network is better, the packet loss and delay are relatively low, the simulation of this stage has a good test effect on the logical function of the network communication protocol.

Through the evaluation of the combat effectiveness of the offensive and defensive combat environment based on the distributed interactive simulation environment, the formation scale of the participating simulation can't be restricted by the capability of the single terminal calculation, and the operation status of the network communication protocol can be checked, at the same time to a certain extent, the entire formation system exposed to a relatively real network environment, the algorithm can test the robustness of network quality of service. Figure 8.18 shows the monitoring and display of attack and defense combat simulation.

8.2.3 The Simulation and Test System Based on HIL for the CGCS of IPCLab

Broadly speaking, the HIL's simulation system can be equivalent to the semi-physical simulation system. In a narrow sense, it refers to a kind of the

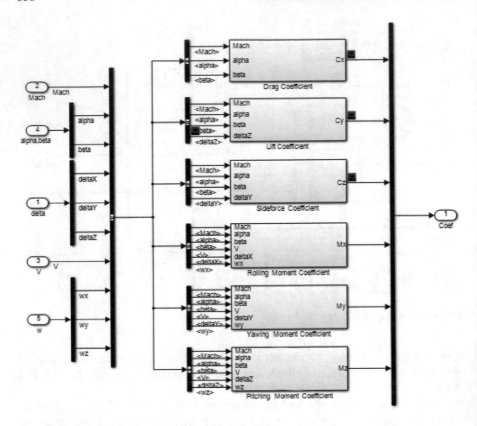

Fig. 8.5 The aerodynamic parameter solving sub-module

semi-physical simulation system composed of the actual controller and the virtual controlled object. The HIL simulation test system structure for the CGCS of IPCLab shown in Fig. 8.19, only the missile node with the computer and turntable simulation, the rest are in kind.

8.2.4 The Embedded Platform for Equivalent Flight Test Verification of the CGCS

The "Open Control Platform—Hardware-In-The-Loop" project, launched by the US Air Force in 2003, has been used in the study of single-aircraft control systems. RMUS (Real-time Multi-UAV Simulator) system and multi-UAV experimental platforms have also been established [144–148], but whether these experimental systems can be used for missile autonomous formation comprehensive validation has not yet been publicly reported.

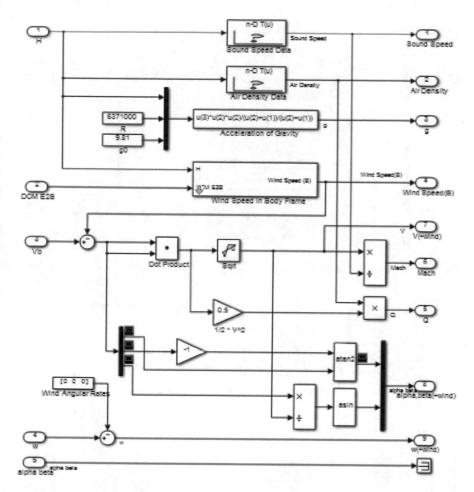

Fig. 8.6 The air environment sub-module

At present, using multiple UAVs to implement equivalent flight test is one of the more cost-effective means of autonomous formation cooperative guidance control key technologies demonstration and validation, the IPCLab of Beihang University, has exploratory developed the equivalent flight test system for missile autonomous formation, and has made important progress. The CGCS of IPCLab embedded equivalent flight test system includes autonomous formation cooperative guidance control system (flight control module, formation guidance module, formation support network module, and emergency control system), ground monitoring system and UAV system. The CGCS of IPCLab embedded flight test verification system uses a hybrid structure of real node and virtual node, the module is based on the VxWorks system development, the system is stable and reliable. The flight data is transmitted back to the ground monitoring system through the wireless support

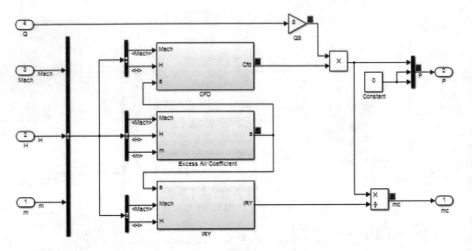

Fig. 8.7 The engine sub-module

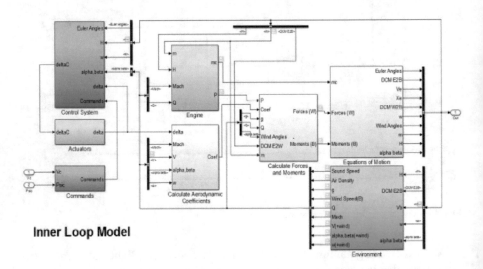

Fig. 8.8 The single-node module

network and stored in the flight test database to provide support for the later analysis. The CGCS of IPCLab embedded equivalent flight test system structure as shown in Fig. 8.20, where the UAV system structure shown in Fig. 8.21.

1. Support network system

The wireless self-organizing support network physical framework for the CGCS of IPCLab embedded equivalent flight test system is shown in Fig. 8.22, the system consists of embedded wireless self-organizing support network protocol,

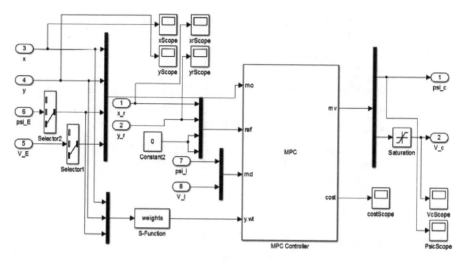

Fig. 8.9 The formation flight controller sub-module

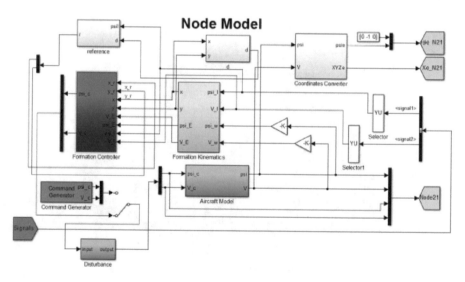

Fig. 8.10 The formation node module

communication terminal and interface circuit. Among them, the communication terminal to complete digital signal processing, network control, security authentication, message processing and other functions. The UAV system according to the structure, power consumption and other characteristics to select the suitable amplifier and antenna.

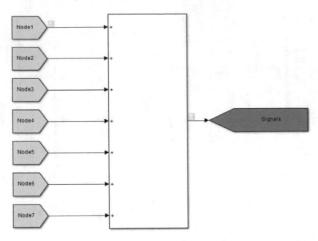

Fig. 8.11 The communication module

Fig. 8.12 The digital simulation system of MAF

(1) Data frame design. The UAV system through the wireless self-organizing support network to transfer the data defined as two frames, which are stored and send in hexadecimal, 10 Byte per frame. Above the data frame is the network parameter frame, which indicates the current network state and the total number of nodes and the ID number of the UAV.

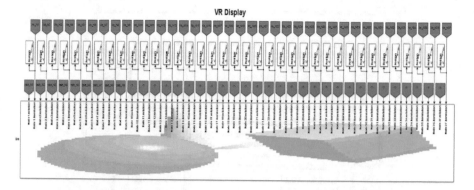

Fig. 8.13 The integrated demonstration sub-module

(2) Protocol implementation. The protocol will be implemented in an embedded system. Select the appropriate embedded system, through the serial port and the radio to exchange data with the upper user. Each layer protocol is programmed in standard C language, each layer protocol exists independently in the embedded system, and the interaction between protocols is the interaction between processes. The communication terminal is the key equipment of the wireless ad hoc network system, and its performance index has a strong constraint on the design of the subsequent upper layer protocol. It should be combined with the actual needs of different applications, flexible use the different performance indicators of radio equipment. These performance indicators include: channel data rate, transmit power, transmit and receive delay, receiver sensitivity, anti-jamming performance, and the size and weight of the radio. It is worth noting that the delay is large when the half-duplex radio station in the send and receive state, when the network nodes need a higher update rate, each radio transceiver state constantly changes, leading to low utilization rate of the

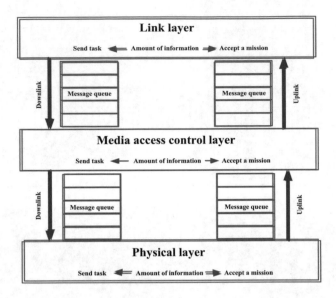

Fig. 8.14 The simulation process of the network communication protocol

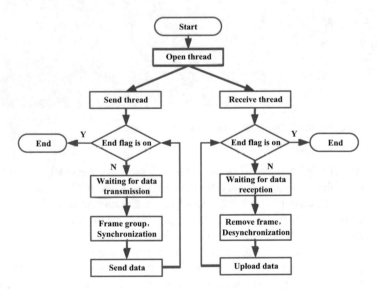

Fig. 8.15 The simulation process of the physical layer protocol

network channel, which seriously restricts the improvement of the network update rate, therefore, it should focus on this index during system design.

At present, there are many alternative digital radio stations for the experiment, and the spread spectrum radio is more suitable for this category, including the direct sequence spread spectrum and frequency hopping spread spectrum two spread spectrum system.

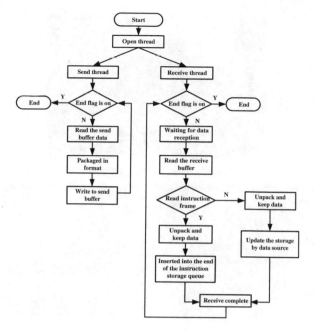

Fig. 8.16 The simulation process of the MAC layer protocol

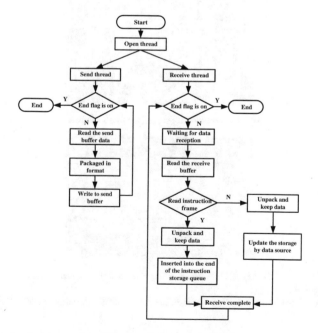

Fig. 8.17 The simulation process of the data link layer protocol

(a) The initial interface of MAF

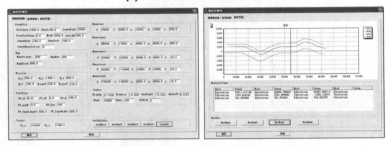

(b) The network data monitoring interface

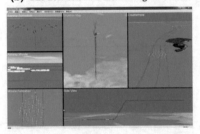

(c) The formation flight path monitoring interface

(d) The formation combat effectiveness evaluation results

Fig. 8.18 The surveillance and display interface for combat simulation of attack and defense

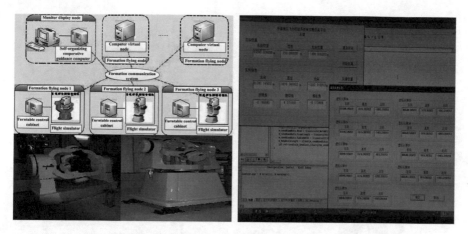

Fig. 8.19 The HIL simulation test system

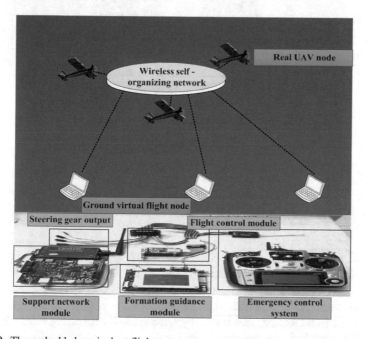

Fig. 8.20 The embedded equivalent flight test system structure

(3) Physical system. The wireless self-organizing network physical system used for
    the CGCS of IPCLab embedded flight test verification system includes the
    following aspects:

  (1) The embedded development environment is built to complete the hardware
      design of embedded system;

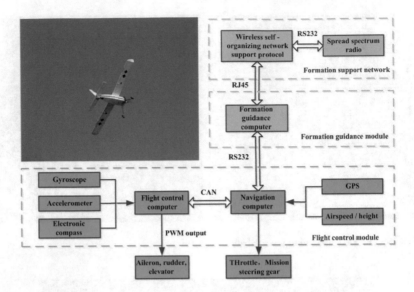

Fig. 8.21 The UAV system structure

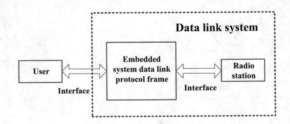

Fig. 8.22 The system framework

(2) According to the special requirements of physical simulation, the wireless self-organizing network protocol based on computer system is modified and the parameter setting is adjusted to obtain the wireless self-organizing network protocol which satisfies the physical bottom of the spread spectrum radio station, and tested by wireless self-organizing network system composed of computer and radio.

(3) Transplant protocol.

2. Embedded member flight control system

The member flight control system uses the PID controller. Mainly use seven PID control channels: Target roll → aileron rudder angle (PID1), Target pitch → elevator angle (PID2), Target course → direction rudder angle (PID3), Target course → target roll (PID4), Side offset → target course (PID5), Target height → target pitch (PID6), Target speed → throttle volume (PID7). They form the UAV lateral control loop, the height control loop and the speed control loop.

When the STT turning control is implemented, the yaw angle command is calculated by the target position and its own position, and the steering angle is generated by the change of the yaw angle command; at the same time, the rolling channel remains stable and the roll angle command is zero, the flow chart shown in Fig. 8.23. When the longitudinal motion control is implemented, the PID6 channel gives the pitch angle command with reference to the target height and the current height, and the PID2 channel controls the elevator to achieve the target pitch angle, the flow chart is shown in Fig. 8.24.

The other kinds of motion control process similar to this, not repeat here. The dynamic parameters and physical parameters of UAV are found in literature [4, 153].

Selected $[V \quad \alpha \quad \beta \quad p \quad q \quad r \quad \phi \quad \theta \quad \psi \quad x \quad y \quad h \quad m \quad \delta_a \quad \delta_e \quad \delta_r \quad \delta_t]^T$ as the system state vector, the design of the member flight controller is:

$$\begin{cases} \delta_e^* = \delta_{e0} - k_1 \cdot (n_y - 1) - k_2 \cdot q - k_3 \cdot (\theta - \alpha_0) - k_4 \cdot (h - h^*) \\ \delta_t^* = t_A - k_5 \cdot (V - V_0) \end{cases} \tag{8.1}$$

where, $k_1 = -0.068$, $k_2 = 0.2632$, $k_3 = 2.4003$, $k_4 = 0.006$, $k_5 = 0.46305$. The definition of the relevant variables, the non-linear six-degree-of-freedom kinematics and dynamics modeling methods and the conversion between the coordinate systems are shown in the Ref. [75, 150], stochastic robust analysis and design (SRAD) method for member flight controllers see Ref. [1].

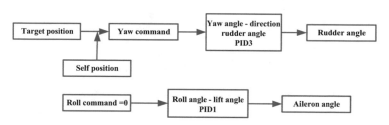

Fig. 8.23 The STT turning control flow

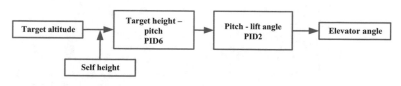

Fig. 8.24 The longitudinal motion control flow

3. **Analysis of the influence of network random delay**

(1) Random delay of network control system

From the delay type, the network random delay can be divided into constant delay, independent random delay and random delay of the Markov characteristic. From the point where the delay occurs, the network random delay can be divided into the sensor observation delay τ_o and the actuator control delay τ_c, as shown in Fig. 8.25. This section mainly uses independent random delay as the main delay type of the networked flight control system, assuming that the sensor observation delay and the actuator control delay obey the mean μ_d, the normal distribution of variance σ_d, i.e.:

$$\tau_o \in N(\mu_d, \sigma_d^2), \tau_c \in N(\mu_d, \sigma_d^2) \tag{8.2}$$

The normal distribution of the network delay as shown in Fig. 8.26.

In Fig. 8.26, the circle position represents the delay size, the abscissa represents the time at which the delay occurs, and the time of the delay occurs is not uniform,

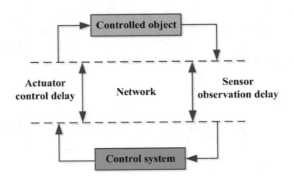

Fig. 8.25   The schematic diagram of sensor and actuator delay

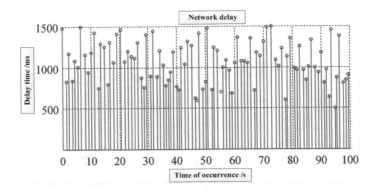

Fig. 8.26   The normal distribution of network latency

and the nonlinear system equation of the UAV is changed from the original $\dot{\bar{X}}(t) = f(\bar{X}(t), u(t))$ to:

$$\dot{\bar{X}}(t) = f(\bar{X}(t - \tau_o), u(t - \tau_c)) \qquad (8.3)$$

where: $\bar{X}(t)$ is the flight state vector, $\dot{\bar{X}}(t)$ is the derivative of the flight state vector, $\bar{X}(t - \tau_o)$ is the flight state vector added to the sensor observation delay, $u(t - \tau_c)$ is the flight control vector added to the actuator control delay, the randomness of network delay provides the conditions for the application of stochastic robust analysis.

(2) Conservatism of robust control for traditional equivalent upper bound network systems

It is assumed that the observer delay and actuator control delay obey the normal distribution with mean μ_d of 500 ms and variance σ_d of 250 ms. The robustness of the control law for UAV control is investigated under the condition of stochastic network delay, and 50 Monte Carlo simulations were performed on the six-degree-of-freedom model of the system with the flight control system and the network random delay, to statistics height of the overshoot and adjust the time, as shown in Fig. 8.27 (height in the figure is m, the time unit is s). The following conclusions can be drawn:

In the case where there is a sensor observation delay τ_o and an actuator control delay τ_c, the maximum overshoot reaches 60%, with respect to the sensor observation delay τ_o and actuator control delay τ_c exists in the case of large.

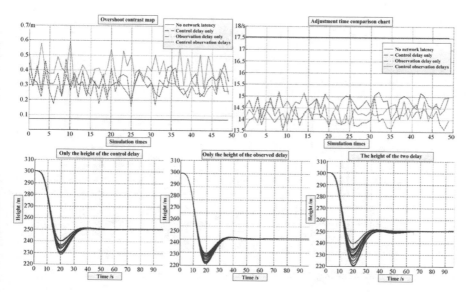

Fig. 8.27 $\mu_d = 500$ ms and $\sigma_d = 250$ ms of the normal distribution delay simulation results

For the sensor observation delay and the actuator control delay exist at the same time, all the network delay in accordance with the upper bound equivalent and then use the traditional robust control method designed control law is too conservative. In the case of both sensor delay τ_o and actuator control delay τ_c, and $\tau_o \in N(500\,\text{ms}, 250\,\text{ms})$, $\tau_c \in N(500, 250)$, if the equivalent delay distribution $\tau_{co} \in N(1000\,\text{ms}, 500\,\text{ms})$ is treated according to the upper bound, from Fig. 8.28, delay distribution such as τ_{co} in both the equivalent end of the sensor or in the actuator end, can control the maximum overshoot value of more than 100%, while the maximum overshoot of the same control law in Fig. 8.27 is not more than 60%. The reason shown in Fig. 8.28, the hollow circle in the figure represents the actuator control delay time, the solid circle represents the sensor observation delay time. After 37.5 s an actuator control delay 670 ms occurred, 38 s occurred with a sensor observed delay 340 ms, that is, in the [37.5 s, the 38.17 s] actuator has a delay, in the [38 s, and the 38.34 s] observer has a delay, while [38.17 s, 38.34 s] is the part of the control delay masking observation delay, this part belongs to the negligible part, and should not be included in the uncertainty Δ of the robust control and then increase the upper bound of uncertainty. The control delay is the direct cause of the decline in flight control quality, when the control delay mask part of the observation delay, it can be considered that this part of the observation delay has no effect on the decline of the flight control quality, so the network delay all in accordance with the upper bound equivalent, and then use the traditional robust control method designed control law is too conservative.

In order to evaluate the quality of system control, the average command deviation within 100 s of simulation time is taken as the measure of system control quality and the stochastic robust analysis of system is as follows:

$$AID = \frac{1}{T_0} \int_0^{T_0} |h - h^*| dt, \quad \text{among } T_0 = 100s \tag{8.4}$$

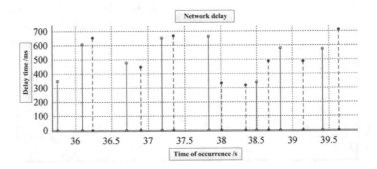

Fig. 8.28 The schematic diagram of equivalent upper bound robust control design

4. **Stochastic robust design for members networked flight control system**

The stochastic robust design schematic diagram for flight control system is shown in Fig. 8.29. The stochastic robustness of the system is 50 times Monte Carlo simulation to obtain the average AID, this index is used to measure the parameters of the particle swarm optimization algorithm, and the parameters of the controller are optimized by particle swarm optimization algorithm.

The controller structure is controlled by PID, using particle swarm to optimize the controller parameters, the number m of particles in the particle group is taken as 30, the dimensionality n of the solution vector is 5, the number k of iterations is 20, the controller parameter $k_1 \ldots k_5$ is optimized, and the stochastic robust design result of the network random delay $\tau_1 \sim N(\mu_1, \sigma_1^2)$ is shown in Fig. 8.30 (The height unit is m and the time unit is s). In this part, the cost convergence curve of the standard particle swarm algorithm is given, the minimum value of AID is 6.26927, the maximum AID obtained in Monte Carlo simulation is 6.33, and the feedback coefficient of the final control system is $K_1^N = \left[k_{11}^N, k_{12}^N, k_{13}^N, k_{14}^N, k_{15}^N \right]$ [4]:

$$k_{11}^N = -0.267848, k_{12}^N = 0.032116, k_{13}^N = 2.63188,$$
$$k_{14}^N = 0.00724909, k_{15}^N = -0.290237$$

5. **Gain scheduling method for flight control systems with random large latency**

In the case of large random variation of network, the single control law with stochastic robust analysis and design can't meet the control quality requirements under the various network delay, so based on the scheduling of different network status to adapt the state feedback gain is very necessary. The gain scheduling method is divided into traditional gain scheduling, gain scheduling based on LPV (linear variable parameter) and robust gain scheduling based on loop forming, the latter two methods have strict mathematical proof and quantifiable control quality evaluation standard, but the gain scheduling structure is complex, is not conducive to the realization and on-line adjustment; gain scheduling design process is complex and difficult for designers to master.

In this section, firstly, the stochastic robustness analysis and design method are used to overcome the shortcomings of the existing network control system design method. Secondly, the linear interpolation gain scheduling strategy is used to enlarge the control range, so that the control system can maintain better control quality under different network conditions; the control strategy is simple and does not include the complicated mathematical operation, which is convenient for real-time requirement of the hardware. The proposed control strategy can reduce the update rate of UAV wireless data link, and thus enhance the robustness of UAV formation flight.

(1) Stochastic robust gain design for three-stage network delay

Assuming that the network delay of the UAV control system can be divided into three levels, one type of network delay belongs to the short time delay, denoted as τ_1,

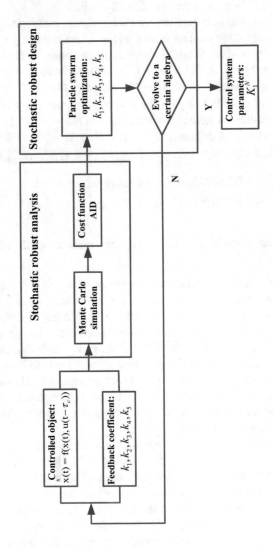

Fig. 8.29 The stochastic robust design schematic diagram

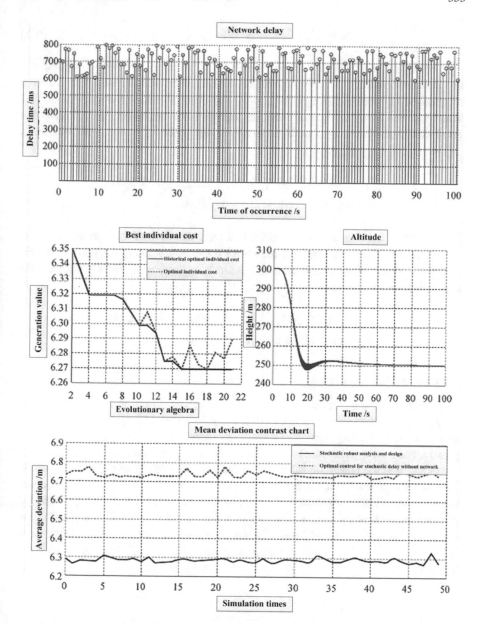

Fig. 8.30 Stochastic robust design for systems with delay τ_1

obey the mean μ_1 is 700 ms, the variance σ_1 is 100 ms standard normal distribution; one type of network delay belongs to middle time delay, denoted as τ_2, obey the mean μ_2 is 700 ms, the variance σ_2 is 100 ms standard normal distribution; one kind of network delay belongs to the long delay, denoted as τ_3, obey the mean μ_3 is 1100 ms, the variance σ_3 is 100 ms standard normal distribution, namely:

$$\tau_1 \sim N(\mu_1, \sigma_1^2), \tau_2 \sim N(\mu_2, \sigma_2^2), \tau_3 \sim N(\mu_3, \sigma_3^2) \tag{8.5}$$

The stochastic robust design results of the network random delay $\tau_1 \sim N(\mu_1, \sigma_1^2)$ are described earlier and will not be described here. For the stochastic robust design of random network delay $\tau_2 \sim N(\mu_2, \sigma_2^2)$, the controller uses the traditional PID structure, and uses the particle swarm optimization algorithm to optimize the controller parameters, the obtained feedback coefficient of the control system is $K_2^N = \left[k_{21}^N, k_{22}^N, k_{23}^N, k_{24}^N, k_{25}^N \right]$:

$$k_{21}^N = -0.303185, k_{22}^N = 0.0232822, k_{23}^N = 2.5781,$$
$$k_{24}^N = 0.00706601, k_{25}^N = -0.269634$$

Similarly, for the stochastic robust design of the network random delay $\tau_3 \sim N(\mu_3, \sigma_3^2)$, the controller still uses the PID structure, using the particle swarm optimization algorithm to optimize the controller parameters, the feedback coefficient of the control system is $K_3^N = \left[k_{31}^N, k_{32}^N, k_{33}^N, k_{34}^N, k_{35}^N \right]$:

$$k_{31}^N = -0.221949, k_{32}^N = 0.032926, k_{33}^N = 2.44389,$$
$$k_{34}^N = 0.00662643, k_{35}^N = -0.231925$$

(2) Gain scheduling strategy

Application of the above-mentioned stochastic robust design of the controller can only guarantee the network random delay within a certain range of robust performance to meet the requirements, when the random delay of the network changes greatly, the robust performance of the controller may not meet the requirement of the index, and the control system adopts the gain scheduling strategy is necessary. Here the linear interpolation method is used to schedule the gain of the network flight control system. Assume that the network delay between τ_1 and τ_i, when the random network delay in $\tau_1 \sim N(\mu_1, \sigma_1^2)$, $\tau_2 \sim N(\mu_2, \sigma_2^2) \ldots \tau_i \sim N(\mu_i, \sigma_i)$, using the stochastic robust analysis and design method to obtain the control law $K_1^N, K_2^N, \ldots, K_i^N$, and the controller currently measured network delay is τ_N, then the actual use of the controller feedback coefficient K_N is:

$$K_N = \begin{cases} \frac{\tau_N - \tau_1}{\tau_2 - \tau_1} \cdot \left(K_2^N - K_1^N \right) + K_1^N, & \tau_1 \leq \tau_N \leq \tau_2 \\ \frac{\tau_N - \tau_2}{\tau_3 - \tau_2} \cdot \left(K_3^N - K_2^N \right) + K_2^N, & \tau_2 \leq \tau_N \leq \tau_3 \\ \quad \vdots \\ \frac{\tau_N - \tau_{i-1}}{\tau_i - \tau_{i-1}} \cdot \left(K_i^N - K_{i-1}^N \right) + K_{i-1}^N, & \tau_{i-1} \leq \tau_N \leq \tau_i \end{cases} \tag{8.6}$$

Assuming the network delay between 600 and 1200 ms, the controller measured the network delay is τ_N, the actual use of the controller feedback coefficient K_N is:

$$K_N = \begin{cases} \frac{\tau_N - \tau_1}{\tau_2 - \tau_1} \cdot (K_2^N - K_1^N) + K_1^N, & \tau_1 \leq \tau_N \leq \tau_2 \\ \frac{\tau_N - \tau_2}{\tau_3 - \tau_2} \cdot (K_3^N - K_2^N) + K_2^N, & \tau_2 \leq \tau_N \leq \tau_3 \end{cases} \tag{8.7}$$

Figure 8.31 shows the overall flow chart for gain scheduling.

(3) Monte Carlo simulation of nonlinear closed-loop system

Figure 8.32 shows the height directive and the network random delay satisfies the following normal distribution:

$$\begin{cases} h^* = 250 \text{ m}, \tau_1 \sim N(\mu_1, \sigma_1^2), & t \leq 50 \text{ s} \\ h^* = 200 \text{ m}, \tau_2 \sim N(\mu_2, \sigma_2^2), & 50 \text{ s} < t \leq 100 \text{ s} \\ h^* = 150 \text{ m}, \tau_3 \sim N(\mu_3, \sigma_3^2), & 100 \text{ s} < t \leq 150 \text{ s} \end{cases} \tag{8.8}$$

The control law gain scheduling strategy is based on 50 Monte Carlo simulation results, the highly controlled quality meets requirements.

Figure 8.33 shows the height command for Eq. (8.9) and the uniform distribution of the network random delay.

$$\begin{cases} h^* = 250 \text{ m}, \tau_1 \sim U(a, b), & t \leq 50 \text{ s} \\ h^* = 200 \text{ m}, \tau_2 \sim U(a, b), & 50 \text{ s} < t \leq 100 \text{ s} \\ h^* = 150 \text{ m}, \tau_3 \sim U(a, b), & 100 \text{ s} < t \leq 150 \text{ s} \end{cases} \tag{8.9}$$

The control law of gain scheduling strategy is achieved across 50 times Monte-Carlo simulation results. Where $a = 600$ ms, $b = 1200$ ms is the worst case of gain scheduling, but the control quality still meets the design requirements.

In the light of the current problems of multi - variable flight control system design by using the network control method, this section introduces the stochastic robustness analysis and design method of network flight control system, the method to some extent alleviate the network delay of flight control system quality deterioration, and can enhance the stability of distributed system. Compared with the previous for linear time-delay systems rely on mathematical methods to obtain the control law, the design process and make full use of the computer data processing capability, on the basis of the linear system LQR method, stochastic robust analysis and design of nonlinear systems using Monte Carlo simulation and modern optimization algorithms, and the engineering and operability are strong [1, 4].

6. Ground monitoring system

The ground monitoring system functions for the CGCS of IPCLab embedded test flight verification system include electronic map display, telemetry data monitoring, flight attitude meter, location and attitude remote control, telemetry data recording and playback, route editing, map loading management, etc., it can carry out rapid detection and fault diagnosis (see Fig. 8.34).

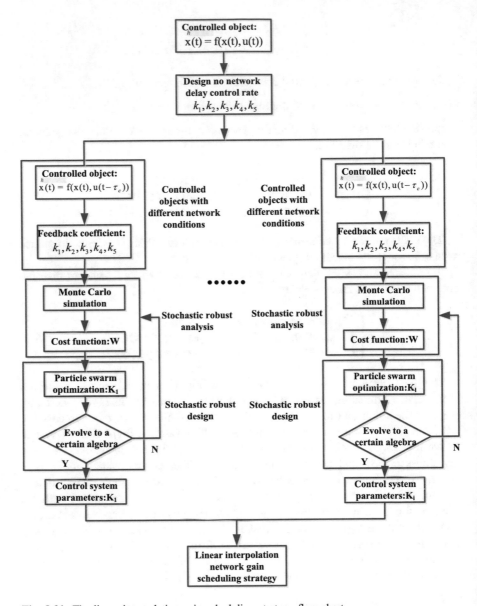

Fig. 8.31 The linear interpolation gain scheduling strategy flow chart

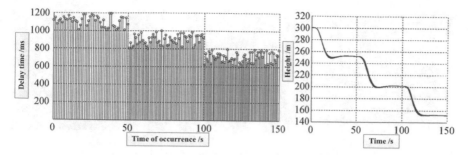

Fig. 8.32 The normal distribution network delay and closed loop nonlinear Monte Carlo simulation verification

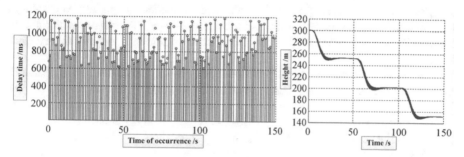

Fig. 8.33 The uniformly distributed network delay and closed-loop nonlinear Monte Carlo simulation

Fig. 8.34 Ground monitoring station for the CGCS of IPCLab

Fig. 8.35 The "Lead-Follow" mode autonomous formation test flight

8.3 The Equivalent Flight Test Verification

At present, the CGCS of IPCLab embedded equivalent flight test verification system has been completed in the field test, the aim of the equivalent flight test verification of autonomous formation cooperative guidance control theory and technology is realized.

Autonomous formation flight test scene for the CGCS of IPCLab embedded equivalent flight test verification system "Lead-Follow" mode as shown in Fig. 8.35.

The ground monitoring system can telemetry and remotely control any UAV, as shown in Fig. 8.36. The UAV's real-time gesture display window is shown in Fig. 8.37. Flight test results show that the missile can well follow the leader to complete the specified task of formation flight, after emergence of gust disturbance also can quickly restore the formation, verify the equivalent test system design is reasonable and reliable work.

The performance of the supporting network is also tested in the equivalent flight test. The test results show that the support network can provide stable data transmission channel, network delay can meet the requirements of the small UAV formation flight, in the process of communication, there is basically no data collision caused by time slot synchronization. At the same time in the communication distance transfinite and external interference, etc., it can still maintain the stability of the network structure. The temporary loss of the node in the communication after

Fig. 8.36 The flight control command

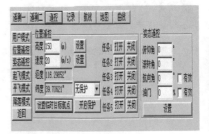

Fig. 8.37 The flight attitude
monitoring window

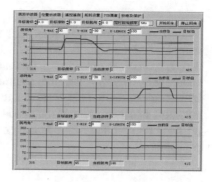

Fig. 8.38 The network
monitoring window

the recovery of the situation can be successfully rejoined, and the loss of any node in the network will not affect the normal network jobs.

In order to real-time detect the wireless self-organizing network communication with the free network monitoring station to listen, and according to the listening data to determine the current network status and the status of the communication nodes, monitoring situation as shown in Fig. 8.38.

Through the CGCS of IPCLab from the three levels of digital simulation, hardware in the loop (HIL) simulation test and embedded equivalent flight test verification to study the theory and technology of missile autonomous formation cooperative guidance control, and achieved the stage results. However, in the practical engineering application, when the missile autonomous formation execute the task, the information of the formation depends mainly on the wireless self-organization support network transmission, When the formation of the low altitude penetration flight will have to face the problem of multipath effect and absorption of electromagnetic wave, and the influence of the Doppler effect on communication data in high speed flight, especially when the enemy electromagnetic suppression and confrontation, communication ability and system security support network formation may be seriously affected. Therefore, it is a development direction that is highly worthy of attention in this field, which enhances the low intercept, anti-jamming and strong fault tolerance of missile autonomous formation system. The research contents mainly though hard kill tasks as the starting point of autonomous missile formation, but the basic theory and design implementation technique of the book is also applicable to more diversified cooperative operations research. For example, in the national defense field including military reconnaissance, early warning, communications relay, electronic warfare, the temptation of harassing and battlefield assessment; in the civil field including geodesy, meteorological observation, environmental monitoring, resource exploration, forest fire, artificial rainfall and traffic management and other aspects have broad application prospects.

References

1. Wu, S.: Stochastic Robustness Analysis and Design for Guidance and Control System of Winged Missile. National Defense Industry Press, Beijing (2010)
2. Liu, X.: Study on Self-organizing Networked Control System of Multi-missiles Formation. Beihang University, Beijing (2009)
3. Mu, X.: Methods of Autonomous Formation Control and Cooperative Route Planning for Multi-missiles. Beihang University, Beijing (2009)
4. Sun, J.: Study on Cooperative Guidance and Control of Vehicles Formation. Beihang University, Beijing (2012)
5. Wang, P.K.: Navigation strategies for multiple autonomous mobile robots moving in formation. J. Robotic Syst. **8**(2), 177–195 (1991)
6. Jaydev, P.D., Jim, O., Vijay, K.: Controlling formations of multiple mobile robots In: IEEE International Conference on Robotics and Automation, pp. 2864–2869. Belgium (1998)
7. Yamaguchi, J., Burdick, W.: Asymptotic stabilization of multiple nonholonomic mobile robot forming group formations. In: IEEE International Conference on Robotics and Automation, pp. 3573–3580. Belgium (1998)
8. Tanner, H.G., Pappas, G.J., Kumar, V.: Leader to formation stability. IEEE Trans. Robot. Autom. **20**(3), 443–455 (2004)
9. David, J.N., Gaurav, S.S.: Negotiated formations. In: International Conference on Intelligent Autonomous Systems, pp. 181–190. Amsterdam (2004)
10. Das, A.K., Fierro Kumar, R., et al.: A vision base formation control framework. IEEE Trans. Robot. Autom. **18**(5), 813–825 (2002)
11. Reynolds, C.W.: Flocks, herds, and schools: a distributed behavioral model. Comput. Graph. **21**(4), 25–34 (1987)
12. Jadbabaie, A., Lin, J., Morse, A.S.: Coordination of groups of mobile autonomous agents using nearest neighbor rules. IEEE Trans. Autom. Control **48**(6), 988–1001 (2003)
13. Balch, T., Hybinette, M.: Social potentials for scalable multi-robot formations. In: Proceedings IEEE International Conference on Robotics and Automation, vol. 1, pp. 73–80. SanFransisco (2000)
14. Kostelnik, P., Samulka, M., Janosik, M.: Scalable multi-robot formations using local sensing and communication. In: Proceedings of the 3rd International Workshop on Robot Motion and Control, pp. 319–324. Poznan (2002)
15. Lewis, M.A., Tan, K.H.: High precision formation control of mobile robots using virtual structures. Auton. Robots **4**(4), 387–403 (1997)
16. Fax, J.A., Murray, R.M.: Information flow and cooperative control of vehicle formations. IEEE Trans. Autom. Control **49**(9), 1465–1476 (2004)

© National Defense Industry Press, Beijing and Springer Nature Singapore Pte Ltd. 2019
S. Wu, *Cooperative Guidance & Control of Missiles Autonomous Formation*,
https://doi.org/10.1007/978-981-13-0953-3

17. Saber, O., Murray, R.: Distributed cooperative control of multiple vehicle formations using structural potential functions. In: Proceedings of the 15th IFAC World Congress. Barcelona (2002)

18. Saber, O., Murray, R.: Graph rigidity and distributed formation stabilization of multi-vehicle systems. In: 41st Conference on Decision and Control, pp. 2965–2971. LasVegas (2002)

19. Lawton, J.R., Beard, R.W., Young, B.J.: A decentralized approach to formation maneuvers. Robot. Autom. IEEE Trans. **19**(6), 933–941 (2003)

20. Saber, O.: Flocking for multi-agent dynamic systems: algorithm and theory. IEEE Trans. Autom. Control **3**(51), 401–420 (2006)

21. Alexander, F.J., Richard, M.M.: Graph laplacians and stabilization of vehicle formations. In: Proceedings of 15th IFAC Conference, pp. 227–241. Barcelona (2002)

22. Yohnnes, K., Gary, B.: Formation stability with limited information exchange between vehicles. In: Proceedings of the American Control Conference, pp. 290–295. Penve (2003)

23. Leonard, N.E., Fiorelli, E.: Virtual leaders, artificial potentials and coordinated control of groups. In: Proceedings of the 40th IEEE International Conference on Decision and Control, pp. 2968–2973. Orlando (2001)

24. Ogren, P., Fiorelli, E., Leonard, N.E.: Formations with a mission: Stable coordination of vehicle group maneuvers. In: Proceedings of the 15th International Symposium on Mathematical Theory of Networks and Systems. Notre Dame (2002)

25. Baras, J.S., Tan, XB., Hovareshti, P.: Decentralized control of autonomous vehicles. In: Proceedings of the 42nd IEEE Conference on Decision and Control, pp. 1532–1537. Hawaii (2003)

26. Williams, A., Lafferriere, G., Veerman, J.: Stable motions of vehicle formations. In: Decision and Control, 2005 and 2005 European Control Conference. CDC-ECC '05. 44th IEEE Conference on, pp. 72–77 (2005)

27. Yang, T.C.: Networked control system: a brief survey. IEEE Proc. Control Theory Appl. **153**(4), 403–412 (2006)

28. Ant saklis, P., Baillieul, J.: Guest editorial special issue on networked control systems. IEEE Trans. Autom. Control **49**(9), 1421–1423 (2004)

29. Ant saklis, P., Baillieul, J.: Special issue on technology of networked control systems. Proc. IEEE **95**(1), 528 (2007)

30. Nilsson, J.: Real-time control systems with delays. Lund Institute of Technology, Sweden (1998)

31. Zhang, W., Michael, S.B., Stephen, M.P.: Stability of networked control systems. IEEE Control Syst. Magazine **21**(1), 84–99 (2001)

32. Wittenmark, B.: Computer-controlled systems: theory and design, Englewwood liffs. Prentice-Hall, N.J., USA (1990)

33. Luck, R., Ray, A.: An observer-based compensator for distributed delays. Automatica **26** (5), 903–908 (1990)

34. José, Y., Pau, M., Josep, M.F.: Control loop performance analysis over networked control systems, In: IEEE 28th Annual Conference of the Industrial Electronics Society, pp. 2880–2885 (2002)

35. Battilotti, S.: Control over a communication channel with random noise and delays. Automatica **44**(2), 348–360 (2008)

36. Feng, L.L, James, M., Dawn, T.: Optimal controller design and evaluation for a class of networked control systems with distributed constant delay. In: Proceedings of American Control Conference Anchorage, pp. 3009–3014 (2002)

37. Qiu, Z., Zhang, Q., Lian, Z., Liu, M.: Exponential stability of state feedback networked control system with time-delay and data packet dropout. Inf. Control **34**(5), 567–571 (2005)

38. Ray, A.: Output feedback control under randomly varying distributed delays. J. Guidance Control Dyn. **17**(4), 701–711 (1994)

39. Nilsson, J., Wittenmark, B.: Stochastic analysis and control of real-time systems with random time delays. Automatica **34**(1), 57–64 (1998)
40. Gao, H.J., Chen, T.W., Lam, J.: A new delay system approach to network-based control. Automatica **44**(1), 39–52 (2008)
41. Montestluque, L.A., Antsaklis, P.J.: State and output feedback control in model-based networked control systems. In: Proceedings of 41st IEEE Conference on Decision and control, pp. 1620–1625. Lasvegas, NV (2002)
42. Xu, S., Lam, J., Chen, T.: Robust H∞ control for uncertain discrete stochastic time delay systems. Syst. Control Lett. **51**(3/4), 203–215 (2004)
43. Song, S.H., Kim, J.K., Yim, C.H., et al.: H∞ control of discrete time linear systems with time varying delays in state. Automatica **35**(9), 1587–1591 (1999)
44. Pan, Y.J., Horacio, J.M., Chen, T.W.: Remote stabilization of networked control systems with unknown time varying delays by LMI techniques. In: Proceedings of IEEE Conference on Decision and Control, 1589–1594. Spain (2005)
45. Xia, Y.Q., Shi, P., Liu, G.P., et al.: Output feedback control of discrete systems with time 2 varying delay. J. Guidance Control Dyn. **17**(4), 701–711 (1994)
46. Xie, G.M., Wang, L.: Stabilization of networked control systems with time varying network induced delay. In: Proceedings of IEEE Conference on Decision and Control. pp. 3551–3556. Bahamas (2004)
47. Yu, M., Wang, L., Chu, T., et al.: Stabilization of networked control systems with data packet dropout and network delays via switching system approach. In: Proceedings of the 43rd IEEE Conference on Decision and Control, pp. 3539–3544. Bahamas (2004)
48. Zhang, X.M., Zheng, Y.F., Lu, G.P.: Stochastic stability of networked control systems with network induced delay and data dropout. In: Proceedings of the 45th IEEE Conference on Decision and Control, pp. 5006–5011. San Diego (2006)
49. Zhang, L.Q., Shi, Y., Chen, T.W., et al.: A new method for stabilization of networked control systems with random delays. In: 2005 American Control Conference, pp. 633–637 Oregon (2005)
50. Costa, O.L.V., Fragoso, M.D.: Stability results for discrete time linear systems with markovian jumping parameters. J. Math. Anal. Appl. **179**(2), 154–178 (1993)
51. Lin, A., Arash, H., Jonathan, P.H.: Control with random communication delays via a discrete time jump system approach. In: Proceedings of the American Control Conference. pp. 2199–2204. Chicago (2000)
52. Sun, M.H., Lam, J., Xu, ShY, et al.: Robust exponential stabilization for markovian jump systems with mode-dependent input delay. Automatica **43**(10), 1799–1807 (2007)
53. Lee, S., Lee, K.C., Kim, H., Lee, M., et al.: Remote control for guaranteeing QoC of networked control system via profibus token passing protocol. In: The 29th Annual Conference of the IEEE, pp. 1425–1430. Berlin (2003)
54. Walsh, G.C., Ye, H., Bushnell, L.: Stability analysis of networked control system. In: Proceedings of the American Control Conference, pp. 2876–2880. California (1999)
55. Walsh, G.C., Ye, H., Bushnell, L.: Scheduling of networked control system. IEEE Control Syst. Magazine **21**(1), 57–65 (2001)
56. Walsh, G.C., Beldiman, O., Bushnell, L.G.: Asymptotic behaviour of nonlinear networked control systems. IEEE Trans. Autom. Control **46**(7), 1093–1097 (2001)
57. Zuberi, K.M., Shin, K.G.: Scheduling messages on controller area network for real-time CIM applications. IEEE Trans. Control Syst. Technol. **3**(2), 310–314 (1997)
58. Fan, W., Cai, H., Chen, Q., Weili, H.: Stability of networked control systems with time-delay. Control Theory Appl. **21**(6), 880–884 (2004)
59. Hong, S.H.: Scheduling algorithm of data sampling times in the integrated communication and control systems. IEEE Trans. Control Syst. Technol. **3**(2), 225–231 (1995)
60. Kim, Y.H., Kwon, W.H., Park, H.S.: Stability and a scheduling method for network-based control systems. In: Proceedings of the International Conference on Industrial Electronics, Control, and Instrumentation, pp. 934–939. Las Vegas (1996)

61. Dacic, D.B., Nesie, D.: Quadratie stabilization of linear networked control systems via simultaneous protocol land controller design. Automatiea **43**, 145–155 (2007)
62. Hristu Varsakelis, D.: Stabilization of networked control systems with access constraints and delays. In: Proceedings of the 45th IEEE Conference on Decision and Control, pp. 1123–1128. San Diego (2006)
63. Li, Z., Fang, H.J.: A novel controller design and evaluation for networked control systems with time-variant delays. J. Franklin Institute **343**(2), 161–167 (2006)
64. Sentang, W., Wu, Z.: Study on cooperative guidance and control of vehicles formation. Commun. CAA. **176**(3), 40–50 (2014)
65. Liu, X., Sentang, W., Xiaomin, M., et al.: Autonomous formation and cooperative guidance of multi-UAV: concept, design and simulation. J Syst. Simul. **20**(19), 5075–5085 (2008)
66. Unmanned Aircraft Systems: (UAS) Roadmap, 2005–2030. Office of the Secretary of Defense, Washington, D. C., USA (2005)
67. Unmanned Aircraft Systems: (UAS) Roadmap, 2000–2025. Office of the Secretary of Defense, Washington, D. C., USA (2001)
68. Unmanned Aircraft Systems: (UAS) Roadmap, 2002–2027. Office of the Secretary of Defense, Washington, D. C., USA (2002)
69. Unmanned Systems. (UAS) Roadmap, 2007–2032. Office of the Secretary of Defense, Washington, D. C., USA (2007)
70. Huang, H. (ed.): Terminology for Specifying the Autonomy Levels for Unmanned Systems, Version 1.0. NIST Special Publication 1011, National Institute of Standards and Technology, Gaithersburg, MD (2004)
71. Huang, H.M., Albus, J., Messinan, E., et al.: Specifying autonomy levels for unmanned systems: interim report. In: Proceedings of the 2004 SPIE Defense and Security Symposium Conference, pp. 386–397 (2004)
72. Larry, A.Y., Jeffrey, A.Y., Mark, D.G.: System analysis applied to autonomy: Application to high-altitudelong-endurance remotely operated aircraft. In: Proceedings of the AIAA Infotech Aerospace Conference, pp. 22–31 (2005)
73. Suresh, M., Ghose, D.: Role of information and communication in redefining unmanned aerial vehicle autonomous control levels. Proc. Inst. Mechanical Eng. Part G J. Aerospace Eng. **224**, 171–197 (2010)
74. Du, Y.: Analysis and Design for Wireless Networked Systems of Vehicles Autonomous Formation. Beihang University, Beijing (2013)
75. Sentang, W.: Flight Control System. Beihang University Press, Beijing (2013)
76. Xu, R.: Study on Stochastic Robustness Design and Applications. Beihang University, Beijing (2003)
77. Sentang, W.: Theory and Applications of System with Random Changing Structure. Science Press, Beijing (2007)
78. DoD Positioning navigation and timing executive committee. Global positioning system standard positioning service performance standard (2008)
79. Kaplan, E.D., Hegarty, C.: Understanding GPS Principles and Applications, 2nd edn, pp. 1–14, 67, 87–100. Artech House Publishers, Boston (2005)
80. Bisnath, S., Gao, Y.: Precise point positioning–a powerful technique with a promising future. GPS World **20**(4), 43–50 (2009)
81. Van Graas, F., Soloviev, A.: Precise velocity estimation using a stand-alone GPS receiver. Navig. J. Inst. Navig. **51**(4), 283–292 (2004)
82. Misra, P., Enge, P.: Global Positioning System–signals, Measurements, and Performance, vol. 32, pp. 124–130, 196–198. Ganga-Jamuna Press, Lincoln (2001)
83. Alighanbari, M.: Task Assignment Algorithms for Teams of UAVs in Dynamic Environments. Massachusetts Institute of Technology, Cambridge (2004)
84. Jennings, N.R.: Controlling cooperative problem solving in industrial multi-agent systems using joint intention. Artif. Intell. **75**(2), 195–240 (1995)

85. Manathara, J.G., Sujit, P.B., Beard, R.W.: Multiple UAV coalition for a search and prosecute mission. J. Int. Robot Syst. **7**, 1–34 (2010)
86. Costa, D.: An evolutionary tabu search algorithm and the NIIL scheduling problem. INFOR **33**, 161–178 (1995)
87. Gendreau, M., Herts, A., Laporte, G.: A tabu search heuristic for the vehicle routing problem. Manag. Sci. **40**, 1276–1290 (1994)
88. Clerc, M., Kennedy, J.: The particle swarm-explosion, stability and convergence in a multidimensional complex space. IEEE Trans. Evol. Comput. **6**(2), 58–73 (2006)
89. Liang, J.J., Suganthan, P.N.: Dynamic multi-swarm particle swarm optimizer. Swarm Intell. Symp. **6**(2), 124–129 (2005)
90. Kennedy, J., Eberhart, R.C.: A discrete binary version of the particle swarm algorithm. In: IEEE Conference on Systems, Man and Cybernetics, pp. 4104–4109. Orlando (1997)
91. Sun, Jian, Sentang, Wu: Methods of cruise missiles route planning based on modified PSO. Beijing J. Beihang Uni. **37**, 1228–1232 (2011)
92. Nan, Ying, Peng, Yun: An effective algorithm of terrain avoidance/terrain following. Flight Dyn. **24**(4), 73–75 (2006)
93. Sun, J., Bao, Y., Sentang, W.: Cooperative route plan of initial stage for multiple missiles formation. In: 2011 Chinese Control and Decision Ofference, pp. 2487–2491. Guilin
94. Shannon, T., Anthony, C.: Eric J.3D Trajectory optimization for terrain following and terrain masking. In: AIAA Guidance, Navigation, and Control Conference and Exhibit, pp. 2006–6102. Colorado (2006)
95. Nikolos, I.K., Zografos, E.S., Brintaki, A.N.: UAV path planning using evolutionary algorithm. Stud. Comput. Intell. **70**, 77–111 (2007)
96. Chen, P.: Study on Guidance Approach of Multi-missiles Autonomous Formation. Beihang University, Beijing (2009)
97. McLain, Timothy W., Beard, Randal W.: Coordination variables, coordination functions, and cooperative-timing missions [J]. Journal of Guidance, Control, and Dynamics. **28**(1), 150–161 (2005)
98. Ren, Wei, Beard, R.W., Atkins, E.M.: Information consensus in multivehicle cooperative control. Control Syst. Magazine IEEE **27**(2), 71–82 (2007)
99. Jeon, In-Soo, Lee, Jin-Ik, Tahk, Min-Jea: Impact-time-control guidance law for anti-ship missiles. Control Syst. Technol. IEEE Trans. **14**(2), 260–266 (2006)
100. Shiyu, Zhao, Rui, Zhou: Cooperative guidance for multi-missile Salvo attack. Chinese J. Aeronaut. **06**, 533–539 (2008)
101. Chen, P., Sen-tang, W.U.: Study on capture probability of anti-ship missiles at hand-over point. Flight Dyn. **26**, 37–40 (2008)
102. Wu, S.T., Zhang, S.X., Zhang, M.: Method of Nonlinear Gauss-Hermite Filtering and anti-interference tracking of Maneuvering targets. Control Decis. **17**(5), 559–562 (2002)
103. Jiang, Z.: Stochastic Robustness Design Method for System With Random Changing Structure With Applications. Beihang University, Beijing (2009)
104. Zhi-chao, J.I.A.N.G., Sen-tang, W.U., Xian-zhe, J.I.N.: Design of robust flight control systems for a flex-wing vehicle. Control Decision **7**, 022 (2008)
105. Sentang, W., Renniu, X.: Stochastic robustness design for overload control of anti-ship missiles. Aerospace Control **21**(83), 1–6 (2003)
106. Bao, Y.: Research on Guidance and Control Methods of Intermittent Gliding Missile. Beihang University, Beijing (2011)
107. Hu, N.: Research on Cooperative Guidance of Vehicles Formation. Beihang University, Beijing (2013)
108. Zhou, X., Yan, J., Xie, Y., Zhai, H.: Task distributed algorithmic for multi-UAV based on auction mechanism. J Naval Aeronautical Astronautical Uni. **27**(3), 308–312 (2012)
109. Olfati-Saber, R., Murray, R.M.: Consensus problems in networks of agents with switching topology and time-delays. Autom. Control IEEE Trans. **49**(9), 1520–1533 (2004)

110. Beckers, R., Deneubourg, J.L., Goss, S.: Trails and U-turns in the selection of a path by the ant Lasius-Niger. J Theor. Biol. **159**(4), 397–415 (1992)
111. Hou, X.G., Siveter, D.J., Aldridge, R.J., Siveter, D.J.: Collective behaviour in an early Cambrian arthropod. Science **322**(5899), 224 (2008)
112. John, T., Emlen, Jr: Flocking behavior in birds. Auk **69**(2), 160–170 (1952)
113. Ballerini, M., Cabibbo, N., Candelier, R., et al.: Interaction ruling animal collective behavior depends on topological rather than metric distance: evidence from a field study. PNAS **105**(4), 1232–1237 (2008)
114. Giardina, Irene: Collective behavior in animal groups: Theoretical models and empirical studies. HFSP J. **2**(4), 205–219 (2008)
115. Warburton, K., Lazarus, J.: Tendency-distance models of social cohesion in animal groups. J. Theor. Biol. **150**, 473–488 (1991)
116. Gerard, R.W.: Synchrony in flock wheeling. Science **97**, 160–161 (1943)
117. Loren, K.: On Aggression. Harcourt Brace, San Diego (1963)
118. Meyn, L.: A closed-form solution to multi-point scheduling problems.. In: AIAA Modeling and Simulation Technologies Conference, pp. 7911, 1–12 (2010)
119. U.S. Air Force. Air Combat Command Manual, pp. 3–3. WashingtonDC (1992)
120. U.S. Army. Field Manual, pp. 7–7j. Washington DC (1986)
121. Balch, T., Arkin, R.C.: Behavior-based formation control for multi-robot teams. IEEE Trans Robot. Autom. **14**(6), 926–939 (1998)
122. Nilsson, J.: Real-time Control Systems With Delays. Department of Automatic Control, Lund Institute of Technology, Sweden (1999, Jan)
123. Wei, Zhen, Li, Changhong, Xie, Jianying: Online delay-evaluation control for networked control systems. Control Decis. **18**(3), 545–549 (2003)
124. Chan, H., Ozguner, U.: Closed-loop control of systems over a communication network with queues. Int. J. Control **62**, 494–510 (1994)
125. Beldiman, O.V.: Networked Control System. Duke University, USA. (2001)
126. Kawka, P.A., Alleyne, A.G.: Stability and feedback control of wireless networked systems. In: 2005 American Control Conference, p. 2953. Portland, OR, USA (2005)
127. Xiao, L., Hassibi, A., How, J.P. Control with random communication delays via a discrete-time jump system approach. In: Proceedings of the American Control Conference, p. 2199. Chicago (2000)
128. Kang, Jun, Dai, Guanzhong: Research on Synthesizing control of networked control systems. Control Decis. **24**(2), 181–185 (2009)
129. Zhixun Yu, Huitang Chen, Yuejian Wang. Research on markov delay characteristic-based closed loop network control system.[J]. Control Theory and Applications,2002,19 (2):263-267
130. Tanner, H.G., Jadbabaie, A., Pappas, G.J.: Flocking in fixed and switching networks. IEEE Trans. Autom. Control **52**(5), 863–868 (2007)
131. Yook, J.K., Tilbury, D.M., Soparkar, N.R.: Trading computation for bandwidth: reducing communication in distributed control systems using state estimators. IEEE Trans. Control Syst. Technol. **10**(4), 503–518 (2002)
132. Qiongjian, F., Yang, Z., Feng, M., Chen, C.: Formation flight hardware-in-loop testbed system for multi-UAV. Chinese J. Sci. Instrum. **30**(3), 503–507 (2009)
133. Hogg, R.W., Rankin, A.L., Roumeliotis, S.I., et al.: Algorithms and sensors for small robot path following. In: Proceedings 2002 IEEE International Conference on Robotics and Automation, pp. 3850-3857. Washington D C (2002)
134. Chatterjee, M., Das, S.K., Turgut, D.: WCA: a weighted clustering algorithm for mobile Ad-Hoc networks. IEEE J. Clustering Compu. **5**, 193–204 (2002)
135. Xu, Y., Heidemann, J., Estrin. D.: Geography informed energy conservation for Ad-Hoc routing. In: Proceedings of the Seventh Annual ACM/IEEE International Conference on Mobile Computing and Networking (ACM Mobicom) pp. 70–84 (2001)

136. Heinzelman, W.R., Chandrakasan, A., Balakrishnan, H.: Energy-efficient communication protocol for wireless microsensor networks. In: Proceedings of the Hawaii International Conference on System Sciences (2000)

137. Younis, O., Fahmy, S., Distributed clustering in ad-hoc sensor networks: a hybrid, energy-efficient approach. In: Proceedings of 13th Joint Conference on IEEE Computer and Communications Societies (INFOCOM) (2004)

138. Bin, Y., Jinwu, X., Jianhong, Y., Debin, Y.: A novel weighted clustering algorithm in mobile ad hoc networks using discrete particle swarm optimization (DPSOWCA). Int. J. Manage. 20(2), 71–84 (2010)

139. Xu,J.-Y., Feng, Y.-P.: A concave hull algorithm for scattered data and its applications. In: Proceedings of 3rd International Congress Image Signal Process (CISP2010), pp. 2430–2433 (2010)

140. Fang, Q., Gao, J., Guiba, L.: Locating and bypassing routing holes in sensor networks. Mobile Netw. App. 11, 187–200 (2006)

141. McLurkin, J., Demaine, E.D.: A distributed boundary detection algorithm for Multi-Robot systems. In: IEEE International Conference Intelligent Robots and Systems (IROS2009), pp. 4791–4798 (2009)

142. Fekete, S.P., Kaufmann, M., Kroeller, A., Lehmann, N.: A new approach for boundary recognition in geometric sensor networks. In: Proceedings of 17th Canadian Conference on Computational Geometry, pp. 82–85 (2005)

143. Kroeller, A., Fekete, S.P., Pfisterer D., Fischer, S.: Deterministic boundary recognition and topology extraction for large sensor networks. In: Proceedings of 7th annual ACM-SIAM symposium on Discrete algorithm, pp. 1000–1009 (2006)

144. Selcuk B., Georgios E.F., George J.P.: Experimental cooperative control of fixed-wing unmanned aerial vehicles. In: Proceedings of the 43rd IEEE Conference on Decision and Control, pp. 4292–4298 (2004)

145. Seanor, B., Campa, G., Gu Y., et al.: Formation flight test results for UAV research aircraft models. In: AIAA 1st Intelligent Systems Technical Conference, pp. 1–14 (2004)

146. Pruetts, H., Slutzg, J., et al.: Hardware in the loop simulation using open control platform. In: AIAA Modeling & Simulation Technologies Conference and Exhibit, p. 1211 (2003)

147. Nettleton, E., Ridley, M., Sukkarieh, S., et al.: Implementation of a decentralised sensing network aboard multiple UAVs. Telecommun. Syst. 26(2–4), 253–284 (2004)

148. Mueller, E.R.: Hardware-in-loop simulation design for evaluation of unmanned aerial vehicle control systems. In: AIAA Modeling & Simulation Technologies Conference and Exhibit, pp. 65–69 (2007)

149. How, J., Kuwata, Y., King E.: Flight demonstrations of cooperative control for UAV teams. In: AIAA 3rd Unmanned Unlimited Technical Conference, Workshop and Exhibit, pp. 64–90 (2004)

150. Xiangyu, J., Sentang, W., Xing, L., et al.: Research and design on physical multi-UAV system for verification of autonomous formation and cooperative guidance. In: International Conference on Electrical and Control Engineering, vol. 1 pp. 1570–1576 (2010)

151. Xiangyu, J.: Design and Simulation on Multi-UAV System for Verification of Autonomous Formation and Cooperative Guidance. Beihang University, Beijing (2010)

152. Xiaomin, M., Sentang, W.: Modeling and simulation for cooperative control system of winged missile autonomous formation. Flight Dyn. 28(4), 59–63 (2010)

153. Gu, Y., Seanor, B., Campa, G., Napolitano, M.R., Rowe, L., Gururajan, S., Wan, S.: Design and flight testing evaluation of formation control laws. IEEE Trans. Control Syst. Technol. 14(6), 1105–1112 (2006)

Printed in the United States
By Bookmasters